# HEAT TRANSFER AND TURBULENT BUOYANT CONVECTION

# SERIES IN THERMAL AND FLUIDS ENGINEERING

**JAMES P. HARTNETT** and **THOMAS F. IRVINE, JR.**, Editors
**JACK P. HOLMAN,** Senior Consulting Editor

| | |
|---|---|
| Chang | • **Control of Flow Separation: Energy Conservation, Operational Efficiency, and Safety** |
| Chi | • **Heat Pipe Theory and Practice: A Sourcebook** |
| Eckert and Goldstein | • **Measurements in Heat Transfer, 2nd edition** |
| Edwards, Denny, and Mills | • **Transfer Processes: An Introduction to Diffusion, Convection, and Radiation** |
| Hsu and Graham | • **Transport Processes in Boiling and Two-Phase Systems, Including Near-Critical Fluids** |
| Kearns | • **Fluidization Technology** |
| Lu | • **Introduction to the Mechanics of Viscous Fluids** |
| Moore and Sieverding | • **Two-Phase Steam Flow in Turbines and Separators: Theory, Instrumentation, Engineering** |
| Richards | • **Measurement of Unsteady Fluid Dynamic Phenomena** |
| Spalding and Afgan | • **Heat Transfer and Turbulent Buoyant Convection: Studies and Applications for Natural Environment, Buildings, Engineering Systems** |
| Sparrow and Cess | • **Radiation Heat Transfer, augmented edition** |
| Tien and Lienhard | • **Statistical Thermodynamics** |

# HEAT TRANSFER AND TURBULENT BUOYANT CONVECTION

Studies and Applications for
- Natural Environment
- Buildings
- Engineering Systems

**VOLUME II**

EDITORS

**D. Brian Spalding**
Imperial College of Science and Technology, London

AND

**N. Afgan**
University of Belgrade

**HEMISPHERE PUBLISHING CORPORATION**
Washington

IN ASSOCIATION WITH
**McGRAW-HILL INTERNATIONAL BOOK COMPANY**
New York   St. Louis   San Francisco   Auckland   Bogotá   Düsseldorf
Johannesburg   London   Madrid   Mexico   Montreal   New Delhi
Panama   Paris   São Paulo   Singapore   Sydney   Tokyo   Toronto

**HEAT TRANSFER AND TURBULENT BUOYANT CONVECTION**

Copyright © 1977 by Hemisphere Publishing Corporation. All rights reserved. Printed in the United States of America. No part of this publication may be reproduced, stored in a retrieval system, or transmitted, in any form or by any means, electronic, mechanical, photocopying, recording, or otherwise, without prior written permission of the publisher.

1 2 3 4 5 6 7 8 9 0    D O D O    7 8 3 2 1 0 9 8 7

**Library of Congress Cataloging in Publication Data**

Main entry under title:

Heat transfer and turbulent buoyant convection.

   (Series in thermal and fluids engineering)
   Papers presented at a seminar of the International Center for Heat and Mass Transfer at Dubrovnik, Yugoslavia, in Aug. 1976.
   Includes index.
   1. Heat—Transmission—Congresses.  2. Heat—Convection—Congresses.  3. Turbulence—Congresses.
I. Spalding, Dudley Brian. II. Afgan, Naim. III. International Center for Heat and Mass Transfer.
QC319.8.H42    536'.25    77-1868
ISBN 0-07-059926-2 (v. II)

**A publication of the International Centre for Heat and Mass Transfer**
Belgrade

**INSTITUTIONAL MEMBERS**

American Geophysical Union
American Institute of Chemical Engineers
American Society of Mechanical Engineers
Associazione Termotecnica Italiana
Canadian Society for Chemical Engineering
Canadian Society for Mechanical Engineering
Egyptian Society of Engineers
Indian National Committee for Heat and Mass Transfer
Institution of Chemical Engineers, London
Institution of Engineers of Australia
Institution of Mechanical Engineers, London
Israel Institute of Chemical Engineers
Koninklijk Instituut van Ingenieurs, Netherlands
National Committee for Heat and Mass Transfer of the Academy
    of Sciences of the USSR
Society of Chemical Engineers of Japan
Societé Française des Thermiciens
Verein Deutscher Ingenieure
Yugoslav Society of Heat Engineers

# CONTENTS

Preface    xiii

## INTERACTIONS OF TURBULENCE AND BUOYANCY    1

Theoretical Considerations of Turbulent Buoyant Flow
*James H. Stuhmiller*    3

A Study of the Interactions of Turbulence and Buoyancy in a Plane Vertical Buoyant Jet
*Nikolas E. Kotsovinos*    15

'Bursting Phenomena' in Turbulent Boundary Layer on a Horizontal Flat Plate Heated from Below
*Nobuhide Kasagi and Masaru Hirata*    27

Influence of Buoyancy on the Turbulence Intensities in Horizontal and Vertical Jets
*M. S. Hossain and W. Rodi*    39

Effect of Thermogravitational Convection on Distribution of Statistical Characteristics in Turbulent Non-Isothermal Incompressible Liquid Flows in Tubes
*O. G. Martynenko, I. A. Vatutin, and I. V. Skutova*    53

A Second-Order Model for Buoyancy Driven Mixed Layers
*Otto Zeman and John L. Lumley*    65

## ONE-DIMENSIONAL MIXING IN TURBULENT STRATIFIED FLUIDS    77

Theories of Turbulent Thermal Convection
*Robert R. Long*    79

A Theoretical Analysis of the Thermally Stratified Layer in the Lake
*Takuma Kawahara*    93

On the Nonlinear Temperature Wave Propagation into the Depth of Ocean
*Branislav S. Bačlić*    105

The Development and Erosion of the Thermocline
*D. B. Spalding and U. Svensson*    113

Development of the Turbulent Mixed Region in a Stratified Medium
*O. F. Vasiliev, B. G. Kuznetsov, Y. M. Lytkin, and G. G. Chernykh*    123

Warm-Water Spreading in Cooling Ponds of Steam Power Plants
*A. Zukauskas*    139

Indented bracketed numbers indicate out of sequence articles to be found at the end of volume II.

## MECHANIC AND HEAT TRANSFER OF LAYERS — 157

A Numerical Study of the Transport and Deposition of Airborne Material
*David A. H. Jacobs* — 159

Spreading of Buoyant Discharges
*J.-C. Chen and E. J. List* — 171

Temperature Modeling in Rivers with Thermal Discharges
*R. Viskanta and R. G. Hills* — 183

Comparison of Laboratory Experiments on Penetrative Convection with Measurements in Nature
*Frank-Dietrich Heidt* — 199

Experimental Study of a Turbulent Stratified Boundary Layer Developing on a Rough Plate
*J.-P. Schon, C. Rey, P. Mery, and J. Mathieu* — 211

Laboratory Experiments on Shear Generated Turbulence in Stratified Fluid and Their Relation to Oceanic Microstructure
*F. K. Browand and G. C. Koop* — 221

## BUOYANT PLUMES — 233

Turbulent Buoyant Plumes
*W. Douglas Baines* — 235

Turbulence in a Two-Dimensional Buoyant Water Plume Discharging into a Moving Stream
*Frederick L. Test, Alexander J. Patton, and Warren M. Hagist* — 251

Analysis of the Zone of Flow Establishment for Buoyant Turbulent Jets in Cross Flow
*Latif M. Jiji and Joseph Hoch* — 263

Calculation of Three-Dimensional Heated Surface Jets
*J. J. McGuirk and W. Rodi* — 275

Influence of Buoyancy on Dispersion in Open Channel-Flow
*Jost Grimm-Strele and William W. Sayre* — 289

Vertical Buoyant Air Jets
*P. H. Oosthuizen* — 303

The Sinking of Thermal Plumes in Winter in Great Lakes Coastal Waters
*G. K. Sato and C. H. Mortimer* — 313

A Line-Impulse Model for Buoyant Jets in a Cross Flow
*Vincent H. Chu* — 325

The Effect of Thermal Instability on Laminar Channel Flow
*Simon Ostrach and Yasuhiro Kamotani* — [729]

Averaging Times for Turbulent Quantities in the Stationary and Horizontal Homogeneous Atmospheric Surface Layer
*J. M. Bessem* — 339

CONTENTS

Turbulent Buoyant Jets
*N. E. Kotsovinos and E. J. List* — 349

The Structure of Turbulent Diffusion in an Axi-Symmetrical Thermal Plume
*Hidendi Nakagome and Masaru Hirata* — 361

Experimental Study of Heat and Mass Transfer in a Vertical Air Layer with Blowing Foreign Gases
*G. V. Tsiklauri, V. G. Puzach, and L. M. Soloviev* — [695]

Experimental Correlations Between Characteristics of Modelled (Atmospheric) Turbulence and the Variances in the Distribution of Chimney-Plumes
*Michael Zeuch* — 373

## BUOYANT FLOW IN DUCTS — 381

Turbulent Flow and Heat Transfer in Pipes Under Considerable Effect of Thermogravitational Forces
*B. S. Petukhov* — [701]

The Stability of a Fluid in a Vertical Rectangular Duct with an Adverse Temperature Gradient
*Orhan Kural* — 383

The Influence of Density Differences on the Mixing of Fluid Injected into Turbulent Pipe Flow
*G. A. L. Delvigne* — 391

Experimental Study of the Effect of Thermogravitation upon Turbulent Flow and Heat Transfer in Horizontal Pipes
*B. S. Petukhov, A. F. Poliakov, Yu. L. Shekhter, and V. A. Kuleshov* — [719]

Enhancement of Turbulent Heat Transfer Due to Buoyancy for Downward Flow of Water in Vertical Tubes
*J. D. Jackson and J. Fewster* — [759]

Heat Transfer to a Supercritical Fluid During Turbulent, Vertical Flow in a Circular Duct
*A. Malhotra and E. G. Hauptmann* — 405

## AIR SMOKE MOVEMENTS IN BUILDINGS — 415

Air and Smoke Movements in Building Fires
*Paul Gerhard Seeger* — [777]

Convection Exchanges Inside a Dwelling Room in Winter
*J. Lebrun and D. Marret* — 417

Numerical Prediction of Three-Dimensional Turbulent Buoyant Flow in a Ventilated Room
*B. H. Hjertager and B. F. Magnussen* — 429

Numerical Study of Problems on High-Intensive Free Convection
*B. M. Berkovsky and V. K. Polevikov* — 443

Accuracy of the Finite Difference Computation of Free Convection
*Jean J. Portier and Ozer A. Arnas* [797]

Experimental Study of Free Convection in a Square Cavity
*G. Burnay, J. Hannay, and J. Portier* [807]

Buoyancy Driven Countercurrent Flows Generated by a Fire Source
*B. J. McCaffrey and J. G. Quintiere* 457

Stability of Free Convection Flow in a Shallow Cavity With and Without Rotation
*Guenter P. Merker and Ulrich Grigull* 473

## FREE CONVECTION IN ENGINEERING EQUIPMENT 485

The Model of Turbulent Free Convection Near a Vertical Heat Transfer Surface
*S. S. Kutateladze* 487

Heat and Mass Transfer Across Gas Filled Enclosed Spaces Between a Hot Liquid Surface and a Cooled Roof
*J. C. Ralph and A. W. Bennett* 497

Flow and Heat Transfer in a Rectangular Cavity and Their Disturbance by Buoyancy
*Dominique Grand* 509

The Structure of Turbulent Free-Convective Flows from Heat Transfer Surfaces of Various Orientation
*A. G. Kirdyashkin* 519

A Theoretical Investigation of Buoyancy-Induced Flow Stratification in the Cylindrical Outlet Plenum of a Liquid-Metal-Cooled Fast Breeder Reactor
*Nicholas C. G. Markatos* 529

Experimental and Analytical Studies of Natural Convection in EBR-II
*Ralph M. Singer, Jerry L. Gillette, Wayne K. Lehto, Dale Mohr, Charles C. Price, and John I. Sackett* 545

Study of Temperature Gradients Due to Gas Thermosyphons Induced within the Phenix Nuclear Reactor
*J. L. Boy-Marcotte, P. Chevalier, and M. Jannot* 555

## FREE CONVECTION PHENOMENA IN GAS-LIQUID MIXTURES 567

The Calculation of Free-Convection Phenomena in Gas-Liquid Mixtures
*D. Brian Spalding* 569

Experiments in Turbulent Thermal Convection Driven by Internal Heat Sources
*J. C. Ralph, R. McGreevy, and R. S. Peckover* 587

Maximum Attainable Superheat and Explosive Boiling-Up of Liquids
*V. P. Skripov* 601

Hydrodynamic and Thermal Study of the Stability of Boundary Layer in the Case of Film Boiling
*F. Moreaux, J. C. Chevrier, and G. Beck* 615

The Collapse of Turbulent Shear Layers in Density Stratified Flow
*Vincent H. Chu*   625

Investigation of Hydrodynamics and Heat Transfer at Turbulent Natural Convection in Flat Slots
*M. D. Diev, S. D. Korneev, H. K. Kurbanov, A. I. Leontiev, and B. M. Mironov*   639

## FREE CONVECTION WITH HEAT ADDITION AND COMBUSTION PHENOMENA   649

Buoyant Diffusion Flames
*John De Ris*   [813]

Effect of Discrete Wall Roughness on Free Convective Heat Transfer From a Vertical Tube
*C. V. N. Sastry, V. Narayana Murthy, and P. K. Sarma*   651

On Laminar Free Convection Stagnation Heat Transfer From a Nonisothermal Cylinder With Internal Sources-Sinks
*Ozer A. Arnas*   663

Turbulent Mixing of a Duststream in a Duct
*Harald Gross*   673

Turbulent Natural Convection Diffusion Flames
*Lawrence A. Kennedy*   683

Index   833

Errata   837

This book contains the papers presented and discussed at the August 1976 seminar of the International Centre for Heat and Mass Transfer, at Dubrovnik, Yugoslavia.

The purpose of the seminar was to draw together, from all parts of the world and from all relevant disciplines, scientists who were interested in and knowledgeable about turbulent free-convection processes in the natural environment, in engineering equipment, and in the laboratory.

The preparation of the seminar began in September 1975; papers were supplied to the publisher as camera-ready copy in April 1976; and unbound offprints were supplied to participants at the seminar itself.

The timetable is mentioned so as to explain why the volume has turned out as it has. The speed of the operation was dictated by the desirability, in a rapidly moving subject, of swift publication of up-to-date results; and it was judged that, if the speed could be achieved, the attendant blemishes would be tolerated by readers of the proceedings.

Papers were submitted and accepted from the following countries: Belgium, Canada, England, France, Japan, The Netherlands, Sweden, Turkey, USSR, USA, West Germany, and Yugoslavia; their authors included oceanographers, mechanical engineers, and specialists in several other branches of pure and applied science.

The papers were organised and presented in ten sessions, with the titles: *Interactions of Turbulence and Buoyancy* • *One-Dimensional Mixing in Turbulent Stratified Fluids* • *Mechanics and Heat Transfer of Layers* • *Buoyant Plumes* (in two sessions) • *Buoyant Flow in Ducts* • *Air and Smoke Movements in Buildings* • *Free Convection in Engineering Equipment* • *Free Convection Phenomena in Gas-Liquid Mixtures* • *Free Convection with Heat Addition and Combustion Phenomena.*

This scheme represented a modification of an earlier one somewhat more systematic in conception, but, even so, not all the papers fit perfectly the title of the session during which they were delivered. Conference organisers can propose, but, when authors are prolific on one topic and more negligent of another, their predilections are perforce observed.

The papers do not appear in the volume in exactly the same order in which they were presented at the seminar, for the publication process required that page numbers should be allotted immediately, and not all the papers were available when printing started.

A summary of the scientific content of the proceedings is hardly to be contrived. It was evident that turbulence models, coupled with numerical methods of solving the associated differential equations, are increasingly occupying the attention of researchers and practitioners; but the way in which buoyancy and turbulence interact, in conditions of stratification, has still not been completely worked out, it appears. The invited lecture by James Stuhmiller, with which the volume opens, raised serious doubts as to whether turbulence models as currently conceived are likely to solve the problem; but the exploration of such suggestions remains for the future.

The papers are far from homogeneous in approach or level of sophistication. The discussions showed that national frontiers, and interdisciplinary boundaries, still present significant barriers to the transfer of ideas and skills. However, at least for the participants in the seminar, those barriers had been appreciably eroded by the end of the week's proceedings.

The editors are glad to express their gratitude for help in soliciting and screening papers, and in chairing the sessions, to the following members of the seminar committee: *P. Nakayama,* Jaycor, USA; *A. P. van Ulden,* Koninklijk Nederlands Meteorologisch Institut; *A. A. Zhukauskas,* Academy of Sciences of the Lithuanian SSR, USSR; *W. D. Baines,* University of Toronto, Canada; *O. F. Vasiliev,* Hydrodynamics Institute, USSR; *S. S. Kutateladze,* Academy of Sciences of the USSR; *D. R. Reineke,* Technische Universität Hannover, West Germany; *B. Gebhart,* State University of New York at Buffalo, USA; *M. A. Styrikovich,* Institute of High Temperatures, USSR; and *B. Magnussen,* University of Trondheim, Norway.

It is also a pleasure to thank Colleen King, who handled the bulk of the pre-seminar arrangements, Dr. Kosta Maglic, who saw to the arrangements at the seminar itself, and Bill Begell, who has personally supervised the whole publication process.

*D. Brian Spalding*
*N. Afgan*

# AIR SMOKE MOVEMENTS IN BUILDINGS

# CONVECTION EXCHANGES INSIDE A DWELLING ROOM IN WINTER

J. LEBRUN and D. MARRET

*Thermodynamic Institute, Service of Prof. G. Burnay*
*University of Liege, Val-Benoit, B-4000 Liege, Belgium*

POSITION OF THE PROBLEM

The practical calculation of heat losses in a room generally implies that transmission heat flow density is everywhere proportional to the difference between the corresponding surface temperature and a central reference inside temperature $t_i$. It is then considered that the emission of the heat source is injected in the room at the temperature $t_i$ and is radially transmitted to the different walls (fig. 1). This representation is inexact; indeed :
- the radiation exchanges between surfaces do practically not make intervene the temperature $t_i$
- the convection exchanges are not transmitted radially, but following the peripheric air flows where the air temperature is necessarily variable.

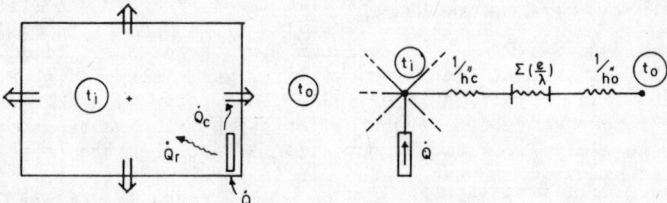

Fig.1 Assumption of radial distribution

In order to arrive at a more elaborated method, it is indispensable to separate the internal exchanges by convection and by radiation, because it is in studying the convection component - where the incertainty is the greatest - that it will be possible to better discern the physical reality of the phenomena [1].

Fig. 2 represents the heat flow densities by convection, radiation and transmission. In the equation of the radiation, the secondary reflexions and the absorption in the air have been neglected. For the convection the air temperatures and thus the real air circulation in the room should be taken into consideration [2].

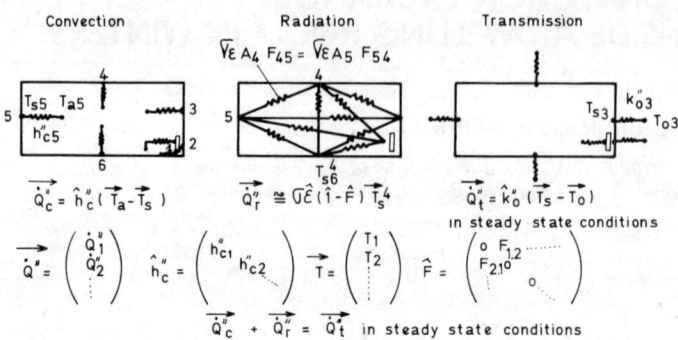

Fig 2 Representation of heat flow densities by convection, radiation and transmission

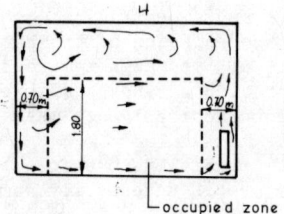

Fig.3 Two-dimensional schema of air circulation

The evolution of the air flows results from mixtures of air different masses and transfers along the surface (fig. 3) : the boundary layer of natural convection along the heat source (zone 1) and the back projection (zone 2) - which is warmed by the radiation of the source - develops a free convection jet; this jet licks the trail above the source (zone 3) and the ceiling (zone 4) where it stretches itself. These zones 3 and 4 are submitted to an exchange by forced convection. Zone 5 represents the totality of the "free" vertical walls along of which develops a descendant free convection layer. The floor (zone 6) can be submitted either the forced convection from the jet which prolongs the boundary layers (5) either to a certain natural convection effect when the temperature of the floor is higher than that of the air in the immediate proximity (for instance, the case of radiation warming of the source). The air infiltrations penetrating through the window are therefore in the zone 3 or 5 according to the heat source is disposed under the window or along an internal wall. The occupied zone is nearly only crossed by a very slow stream induced by the natural convection jet which develops above the source (stream 5 → 3); there can also be a jet on the floor which prolongs the boundary layer issued from 5 and which is eventually reinforced by air infiltrations, when the heat source is disposed along an internal wall, which may cause a risk of convection local discomfort.

The following results provide information on zone 3 which is particularly well studied and some indications - less precise - on the zones 4 and 5. This do not yet allow to predict of what happens on the floor level (zone 6).

As for the thermal comfort aspect, the global comfort is defined by the resultant temperature at about 0.75 m from the floor :

$$t_R = \frac{h_c t_a + h_r t_w}{h_c + h_r} \simeq \frac{1}{2} (t_a + t_w) \text{ in calm air (1)}$$

which $h_c$ and $h_r$ = convection and radiation transfer coefficients of the human body with the environment
$t_a$ = air temperature at about 0.75 m from the floor

$t_W$ = mean radiant temperature (spheric) at about 0.75m from the floor
$t_W$ can be calculated in function of the surface temperatures :

$$T_W^4 \cong \sum F_{h\downarrow} \; T_s^4 \quad (2)$$

$F_{h\downarrow}$ = angle factor between the human body and the surface $\downarrow$
As for the risks of local discomfort, one must first calculate the spheric ($t_W$) and hemispheric ($t_{Wh}$) mean radiant temperatures variations in function of the surface temperatures, and on the other hand one must predict the local values of air temperatures and velocities in the occupied zone, which is possible by the description of the air circulation in the room.

AVAILABLE RESULTS

This communication uses experimental data obtained during a research performed at Liège University with participation of the CSTC (Centre Scientifique et Technique de la Construction). The main aim was to arrive at a better definition of the comfort level in relation to the energy consumption with different heat sources and different qualities of insulation and tightness [3] [4].
A new test room with an exposed wall (so-called frontage) was specially equipped for this purpose [5]. Most tests were performed in steady state conditions and generally with an inside-outside temperature difference of 25 K (standard conditions : $t_{R0.75}$ = 22°C and $t_0$ = - 3°C). The principal measuring points are represented in fig. 4a for the air temperatures and in fig. 4b for the heat flow densities and corresponding surface temperatures with the zones with which are associated; the results of a test have been indicated as an illustration.

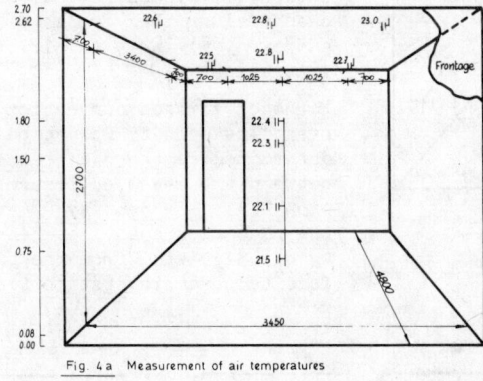

Fig. 4a Measurement of air temperatures

*Explications of the fig. 4a and 4 b*

*Experimental conditions*

- heating system : single panel under the window
- insulation : double glazing and good insulation
- air infiltration rate : n = 0
- inside-outside temperature difference : $t_{R0.75} - t_0 = 25$ K

*Thermal balance*

- *transmission heat flow through the frontage* $Q_F$ = 811 W
- *transmission heat flow through the internal walls* → $Q_i$ = 133 W

                        *Total heat losses*    $Q_t$ = 944 W

- *emission of the heat source (without consideration of connections)*
  $Q_s$ = 931 W
- *accuracy of thermal balance* : + 1.4 %

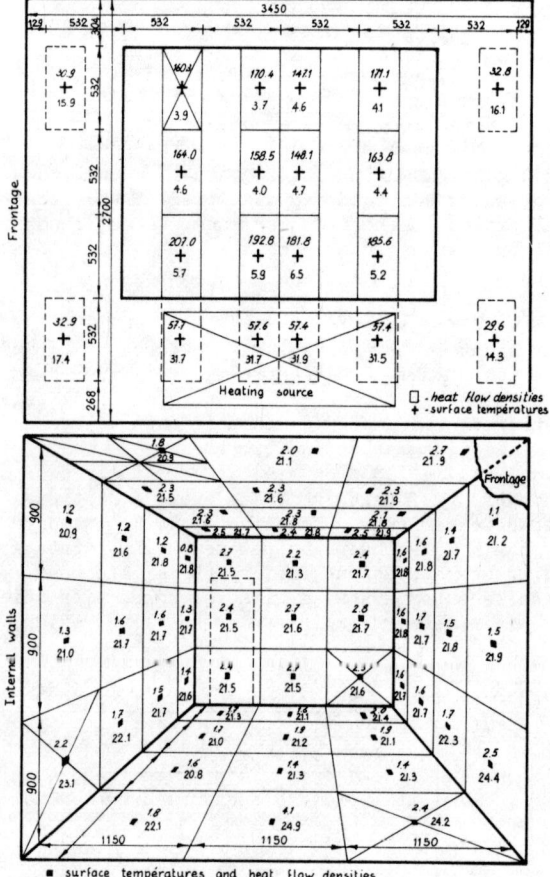

Fig. 4b Distribution of surface temperatures (°C) and transmission heat flow densities (W.m$^{-2}$)

The vertical air temperature profiles in the centre of the room have been established for the four types of the static sources associated with two degrees of insulation for the frontage (single glazing and bad insulation with a resistance of 0.55 m$^2$KW$^{-1}$ - double glazing and good insulation with a resistance of 1.09 m$^2$KW$^{-1}$) and an air renewal rate of 1 volume per hour as well as for the well insulated room without source and without ventilation (fig.5).

The results of the test carried out without source have been extrapolated to standard conditions to allow for a direct comparison with the other results. The most important vertical temperature gradient has been obtained when the jet issued from the source attains the ceiling without having been cooled off by convection exchange along the "frontage" and/or by mixing with air infiltration.

For each surface convection heat flow density $Q_c''$ can be determined as the difference between the measured transmission heat flow density $Q_t''$ and the radiation heat flow density $Q_r''$ calculated from the surface temperature distribution.

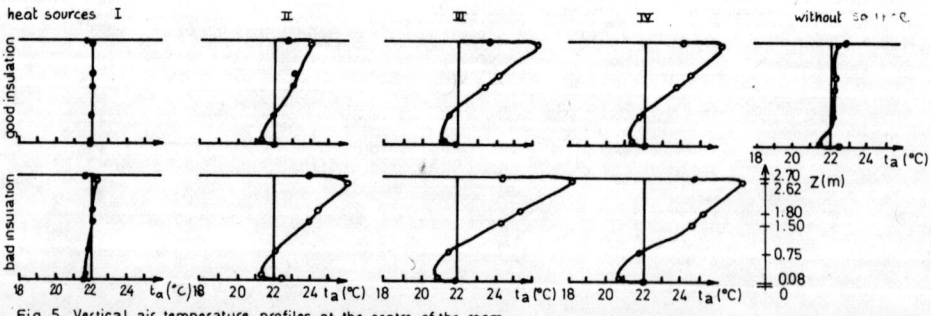

Fig. 5 Vertical air temperature profiles at the centre of the room

i.e. in matrix notations : $\vec{Q}_c'' = \vec{Q}_t'' - \vec{Q}_r''$         (3)

                                       mesured   computed

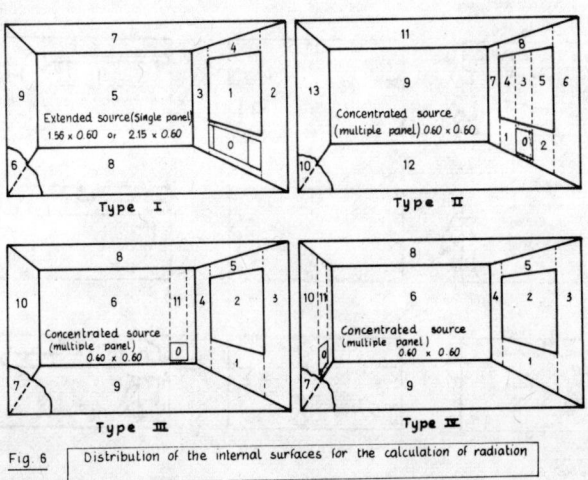

Fig. 6   Distribution of the internal surfaces for the calculation of radiation

For the calculation of radiation heat flow densities, the room has been decomposed in 10 or 14 surfaces taking into account four types of source (fig. 6). The distribution of radiation and convection flows (fig. 7) shows the importance of convection exchanges at the level of the trail above the source and of the ceiling which is precisely connected with vertical temperature gradient.

Fig. 7 shows the room by bringing together the surfaces in two groups, i.e. the total "frontage" and the total internal vertical walls. In each case, the distribution of convection exchanges corresponds roughly to the main air circulation resulting from the natural convection source.

When there is no heat source (last case shown by fig. 7) air circulation in the room is mainly caused by the boundary layer of natural convection developping along the "frontage", to which a negative enthalpy flow corresponds as shown in fig. 7.

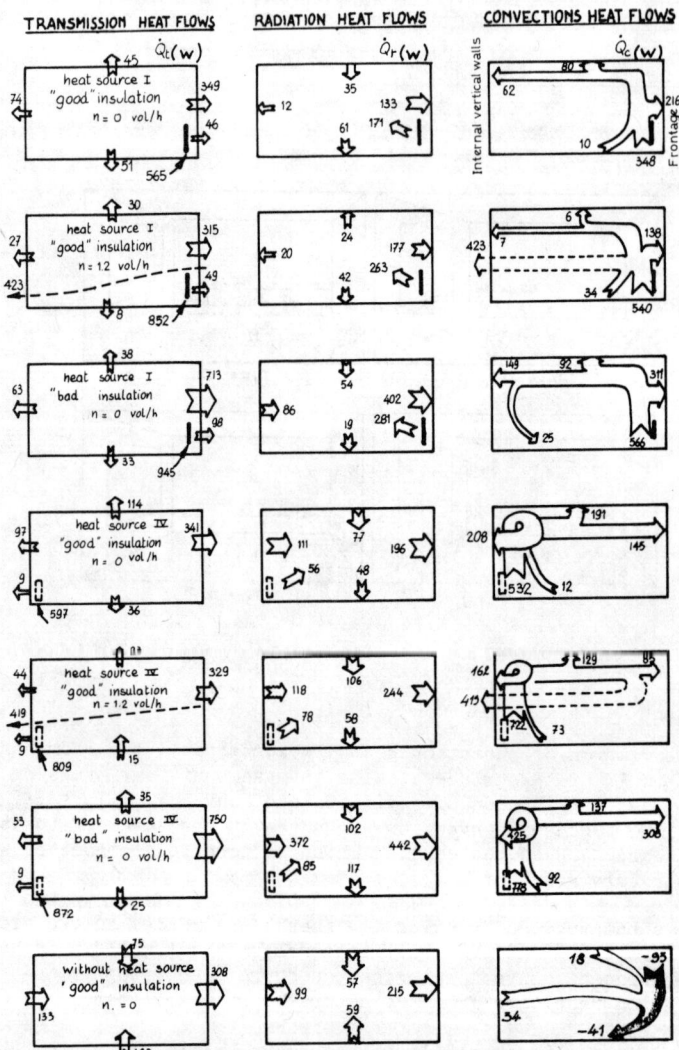

Fig. 7 Distribution of transmission, radiation and convection flows

Air temperatures along the various surfaces can be predicted according to the air enthalpy flow. The latter may be determined taking into account its progressive weakening from the source's starting point (fig. 8) :

$$\dot{Q}_{c1+2} = \dot{Q}_a - \dot{Q}_r - A_2\dot{Q}''_{t2}$$
(mesured) (computed)

$\dot{Q}_{c3} = \dot{Q}_{c1+2} - \dot{Q}_{v3}$ (enthalpy flow at level of surface 3)
(mesured)

$\dot{Q}_{v3}$ = ventilation enthalpy flow through the surface 3

$\dot{Q}_{v3} \cong \dot{Q}_v, \cong \frac{1}{2}\dot{Q}_v, = 0$ respectively for the sources I, II, III and IV (cf fig. 6).

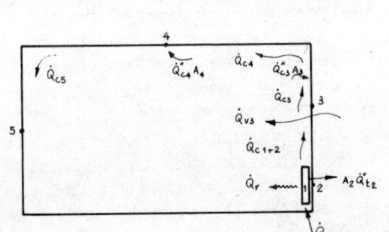

Fig. 8 Two-dimensional schema of convection exchanges

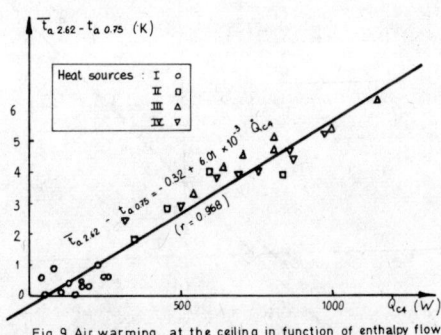

Fig. 9 Air warming at the ceiling in function of enthalpy flow

So the diagram (fig. 9), where the mean air temperature at 0.08 m from the ceiling $\bar{t}_{a2.62}$ has been reported, confirms that the warming of air in the vicinity of the ceiling is in good correlation with the enthalpy flow of the jet reaching that zone : there is no significant difference between the four examined situations.

One can try to physically interpret the convection flow densities observed on the different surfaces in function of the corresponding air temperatures. Air warming in the vinicity of a surface can be evaluated in function of the enthalpy flow of the jet at this level. One can use a relation of type (jet issued from a linear or punctual source [6]):

$$\Delta t_{a_i} \div (\dot{Q}_{c_i})^{2/3} \quad (4)$$

with $\dot{Q}_{c_i}$ = the enthalpy flow of the jet at the level of surface $i$

$$\dot{Q}''_{c_i} = h''_{c_i}(t_{a_i} - t_{s_i}) \quad (5)$$

with $t_{a_i} = \Delta t_{a_i} + t_{a\,0.75}$

and $\Delta t_{a_i} = c_i(\dot{Q}_{c_i})^{2/3} \quad (6)$

Combining (5) and (6), one obtains :

$$\frac{\dot{Q}''_{c_i}}{t_{a0.75} - t_{s_i}} = h''_{c_i}\left(1 + c_i \frac{\dot{Q}_{c_i}}{t_{a0.75} - t_{s_i}}\right)^{2/3} \quad (7)$$

For each of the surfaces 3-4-5 (fig. 8), one has tried to evaluate the convection transfer coefficient $h''_{c_i}$ and the constant $c_i$ by calcula-

ting a regression law of type (7). Zones 3 and 5 correspond here respectively to the trail above the source and to the totality of "free" vertical surfaces wherever the source is in the local :
- at the trail (zone 3), $h''_{c3} \cong 3.2$ W m$^{-2}$ K$^{-1}$ and $c_3 \cong 0.17$ K W$^{-2/3}$ (fig. 10). The relations of convection flow densities in function of temperature differences, taking into account or not air warming, can be compared (fig. 11 a and 11 b). The calculation of this warming is very significant as shown by the thus obtained high correlation (fig. 11 b);

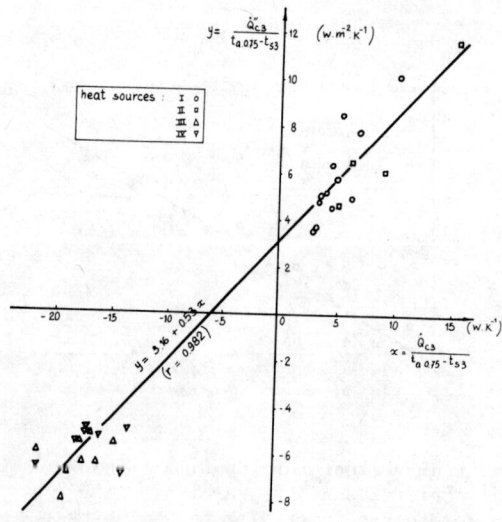

Fig. 10 Prediction of air temperature and heat transfer coefficient along the trail above the source in function of convection power

- the measuring results on the level of the ceiling (zone 4) are less precise and therefore allow no good experimental definition of the transfer coefficient. By taking $h''_{c4} \cong 2.5$ Wm$^{-2}$K$^{-1}$, one finds a constant $c_4 \cong 0.08$ KW$^{-2/3}$. When air warming is not considered, one has a negative correlation between the flow density and the temperature difference, which has no physical significance (fig. 12 a). However the application of air warming does not allow to define a unique transfer coefficient on the level of this surface (fig. 12 b).

- Finally at the level of vertical surfaces (zone 5) by taking $h''_{c5} \cong 2.5$ W m$^{-2}$ K$^{-1}$, one has $c_5 \cong 0.06$ K W$^{-2/3}$. The application of this air warming law allows to find a convection exchange comparable with the one obtained by the relation $N_u \cong 0.13$ $(G_r)^{1/3}$ defined for a vertical sheet in an infinite environment and in a turbulent flow ($G_r > 10^9$) [7] [8] (fig. 13).

The exchanges along the "frontage" with air infiltration have not been considered. The temperature and air velocity profiles have been measured inside the room for certain situations; fig. 14 gives an example of this.

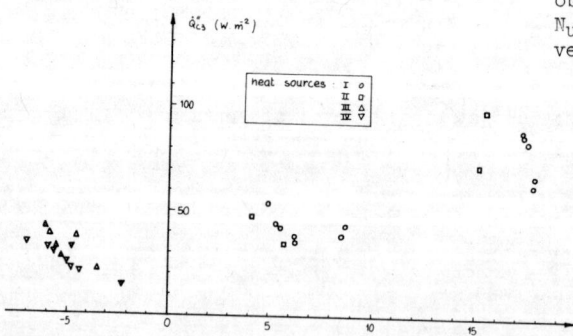

Fig. 11a Convection exchange along the trail in function of air temperature

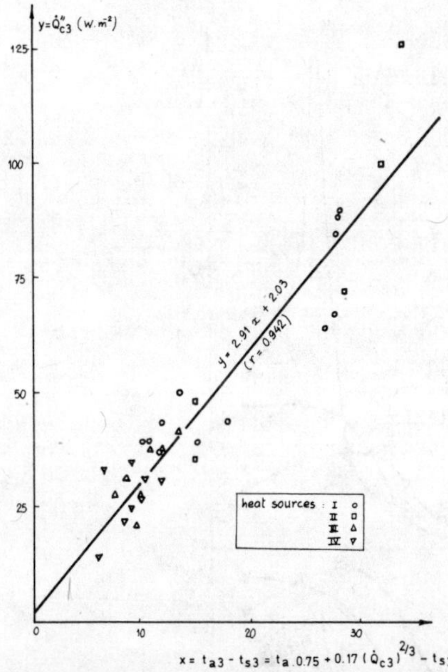

# CONVECTION EXCHANGES INSIDE A DWELLING ROOM IN WINTER

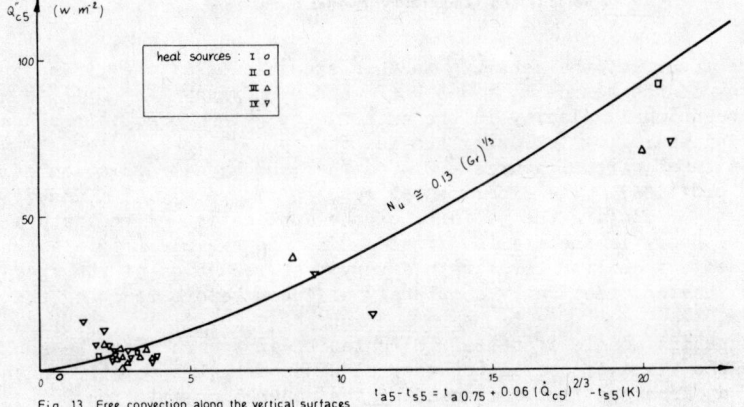

Fig. 11b Convection exchange along the trail in function of air warming

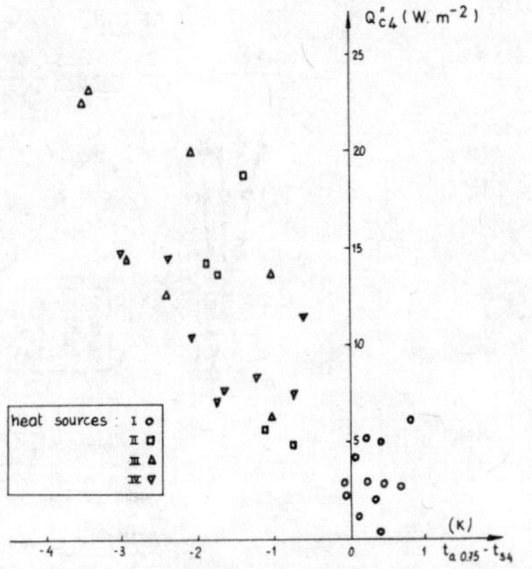

Fig. 12a Convection exchange along the ceiling in function of air temperature

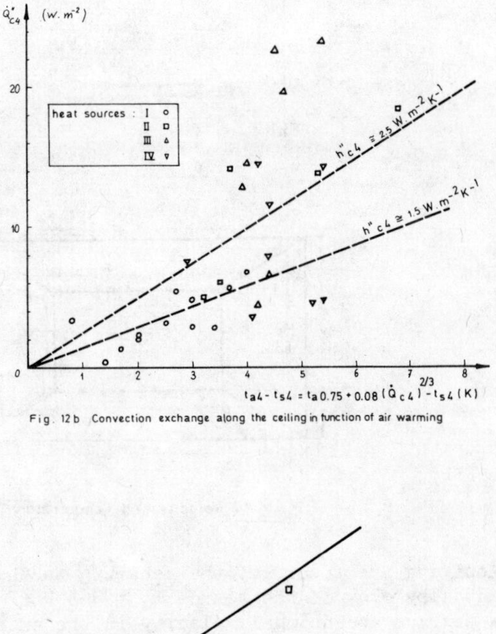

Fig. 12b Convection exchange along the ceiling in function of air warming

Fig. 13 Free convection along the vertical surfaces

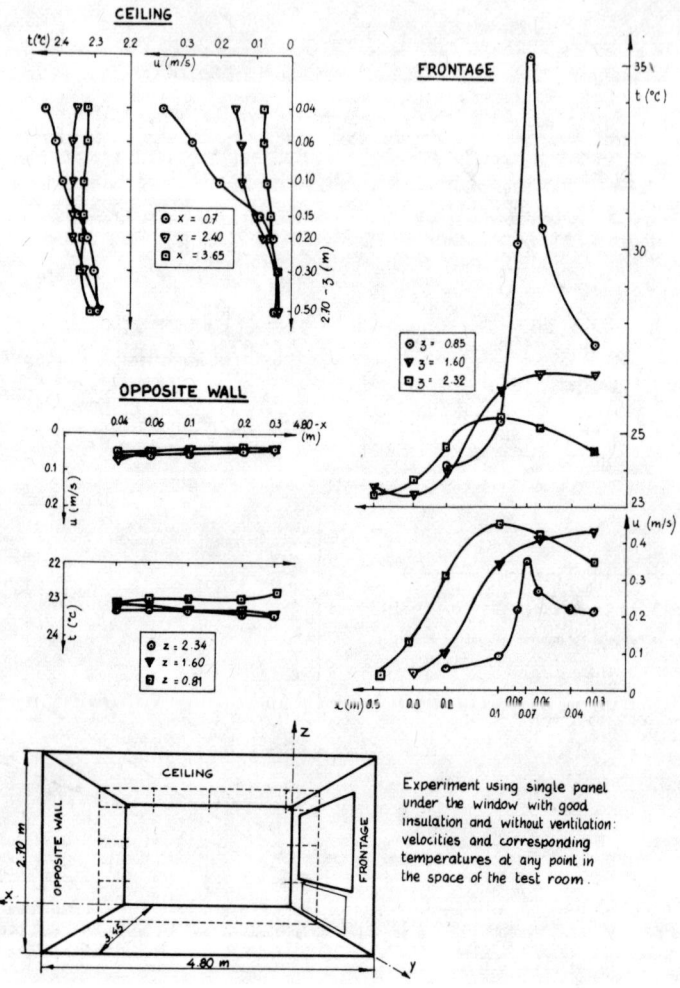

Fig. 14 Velocity and temperature profiles in the room

Applying the theory of the jets [9] to this specific test, one finds for a linear enthalpy flow of the source $\dot{Q}_c$ = 363 W m$^{-1}$ and a buoyancy $\dot{E}_o''$ = 0.122 N m$^{-1}$ s$^{-1}$, a maximum ascensional velocity in the axis of the jet $u_{max}$ = 0.4 m s$^{-1}$ and an evolution of linear flow following the law : $\dot{V}'$ = 0.07 $(z - z_0)$ m$^2$ s$^{-1}$ where $z_0$ is the position of virtual source $\cong$ 0.40 m (defined by extrapolation of flow and warming profiles). The experimental results - $u_{max}$ = 0.45 m and $\dot{V}'$ = 0.06 $(z - z_0)$ m$^2$ s$^{-1}$ - confirm the validity of the application of the natural convection jets theory in the examined field. In this example (fig. 14), the jet stretchs itself at ceiling level with a very fast reduction of its speed and its warming, having practically completely diffused before reaching the opposite wall.

The same computing models of the air circulation in the room can be applied to evaluate the risks of convection thermal discomfort, especially in the vicinity of a "frontage" not compensated by the convection emission of the source. Below at this surface, the maximum air velocity and its cooling can be computed considering the natural convection jet issued from a source disposed more or less at half height of the window; the convection emission of this source includes the convection flow absorbed by the glazing and the ventilation flow.

Yet there are no sufficiently accurate data available to allow for the calculation of the risks of local thermal discomfort at that place. A research is now being carried on that matter.

Acknowledgments

The mentioned research was performed with the participation of the staff of the Thermodynamic Institute; especially J. Hannay for the test room, P. Nusgens for the measuring probes, J.P. Wannyn and L. Laret for data computation. Took part in the interpretation calculations of radiation and convection exchanges, the two students-assistants D. Orban and C. Grégoire.

Reference

[1] LEBRUN, J. MARRET, D. : "Heat exchanges in a test room in steady state winter conditions" I.I.F., Moscou, 1975

[2] LEBRUN, J. : "Exigences physiologiques et modalités physiques de la climatisation par source statique concentrée" Collection des Publications de la Faculté des Sciences appliquées, n° 24, Liège, 1971

[3] LEBRUN, J. MARRET, D. : "Etude du Confort thermique et de la consommation d'énergie dans les conditions d'hiver". Rapport de Recherche IC-IB, Bruxelles, Mai 1975

[4] LEBRUN, J. MARRET, D. NUSGENS, P. WANNYN, J.P. LARET, L. : "Confort thermique et consommation d'énergie dans les conditions d'hiver". Collection des Publications de la Faculté des Sciences appliquées, n° 56, Liège, 1975

[5] HANNAY, J. LARET, L. LEBRUN, J. MARRET, D. NUSGENS, P. : "Experimental study on heat losses due to different heating systems" VIth International Congress of climatistics, Milan, March 1975

[6] BATCHELOR, G.K. : "Heat convection and buoyancy effects in fluids" Quarterly Journal Royal Meteorological Society, n° 80, 1954, p. 339

[7] ECKERT, E.R.G. : "Heat and mass transfer" Mc Graw-Hill, 1959

[8] CHEESEWRIGHT, R. : "Turbulent natural convection from a vertical plane surface" A.S.M.E./C. Transactions, February 1968

[9] SHAO-LIN LEE and EMMONS, H.W. : "A study of natural convection above a line fire" Journal of fluid mechanic, 1961, vol. II, p. 353.

# NUMERICAL PREDICTION OF THREE-DIMENSIONAL TURBULENT BUOYANT FLOW IN A VENTILATED ROOM

B. H. HJERTAGER and B. F. MAGNUSSEN

*The University of Trondheim, The Norwegian Institute of Technology, 7034 Trondheim-NTH, Norway*

ABSTRACT

A numerical study is performed of three-dimensional turbulent isothermal and buoyant flow in a ventilated box-shaped empty room. The Patankar and Spalding algorithm has been used to solve the equations for the three velocity components and continuity. Turbulence is modeled by equations for kinetic energy of turbulence and its rate of dissipation. The mathematical problem thus consists of solution of seven partial differential equations in a three-dimensional space. Predictions are compared with measurements of mean velocities. The agreement is good for the isothermal flow and there is also a fair agreement for the buoyant flow. It is concluded that more measurements of local mean velocities and turbulence velocities are needed to further validate the mathematical model.

NOMENCLATURE

| | |
|---|---|
| B | width of room (m) |
| b | width of nozzle (mm) |
| $C_1, C_2, C_D$ | constants |
| $g_i, g$ | gravitation constant (m/s$^2$) |
| H | height of room (m) |
| h | height of nozzle (mm) |
| k | kinetic energy of turbulence $0.5(\overline{u_i u_i})$ (m$^2$/s$^2$) |
| L | length of room (m) |
| p | pressure (N/m$^2$) |
| Re | Reynolds number |
| T | mean temperature (K) |
| $T_o$ | average temperature of room (K) |
| U, V, W | mean velocities in x, y, z directions (m/s) |
| $U_i$ | mean velocity in $x_i$ direction (m/s) |
| $u_i$ | fluctuating velocity in $x_i$ direction (m/s) |

| | |
|---|---|
| x, y, z | coordinates |
| $x_i$ | coordinates |

Greek

| | |
|---|---|
| $\delta_{ij}$ | = 0 i ≠ j<br>= 1 i = j |
| $\mu_t$ | turbulent viscosity (kg/ms) |
| $\rho$ | density (kg/m$^3$) |
| $\Delta\rho$ | difference in density between temperature T and $T_o$ (kg/m$^3$) |
| $\sigma_\varepsilon$, $\sigma_k$, $\sigma_\phi$ | effective turbulent Schmidt/Prandtl numbers |
| $\Phi$ | mean temperature difference $T-T_o$ (K) |
| $\phi$ | fluctuating temperature difference (K) |
| $\varepsilon$ | rate of dissipation of turbulence kinetic energy |

Subscripts

| | |
|---|---|
| in | inlet value |
| i, j | coordinate directions |
| o | average temperature of room |

Superscript

| | |
|---|---|
| − | time mean value |

INTRODUCTION

The problem considered

The problem of forced ventilation of rooms gives rise to questions concerning the arrangement of the fresh air supply. It is important to get adequate mixing of the inlet air with the room air, to obtain a uniform temperature and fresh air distribution. At the same time the mixing should not be so intense that people present in the room feel draught. These requirements are often opposite in such a way that intense mixing gives rise to large velocities and velocity fluctuations. It is therefore of great importance to be able to predict the flow and mixing conditions before the instalation of the ventilation equipment. Predictions are often obtained by setting up the flow configuration in a model or a full scale room. Flow vizualasation and measurements are used to get a picture of the flow. Such investigations are often very expensive in terms of man power and experimental equipment and it can be difficult to alter geometry, and inlet conditions. A numerical computation model of the flow and mixing processes in the room will therefore make a helpful tool to predict ventilation of rooms. To alter the flow conditions in the computer model only means to change the boundary conditions. This paper reports on some of the results of the development of a numerical computation procedure for three dimensional turbulent flow.

## Previous work

Although computation procedures for flow have been available for some time, relatively few applications to ventilation problems have appeared. The works of Nielsen [1] and Holmberg et al. [2] are concerned with numerical predictions of two-dimensional isothermal ventilation. These methods cannot be extended to three-dimensional flows. The defect lies in the formulation of the equations. Both methods use the transformed variables, vorticity and stream function. However, to solve flow problems in three-dimensional rooms it is preferable to solve equations for the primitive variables, the three velocity components and pressure.

## Objectives of the present work

The purpose of the present paper is to report on applications of the Patankar and Spalding [3] algorithm in the form presented by Hjertager and Magnussen [4] to predict three-dimensional turbulent flow in a ventilated room. Both isothermal and buoyant flows are considered.

# MATHEMATICAL FORMULATION

## Equations for the hydrodynamics and temperature

The flow in ventilated rooms is turbulent and equations for mean velocities and mean temperature have to be solved for a three-dimensional space. The conservation equations for the hydrodynamics expressed in tensor notation read:

- momentum:

$$\frac{\partial}{\partial x_i}(\rho U_i U_j) = -\frac{\partial \bar{p}}{\partial x_i} + \frac{\partial}{\partial x_i}(-\rho \overline{u_i u_j}) + \Delta \rho g_i \qquad (1)$$

- continuity:

$$\frac{\partial}{\partial x_i}(\rho U_i) = 0 \qquad (2)$$

The equation for the temperature distribution $\Phi = T - T_o$ is:

$$\frac{\partial}{\partial x_i}(\rho U_i \Phi) = \frac{\partial}{\partial x_i}(-\overline{\rho u_i \phi}) \qquad (3)$$

In equations 1 and 3 the correlations $\overline{\rho u_i u_j}$ and $\overline{\rho u_i \phi}$ have to be modeled. The k-ε model of turbulence has been used and will be described in the next section. The buoyancy term in the vertical direction is expressed by:

$$\Delta \rho g = \frac{\rho}{T_o} \Phi g$$

## Turbulence model

The two-equation turbulence model of Launder and Spalding [5] has been applied to model the correlations in equations 1 and 3. The correlations have been expressed

$$-\rho \overline{u_i u_j} = \mu_t \left(\frac{\partial U_j}{\partial x_i} + \frac{\partial U_i}{\partial x_j}\right) - 2/3 \rho k \delta_{ij}$$

and

$$-\rho \overline{u_i \phi} = -\frac{\mu_t}{\sigma_\phi}\left(\frac{\partial \Phi}{\partial x_i}\right)$$

The partial differential equations for the kinetic energy of turbulence k, and its rate of dissipation ε read:

$$\frac{\partial}{\partial x_i}(\rho U_i k) = \frac{\partial}{\partial x_i}\left(\frac{\mu_t}{\sigma_k}\frac{\partial k}{\partial x_i}\right) + \mu_t \left(\frac{\partial U_i}{\partial x_j} + \frac{\partial U_j}{\partial x_i}\right)\frac{\partial U_j}{\partial x_i} - \rho \varepsilon \quad (5)$$

and

$$\frac{\partial}{\partial x_i}(\rho U_i \varepsilon) = \frac{\partial}{\partial x_i}\left(\frac{\mu_t}{\sigma_\varepsilon}\frac{\partial \varepsilon}{\partial x_i}\right) + C_1 \mu_t \, \varepsilon/k \left(\frac{\partial U_i}{\partial x_j} + \frac{\partial U_j}{\partial x_i}\right)\frac{\partial U_j}{\partial x_i}$$

$$- C_2 \rho \varepsilon^2 / k \quad (6)$$

where the turbulent viscosity $\mu_t$ is expressed as

$$\mu_t = \rho C_D k^2 / \varepsilon$$

The constants appearing above have been given the following values

$$\sigma_\phi = 0.7, \quad \sigma_k = 1.0, \quad \sigma_\varepsilon = 1.3,$$

$$C_D = 0.09, \quad C_1 = 1.44, \quad C_2 = 1.92$$

The wall function method of Launder and Spalding [5] has been used for the near wall regions. The problem thus consisted of the solution of the seven partial differential equations for a three-dimensional space with the appropriate boundary conditions.

Solution procedure

The partial differential equations were solved by means of the Patankar and Spalding algorithm [3]. The solution procedure have been discussed by Hjertager and Magnussen [4]. The main features of the method shall briefly be reviewed.

Each of the differential equations are cast into finite-difference form and integrated over a control volume. Displaced grids are used, which means that different control volumes are used for velocity components and the scalar quantities. Implicit formulation of the finite-difference equations insures that convergence is obtained in relatively few iterations compared with explicit formulation. The solution of equations for momentum, temperature, k and ε are succesively obtained by using the Tri-Diagonal Matrix Algorithm (TDMA) in each of the three coordinate directions for each of the equations. The coupling of the momentum equations with the continuity equation is taken into account by solving an equation for the pressure correction. This gives the proper correction to the pressure and velocity components to obey continuity. Because of nonlinearities and interlinkage of the equations iterations with under-relaxation must be performed until a converged solution is obtained.

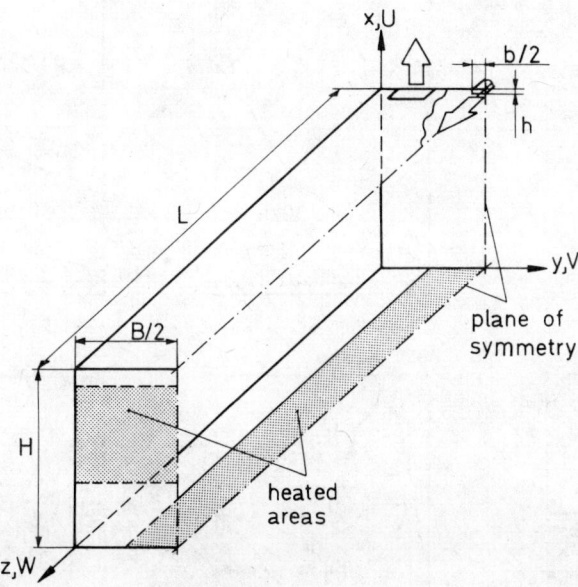

Fig. 1. Geometry of room and notation.

## RESULTS

### The geometry and boundary conditions

The procedure has been used to calculate the isothermal and buoyant flow in a full scale square boxed empty room shown in Fig. 1. The dimensions of the room are: height H = 2.4 m, width B = 2.9 m and lenght L = 5.6 m. The rectangular inlet nozzle is placed under the ceiling with dimensions: height h = 35 mm, width w = 243 mm. The figure also shows the regions of the floor and walls that were heated in the non-isothermal case. Because of symmetry about the plane y = B/2 only the flow in one half of the room has been considered. Nine grid points were used in each of the three coordinate directions with a concentration of points near the walls. The inlet conditions were approximated by uniform velocity and temperature profiles. The inlet velocity was $W_{in}$ = 2.42 m/s and the Reynolds number based on hydraulic diameter of inlet nozzle and inlet velocity was Re = 9800. The outlet velocity was given a value determined from overall continuity.

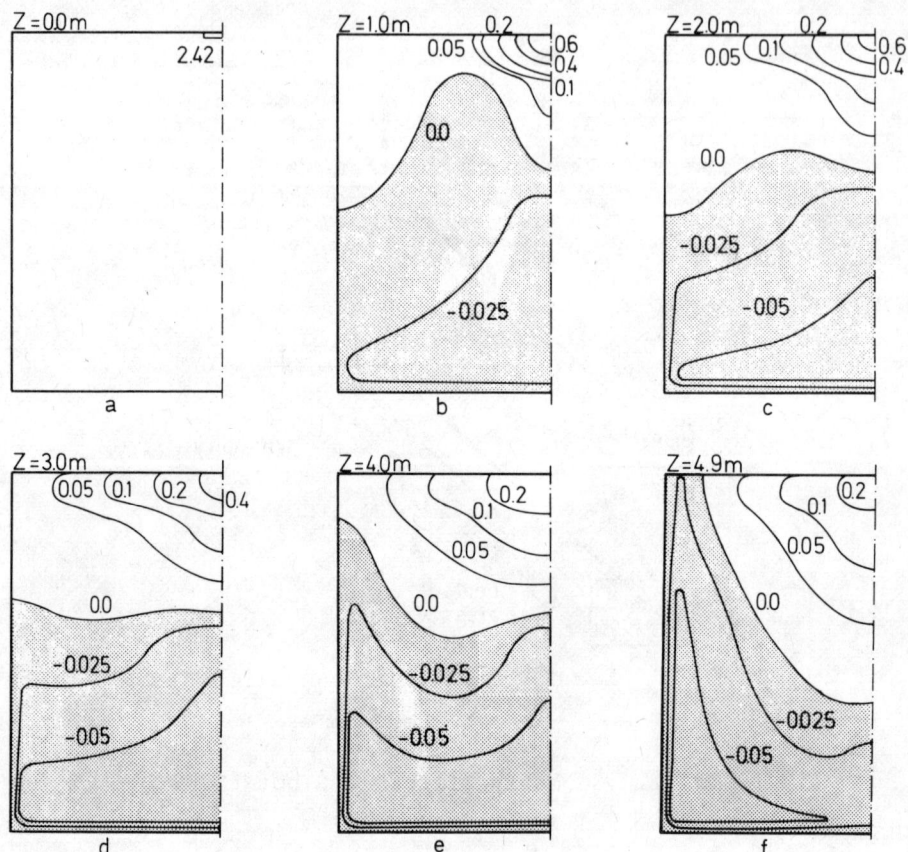

Fig. 2 a-f. Distribution of W-velocity in the xy planes. Numbers denote m/s. Isothermal flow.

Isothermal Flow

Figures 2 to 4 show the results for an isothermal flow calculation. The results are shown as succesive pictures of the flow in the xy-planes. Figure 4 shows comparison of the isovelocity contours of Hestad [6] and the velocity vectors in the symmetry plane. The agreement is good. The main discrepancy is in the recirculating region, which is the same trend as reported by Nielsen [1] and Holmberg et al. [2] for isothermal two-dimensional flows.

The measurements were performed with a thermocouple anemometer. This instrument measures the temperature difference between two spheres, one heated and one unheated. The instrument which was calibrated against known velocities has no directional sensitivity. Therefore the output is supposed to be a measure of the length of the velocity vector. Although the instrument has apparant disadvantages it seems to be fairly much used in ventilation flows in full scale rooms.

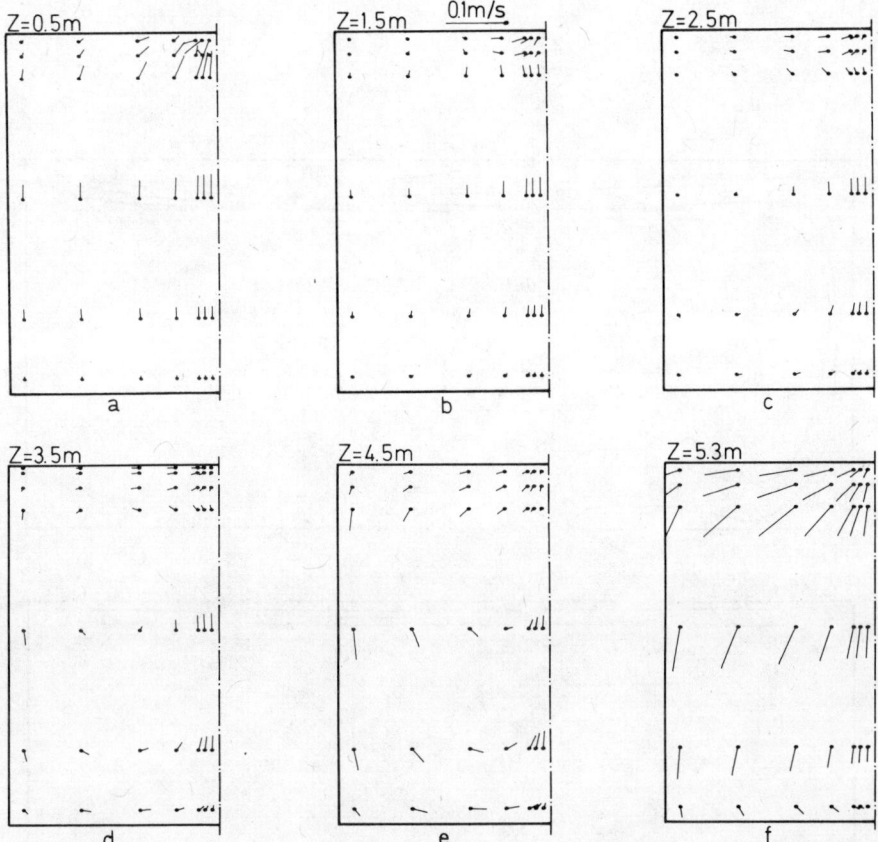

Fig. 3 a-f.  Distribution of velocity vectors in the xy planes. Isothermal flow.

Buoyant flow

Figures 5 to 8 show the results for a buoyant flow calculation. The inlet temperature was 11K lower than the average temperature of the room, $T_o$ = 293K. Heat was supplied at the heated areas shown in figure 1. The total heat was devided equally between the two surfaces. The temperatures of the unheated walls were at the average room temperature. Figure 7 shows a comparison between measurements of Hestad [6] and the predictions for the symmetry plane. The agreement is good along the ceiling and we can see that the point of separation has been well predicted. In the predictions the cold jet falls more rapidly towards the floor than the measurements seem to indicate.

The cause of this discrepancy is difficult to state, due to uncertainties in the experimental method as well as possible discrepancies between the numerical and experimental boundary conditions.

An interesting result in figure 5 is that recirculation in this

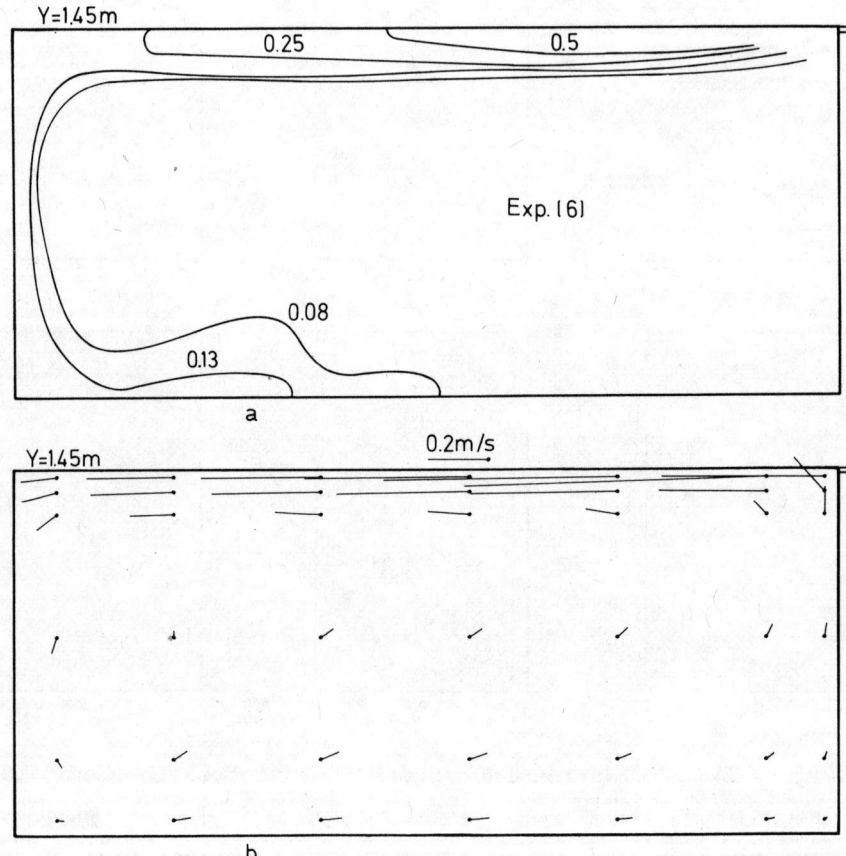

Fig. 4 a-b. Comparison of measured velocity contours of Hestad [6] a and predicted velocity vectors b in the symmetry plane. Numbers denote m/s.

# THREE-DIMENSIONAL TURBULENT BUOYANT FLOW

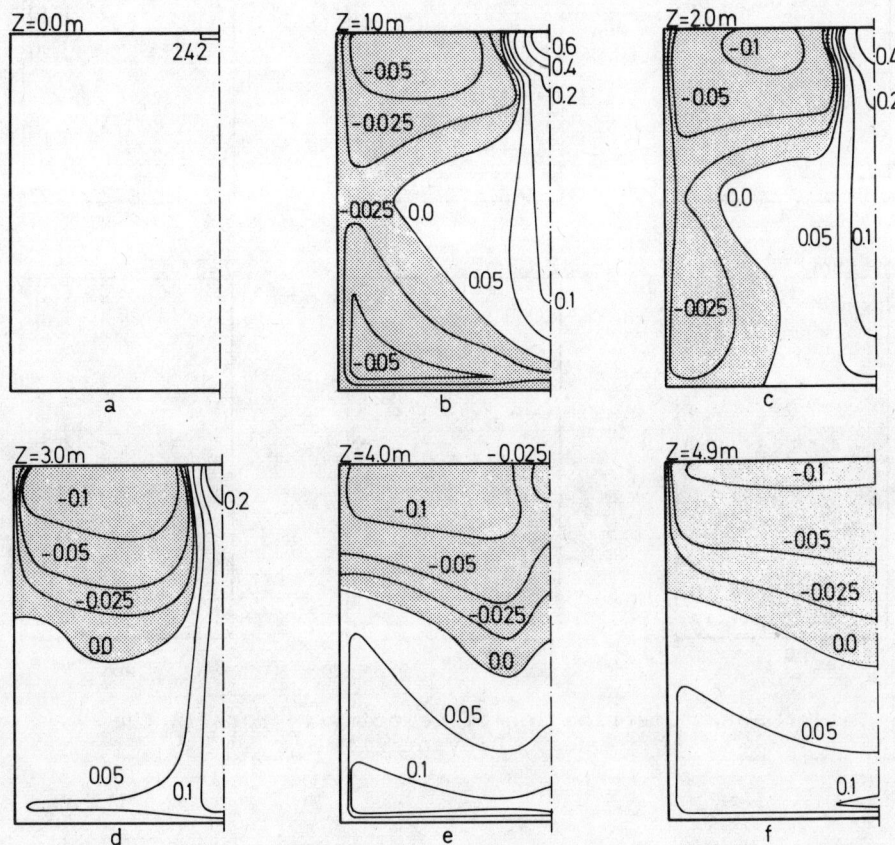

Fig. 5 a-f. Distribution of W-velocity in the xy planes. Buoyant flow. Numbers denote m/s.

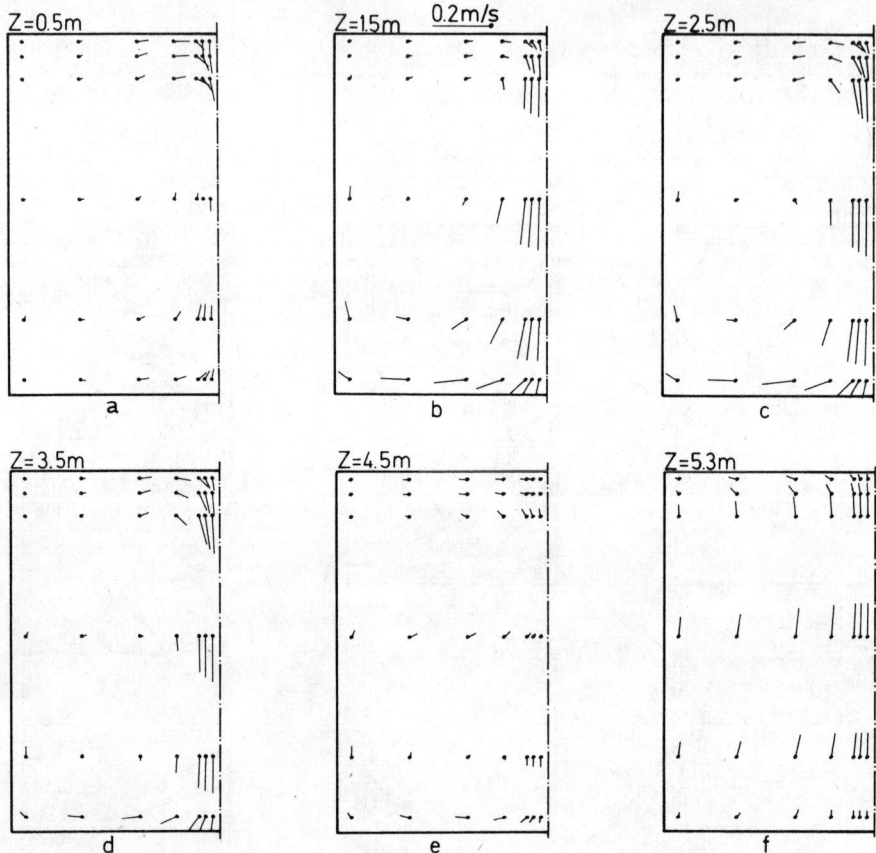

Fig. 6 a-f. Distribution of velocity vectors in the xy planes. Buoyant flow.

case has moved towards the ceiling. This is quite different from the isothermal flow of figure 2 where the recirculation was along the floor. Another feature to note is the circulation set up in the xy-planes (fig. 6).

DETAILS OF COMPUTATION

The calculations were performed on a U-1108 computer. Each iteration took 6.5 seconds for the isothermal flow and 7.5 seconds for the buoyant flow. To obtain converged solutions 120 iterations had to be performed for the isothermal flow and 350 for the buoyant flow. The computation of the buoyant flow was started with the

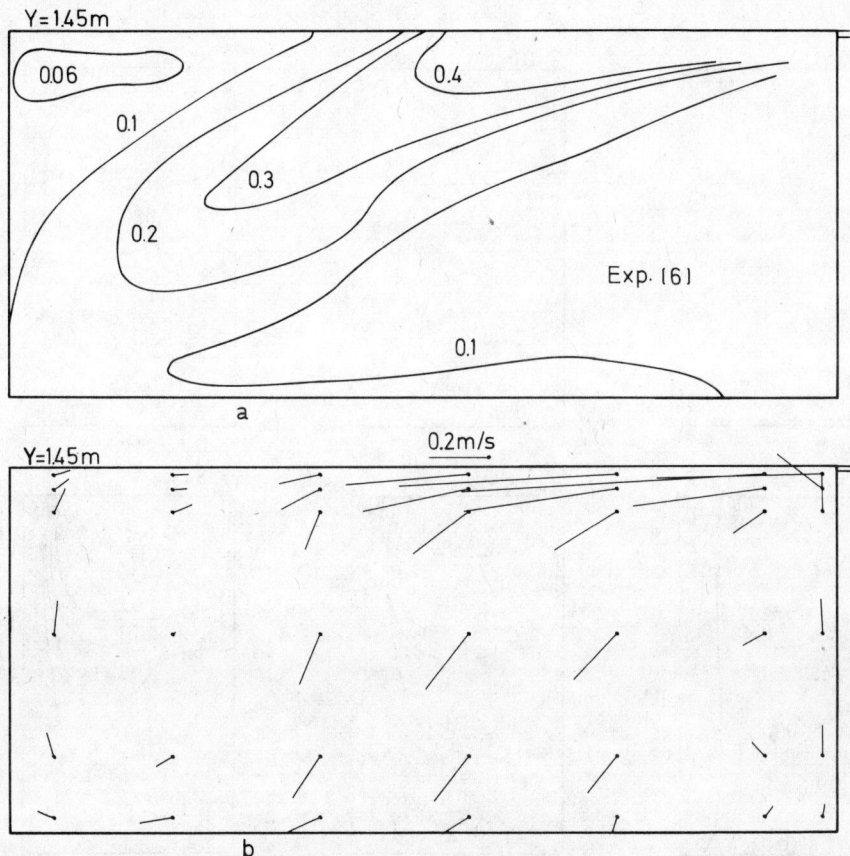

Fig. 7 a-b. Comparison of measured velocity contours of Hestad [6] a and predicted velocity vectors b in the symmetry plane of the buoyant flow. Numbers denote m/s.

solution of the isothermal flow as initial condition. Relaxation factors were gradually increased with numbers of iterations from 0.3 to 0.5 for the velocity components, k, and $\varepsilon$, while for the temperature it was increased from 0.05 to 0.1. For the pressure correction the relaxation factor was 1.0, and for the turbulent viscosity 0.5.

CONCLUSIONS

The predeeding sections have shown that numerical predictions of three-dimensional flow in ventilated rooms can be performed. Comparisons with experiments show that the agreement for isothermal

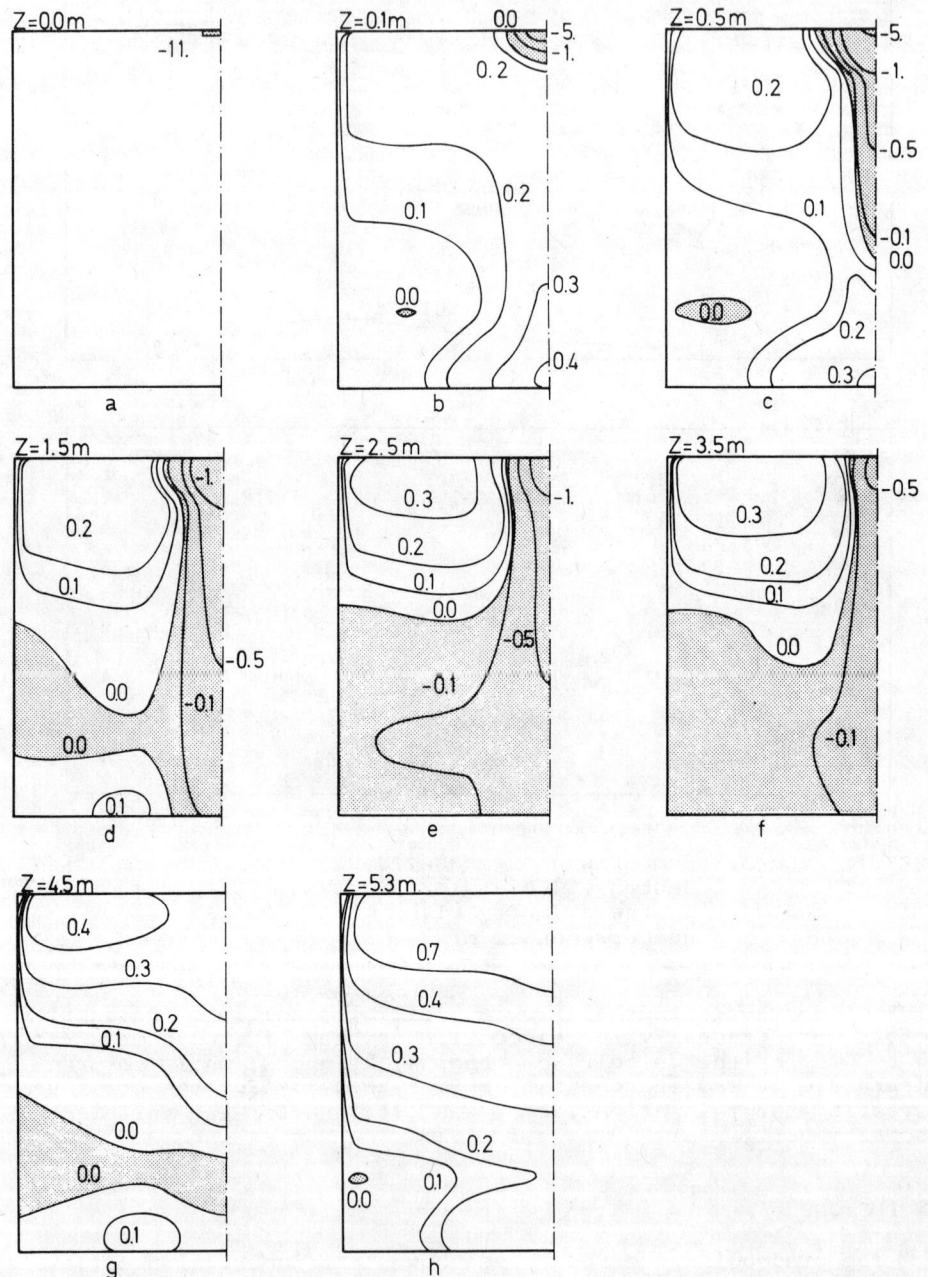

Fig. 8 a-h. Temperature distribution in the xy planes. Buoyant flow. Numbers denote temperature-difference $\Phi = T - T_o$ (K)

flow is good. Flow with buoyancy compares well with experiments in some regions and less in others. The experimental data are at present insufficient to indicate if special measures are needed to improve the mathematical model. More reliable measurements of local mean velocities and velocity fluctuations are therefore needed for the flow in ventilated rooms to further validate the computer model.

ACKNOWLEDGEMENT

This work has been supported by the Royal Norwegian Council for Scientific and Industrial Research.

REFERENCES

1. P. Nielsen (in Danish), 1973, Predictions of air distribution in a forced ventilated room, Ingeniørens Ugeblad No. 5.

2. R. Holmberg, M. Larsson and S.G. Sundkvist (in Swedish), 1975, Calculation of velocity distribution in a ventilated room, VVS, No. 5, pp. 59-66.

3. S.V. Patankar and D.B. Spalding, 1972, A calculation procedure for heat, mass and momentum transfer in three-dimensional parabolic flows, Int. Journal of Heat and Mass Transfer, $\underline{15}$, pp. 1787-1806.

4. B.H. Hjertager and B.F. Magnussen, 1976, Computation of some three-dimensional laminar incompressible internal flows, Proc. of The 1976 Heat Transfer and Fluid Mechanics Institute, Stanford University Press, in press, 15 pp.

5. B.E. Launder and D.B. Spalding, 1974, The numerical computation of turbulent flows, Computer Methods in Applied Mechanics and Engineering, $\underline{3}$, pp. 269-289.

6. T. Hestad, 1975, Private communication, Division of Heating and Ventilating, The Norwegian Institute of Technology.

# NUMERICAL STUDY OF PROBLEMS ON HIGH-INTENSIVE FREE CONVECTION

**B. M. BERKOVSKY and V. K. POLEVIKOV**

*Luikov Heat and Mass Transfer Institute, Minsk, USSR*

## ABSTRACT

The methods to build a difference scheme of the higher order of approximation are presented for multi-dimensional convection equations. This scheme is recommended to calculate the problems on high-rate convection. The scheme obtained was checked on the problem for free convection in the vertical finite layers. Operating parameters varied: $Pr \leq 10^5$, $Ra \leq 10^{10}$, $I \leq H/L \leq 10$. Specific features of development of thermoconvective processes at high Rayleigh numbers are discussed. Criterial relations $Nu = F(Pr, Ra, H/L)$ are built. A good agreement between the known numerical and experimental data is found.

## NOMENCLATURE

p     dimensionality of the problem

t     time

$x_1, x_2, \ldots, x_p$ , space coordinates

$x = (x_1, x_2 \ldots x_p)$ , space point

$h_1, h_2, \ldots, h_p$ , spatial difference-grid steps

$\tau$     time step

$x^{(\pm 1\alpha)} = (x_1, \ldots, x_{\alpha-1}, x_\alpha \pm h_\alpha, x_{\alpha+1}, \ldots, x_p)$, $\check{t} = t - \tau/2$,

$\hat{t} = t + \tau/2$, $y = y(x,t)$, $\check{y} = y(x,\check{t})$, $\hat{y} = y(x,\hat{t})$,

$y^{(\pm 1\alpha)} = y(x^{(\pm 1\alpha)}, t)$, $y_{\bar{x}_\alpha} = (y - y^{(-1\alpha)})/h_\alpha$

$y_{x_\alpha} = (y^{(+1\alpha)} - y)/h_\alpha$, $y^\pm = (y \pm |y|)/2$ , notations for the grid functions

H     layer height

L     layer thickness

$T, \upsilon, \psi, \varphi$     dimensionless temperature, velocity, stream function and vorticity

Pr     ,Prandtl number

Gr and Ra , Grashof and Rayleigh numbers based on layer thickness L

Nu     , Nusselt number

     In contrast to convective motions specific for low Rayleigh numbers, high-rate thermoconvective processes are characterized by boundary layer development at large velocity and temperature gradients. Due to this, utilization of numerical methods in such problems entails great difficulties which may be overcome either by developing computational technique or effective finite-difference algorithms.

     The works on numerical investigation of free convection at Rayleigh numbers of order of Ra $\sim 10^{10}$ are available [1-5]. The monotonic difference schemes of first [1,2], second [3] and third [4,5] orders of accuracy were used. The monotonic schemes (whose specific features are that the convective terms in the governing equations are approximated by asymmetrically directed differences oriented against a convective flow) proved to be most suitable to solve the problems on high-rate free convection, as in contrast to symmetrical approximations these have a stabilizating effect on computations and give stable algorithms at sufficiently large Rayleigh numbers. However, due to great approximation error the numerical results are in good agreement with the experimental ones only at low velocity and temperature gradients. The calculations show that when studying laminar processes at the onset of convective flow turbulization and especially turbulent processes, the use of difference algorithms both of the first and second orders is not successful since the internal memory capacity of electronic computers is often unable to provide a satisfactory accuracy of the results.

     The present paper is a continuation of works [4 - 6] on building up monotonic finite-difference schemes of higher accuracy for free convection equations. The scheme proposed is defined on the same grid pattern as used for the best algorithms of the first and second order. Thus, an increase in the approximation order has not resulted in decreasing the resolution of the method. This scheme may be successfully used to solve free convection problems involving Ra $\sim 10^{10} - 10^{12}$. Unlike the schemes earlier obtained by the authors, it possesses not only a higher order of approximation but more simple formulation too. The scheme is written down for a differential convection equation in a general form in an arbitrary orthogonal curvilinear coordinate system.

# I. HIGHER ACCURACY DIFFERENCE SCHEME

Introduce an orthogonal curvilinear coordinate system $x_1, x_2, \ldots, x_p$ in p-dimensional real space. Consider the convection equation in a general form within the confined domain

$$\frac{\partial u}{\partial t} = \sum_{\alpha=1}^{P} L_\alpha^{(k,\vartheta)} u + f(x,t), \qquad (I)$$

$$L_\alpha^{(k,\vartheta)} u = L_\alpha u - \vartheta_\alpha \frac{\partial u}{\partial x_\alpha}, \quad L_\alpha u = \frac{\partial}{\partial x_\alpha}\left(k_\alpha \frac{\partial u}{\partial x_\alpha}\right)$$

Here $k_\alpha = k_\alpha(x,t) > 0$; $\vartheta_\alpha = \vartheta_\alpha(x,t)$ is some linear function for velocity components defined by a choice of the coordinate system. Let coefficients $k_\alpha, \vartheta_\alpha, f$ together with the solution $u=u(x,t)$ be sufficiently smooth functions, that is characteristic of convection problems.

Construct a finite-difference scheme approximating Eq.(I) with an error $\Theta(\tau^2 + |h|^4)$ on a minimum two-level grid pattern. For this purpose consider a monotonic differential scheme with the second order of accuracy /7/

$$\Lambda_\alpha^{(a,\vartheta)} y = \mathscr{H}_\alpha (a_\alpha y_{\bar{x}_\alpha})_{x_\alpha} - \tilde{\vartheta}_\alpha^{-} a_\alpha^{(+1\alpha)} y_{x_\alpha} - \tilde{\vartheta}_\alpha^{+} a_\alpha y_{\bar{x}_\alpha}, \qquad (2)$$

where $\tilde{\vartheta}_\alpha = \vartheta_\alpha / k_\alpha$, $\mathscr{H}_\alpha = 1/(1 + r_\alpha + r_\alpha^2 + r_\alpha^3)$, $r_\alpha = h_\alpha |\tilde{\vartheta}_\alpha|/2$

Find coefficient $a_\alpha$ by Samarsky's formula /8/ for the schemes approximating the classical heat conduction equation with higher accuracy. This formula has the form

$$a_\alpha = 1 / \left[\frac{1}{6}\left(p_\alpha^{(-1\alpha)} + p_\alpha\right) + \frac{2}{3} p_\alpha^{(-0.5\alpha)}\right], \quad p_\alpha = 1/k_\alpha > 0$$

To write down the higher-accuracy schemes, use is made of the asymptotic expansions which give:

$$\Lambda_\alpha^{(a,\vartheta)} u = L_\alpha^{(k,\vartheta)} u + \frac{1}{12} h_\alpha^2 \left[L_\alpha(p_\alpha L_\alpha u) - 2\vartheta_\alpha \sqrt{p_\alpha} \frac{\partial}{\partial x_\alpha}(\sqrt{p_\alpha} L_\alpha u)\right] + \Theta(|h|^4) \qquad (3)$$

Since $u = u(x,t)$ is the solution to Eq.(I), then the differential operator $L_\alpha u$ may be expressed in an explicit form from (I) and substituted into (3). In this way the higher order derivatives unamenable to difference approximation on a minimum grid pattern may be omitted. Then, introduce into (3) the regularization parameters $R_\alpha = R_\alpha(x,t) > 0$, which allow essential improvement of the desired scheme stability without decreasing its approximation order. So, proceed to transformation:

$$\sum_{\alpha=1}^{p} h_\alpha^2 \left[ L_\alpha(p_\alpha L_\alpha u) - 2\vartheta_\alpha \sqrt{p_\alpha} \frac{\partial}{\partial x_\alpha}(\sqrt{p_\alpha} L_\alpha u) \right] = \sum_{\alpha=1}^{p} h_\alpha^2 \Big\{ L_\alpha \Big[ p_\alpha \Big( \frac{\partial u}{\partial t} - \sum_{\beta \neq \alpha} L_\beta^{(h,\vartheta)} u + \vartheta_\alpha \frac{\partial u}{\partial x_\alpha} - f \Big) \Big] - 2\vartheta_\alpha \sqrt{p_\alpha} \frac{\partial}{\partial x_\alpha}(\sqrt{p_\alpha} L_\alpha u) \Big\} =$$

$$= \sum_{\alpha=1}^{p} h_\alpha^2 \Big\{ L_\alpha \Big[ p_\alpha \Big( \frac{\partial u}{\partial t} - f \Big) \Big] + \frac{\partial}{\partial x_\alpha}\Big( h_\alpha \frac{\partial (p_\alpha \tilde{\vartheta}_\alpha)}{\partial x_\alpha} \Big) h_\alpha \frac{\partial u}{\partial x_\alpha} + \Big( \frac{1}{2} \vartheta_\alpha \frac{\partial p_\alpha}{\partial x_\alpha} +$$

$$+ 2 \frac{\partial \tilde{\vartheta}_\alpha}{\partial x_\alpha} \Big) L_\alpha u - L_\alpha \Big( p_\alpha \sum_{\beta \neq \alpha} L_\beta^{(h,\vartheta)} u \Big) - \vartheta_\alpha \sqrt{p_\alpha} \frac{\partial}{\partial x_\alpha}\Big[ \sqrt{p_\alpha} \Big( \frac{\partial u}{\partial t} - \sum_{\beta \neq \alpha} L_\beta^{(h,\vartheta)} u +$$

$$+ \vartheta_\alpha \frac{\partial u}{\partial x_\alpha} - f \Big) \Big] + (1+R_\alpha)\Big( \frac{\partial u}{\partial t} - \sum_{\beta=1}^{p} L_\beta^{(h,\vartheta)} u - f \Big) \Big\} = \sum_{\alpha=1}^{p} h_\alpha^2 \Big[ -k_\alpha^* L_\alpha^{(h,\vartheta)} u -$$

$$- k_\alpha' L_\alpha^{(h,\vartheta')} u + L_\alpha^{(h,\vartheta)}\Big( p_\alpha \frac{\partial u}{\partial t} \Big) + (1+R_\alpha)\frac{\partial u}{\partial t} - L_\alpha^{(h,\vartheta)}\Big( p_\alpha \sum_{\beta \neq \alpha} L_\beta^{(h,\vartheta)} u \Big) - f_\alpha^* \Big]$$

Here the coefficients are given in the form:

$$k_\alpha^* = \mu_\alpha + \sum_{\beta=1}^{p} R_\beta h_\beta^2 / h_\alpha^2, \quad k_\alpha' = \sum_{\beta=1}^{p} h_\beta^2 / h_\alpha^2 > 0,$$

$$\vartheta_\alpha' = \left[ k_\alpha L_\alpha^{(h,\vartheta)}(p_\alpha \tilde{\vartheta}_\alpha) + \vartheta_\alpha(k_\alpha' - \mu_\alpha) \right] / k_\alpha'$$

$$\mu_\alpha = \tilde{\vartheta}_\alpha^2 - 2\frac{\partial \tilde{\vartheta}_\alpha}{\partial x_\alpha} - \vartheta_\alpha \frac{\partial p_\alpha}{\partial x_\alpha}, \quad f_\alpha^* = L_\alpha^{(h,\vartheta)}(p_\alpha f) + (1+R_\alpha)f$$

Hence it follows that the scheme

$$\lambda y_t = \sum_{\alpha=1}^{p} \Big\{ \Lambda_\alpha^{(a,\vartheta)} \big[ \sigma_\alpha \hat{y} + (1-\sigma_\alpha)\check{y} \big] + B_\alpha \hat{y} \Big\} + F[\check{y}], \quad (4)$$

$$\lambda = 1 + \frac{1}{12}\sum_{\alpha=1}^{p} h_\alpha^2 (1+R_\alpha), \quad \sigma_\alpha = \frac{1}{2} - \frac{h_\alpha^2}{12\tau} p_\alpha.$$

$$B_\alpha \hat{y} = \frac{1}{12} h_\alpha^2 \big( k_\alpha^* \Lambda_\alpha^{(a,\vartheta)} \hat{y} + k_\alpha' \Lambda_\alpha^{(a,\vartheta')} \hat{y} \big),$$

$$F[\check{y}] = \frac{1}{12}\sum_{\alpha=1}^{p} h_\alpha^2 \Big[ \Lambda_\alpha^{(a,\vartheta)}\Big( p_\alpha \sum_{\beta \neq \alpha} \Lambda_\beta^{(a,\vartheta)} \check{y} \Big) + f_\alpha^* \Big] + f$$

has the second order of approximation over $\tau$ and the fourth, over $/h/$.

It is not difficult to notice that (4) is non-economic. Reducing factored scheme (4) to alternating direction algorithms yields an economic scheme

$$\begin{aligned}
(y_{(1)} - \check{y})/(\tau/\lambda) &= (\Lambda_1^{(a,\nu)} \sigma_1 + B_1) y_{(1)} + \Phi[\check{y}], \\
(y_{(2)} - y_{(1)})/(\tau/\lambda) &= (\Lambda_2^{(a,\nu)} \sigma_2 + B_2) y_{(2)}, \\
&\cdots \\
(y_{(p)} - y_{(p-1)})/(\tau/\lambda) &= (\Lambda_p^{(a,\nu)} \sigma_p + B_p) y_{(p)}, \quad y_{(p)} = \hat{y}
\end{aligned} \qquad (5)$$

$$\Phi[\check{y}] = F[\check{y}] + \sum_{\alpha=1}^{p} \Lambda_\alpha^{(a,\nu)} \left[ (1-\sigma_\alpha)\check{y} + \frac{\tau}{\lambda} \sigma_\alpha \sum_{\beta > \alpha} \Lambda_\beta^{(a,\nu)} \sigma_\beta \check{y} \right]$$

The boundary conditions only for $\check{y}$ and $\hat{y}$ are prescribed to solve boundary-value problems. The boundary values of $y_{(1)}$, $y_{(2)}$, ..., $y_{(p-1)}$ are found according to /9/:

$$y_{(\alpha)} = A_p A_{p-1} \ldots A_{\alpha+1} \hat{y}; \quad \alpha = 1, 2, \ldots p-1 \qquad (6)$$

where $A_\alpha = E - \frac{\tau}{\lambda}(\Lambda_\alpha^{(a,\nu)} \sigma_\alpha + B_\alpha)$.

Factored scheme (5) with boundary relations (6) ensures overall approximation of initial differential equation (I) with an error of $\Theta(\tau^2 + |h|^4)$. Based on the general stability theory /10/, the preliminary studies have shown that at $k_\alpha = 1$ higher accuracy scheme (5)-(6) is stable within the energy norm if

$$p \leq 7 - \sum_{\beta \neq \alpha} \frac{h_\alpha^2}{h_\beta^2}, \quad R_\alpha \geq |\mu_\alpha| + \frac{1}{8} h_\alpha^2 \, \mathscr{x}_\alpha \, \mathsf{v}_\alpha^4 + \qquad (7)$$

$$+ \frac{1}{8} \sum_{\beta \neq \alpha} h_\beta^2 \left[ \mathscr{x}_\alpha^{(+1\beta)} (\mathsf{v}_\alpha^{(+1\beta)})^4 + \mathscr{x}_\alpha^{(-1\beta)} (\mathsf{v}_\alpha^{(-1\beta)})^4 \right], \quad \alpha = 1,2,\ldots,p$$

Computations are performed as follows. First, $\Phi[\check{y}]$ is calculated by a known value of $\check{y}$. Then, Eqs (5) are solved in succession with boundary conditions (6) by one-dimensional formulae of elimination. Here, due to a monotonic character of operators $\Lambda_\alpha^{(a,\nu)}$ the elimination algorithm is stable with any $\tau$, /h/.

## 2. STATEMENT OF FREE CONVECTION PROBLEM

Consideration is made of two-dimensional free convective viscous incompressible liquid flow in a bounded vertical layer described by the known system of the differential dimensionless equations /3 - 5/:

$$\frac{\partial T}{\partial t} + v_1 \frac{\partial T}{\partial x_1} + v_2 \frac{\partial T}{\partial x_2} = \frac{1}{\Pr} \left( \frac{\partial^2 T}{\partial x_1^2} + \frac{\partial^2 T}{\partial x_2^2} \right) \qquad (8)$$

$$\frac{\partial^2 \psi}{\partial x_1^2} + \frac{\partial^2 \psi}{\partial x_2^2} + \varphi = 0, \quad v_1 = \frac{\partial \psi}{\partial x_2}, \quad v_2 = -\frac{\partial \psi}{\partial x_1} \qquad (9)$$

$$\frac{\partial \varphi}{\partial t} + v_1 \frac{\partial \varphi}{\partial x_1} + v_1 \frac{\partial \varphi}{\partial x_2} = \frac{\partial^2 \varphi}{\partial x_1^2} + \frac{\partial^2 \varphi}{\partial x_2^2} + Gr \frac{\partial T}{\partial x_1} \qquad (10)$$

At the region boundary $\Gamma$ instability, impermeability and attachment are given

$$v_1|_\Gamma = v_2|_\Gamma = \psi|_\Gamma = 0 \qquad (11)$$

Uniform temperature distribution is assumed to be at the vertical boundaries with horizontal walls thermally insulated:

$$T(0, x_2, t) = 1, \quad T(1, x_2, t) = 0, \qquad (12)$$

$$\frac{\partial T(x_1, 0, t)}{\partial x_2} = \frac{\partial T(x_1, H/L, t)}{\partial x_2} = 0$$

The operating parameters are chosen within $10^{-3} \leq Pr \leq 10^5$, $Ra \leq 10^{10}$, $1 \leq H/L \leq 10$. The range of high Rayleigh numbers $10^5 \leq Ra \leq 10^{10}$ is investigated in more detail.

The problem formulated is one of the most known problems on free convection. Recently, a great number of works reflecting various aspects of thermoconvective processes in vertical liquid layers have been published. The available extensive information allows the above problem to be used to verify different theoretical models and methods. It should be noted that only except the case $H/L = 1$ /3-5/ all investigations of high-rate convection have been carried out experimentally while analytical and numerical methods have been applied at low and moderate Rayleigh numbers ($Ra \leq 10^5$).

Equations (8) - (10) are approximated by scheme (5) - (7). For the vorticity and the thermally insulated wall temperatures the boundary values are approximately calculated $\Theta(\tau^2 + |h|^4)$ by the formulae built with regard for differential equations (8) - (12) by the technique described in the previous section. The first-order derivatives, containing in the coefficients, and also the absolute terms are approximated in a similar way. The steady solutions to the difference problems result from steading at $t \to \infty$. External iteration procedure has been performed on every time row. The basic calculations within $Ra \gtrsim 10^6$ are carried out on a nonuniform grid, whose mesh sizes $h_1$ and $h_2$ have varied from 1/100 to 1/10, depending on the temperature and velocity gradient distribution in a layer.

## 3. DISCUSSION

As a result of the calculations made it has become clear that uniform temperature gradient distribution in an enclosure characteristic of low-intensive convection undergoes disturbance with $Ra \sim 10^4$. With increase in Ra at the vertical boundaries one can observe boundary layer development with velocity and temperature gradients prevailing (Figs I, 2). The isotherms in the central regions take the form of horizontal straight lines with curved end sections.

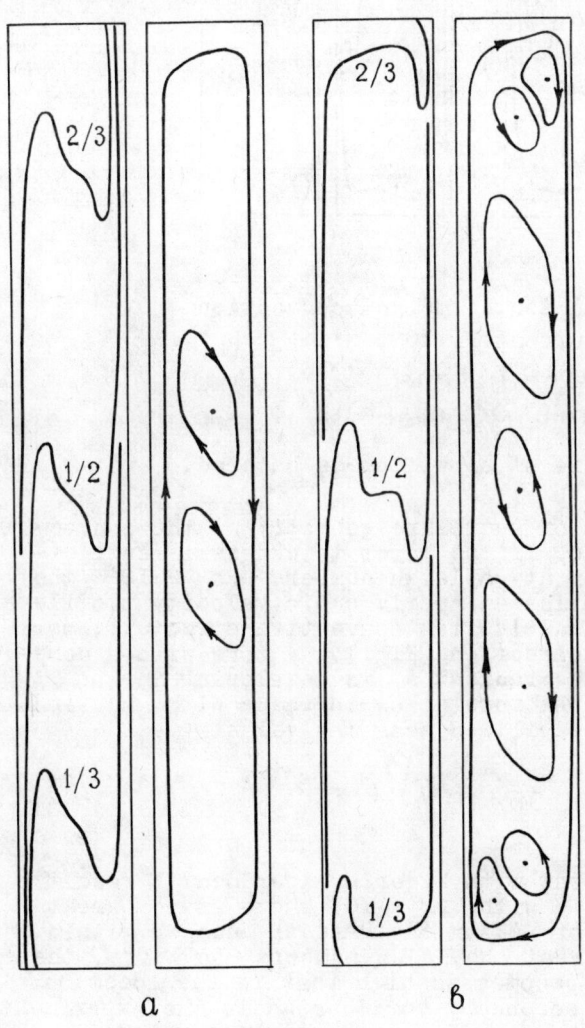

Fig.I. Typical isotherms and convection structure at $Pr = 10^3$, $H/L = 8$.
a - $Ra = 5 \cdot 10^5$, $|\Psi|_{max} \approx 0.1$;
б - $Ra = 5 \cdot 10^6$, $|\Psi|_{max} \approx 0.6$

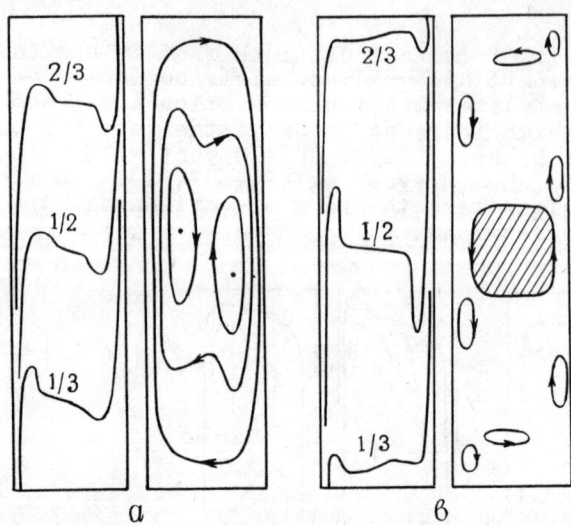

Fig.2. Typical isotherms and convection structure at $Pr=10^3$, $H/L=4$.
$a - Ra = 10^6$, $|\Psi|_{max} \approx 0.05$;
$6 - Ra = 5 \cdot 10^7$, $|\Psi|_{max} \approx 0.5$.

The core of slow flow is hatched.

An essential effect of enclosure geometry on the convection structure is revealed. If the layer thickness is sufficiently small ($H/L \gtrsim 7$), then at Rayleigh numbers $Ra \sim 2 \cdot 10^5$ the primary plate-parallel motion at nearly cubic velocity profile is followed by secondary multivortex convective structure, periodic in the longitudinal direction (Fig.I). A zone with a constant vertical temperature gradient A is developing in the middle of the enclosure. The zone sizes increase with Ra, and the value of $A \cdot H/L$ is stabilized near the value

$$A H / L = 0.55 \pm 10\%. \qquad (13)$$

which is in a good agreement with Elder's experimental result /II/ $A H/L \approx 0.5$, evaluated with $Pr = 10^3$. There is no great dependence of the value of A on the Prandtl number within $H/L \gtrsim 7$. At sufficiently great Rayleigh numbers $Ra \sim 10^6$, the intensity of liquid flow becomes so high that in the location of the shear layers between secondary vortices as in the experiment of /II/, the so-called tertiary flows appear with a direction of circulation opposite to the main flow. Otherwise pattern is observed at $H/L \lesssim 7$ (Fig.2). At Rayleigh numbers $5 \cdot 10^4 \lesssim Ra \lesssim 5 \cdot 10^5$ discrete structures shift takes place, while unlike $H/L \gtrsim 7$, first the primary vortex breaks down into two secondary transverse ones which with Ra increase shift to the side walls releasing the central part of the enclosure where the flow core is forming with velocities being negligible, as compared to the wall ones. At $Ra \sim 5 \cdot 10^6$ each of the secondary vortices breaks down into some longitudinal ones which locate along the side walls. At $Ra \sim 10^7$ tertiary flow formation takes place here as well as in narrow layers. The numerical results on the flow core structure at $H/L \leqslant 7$ under high-rate convection confirm the author's conclusions made for a square enclosure with heat conducting horizontal boundaries /3 - 5/. With Gr increasing, the value of a vertical temperature gradient in a core is found about some quantity which depends on the Prandtl number and is within the (0,I) interval. When Pr increases, the value of this gradient asymptotically falls to zero, and the core becomes isothermal.

In calculations much attention was paid to determining the boundary of laminar convective flow instability. Within the Pr range considered and at $2 \leqslant H/L \leqslant 10$ fully steady-state solutions at the Rayleigh numbers are obtained:

$$Ra \leqslant Ra_c = 6 \cdot 10^8 \frac{\sqrt{Pr}}{(H/L)^3} \pm 30\% \qquad (I4)$$

Within the range of $I \leqslant H/L \leqslant 2$ the critical Rayleigh numbers appeared to be a very weak function of geometrical dimensions, and laminar convection occurred at $Ra \leqslant 10^8 \sqrt{Pr}$. At $Ra > Ra_c$ the numerical calculations give no steady-state solutions. Steady-state oscillations are established, in this case the stream and temperature functions as well as all parameters (temperature gradient in the core, the Nusselt number, etc) oscillate about some mean values, the oscillation amplitude and frequency being increased with Ra. It should be noted that formula (I4) gives somewhat underestimated ($\sim 30$ per cent) values of the critical Rayleigh number as compared to Elder's experimental relation /I2/. This disagreement appears to be attributed to the higher sensitivity of the numerical method of solution to the continuous oscillation onset.

Figure 3 gives the curves for the effect of the Rayleigh number on convective heat transfer at different Prandtl numbers. As we can see calculation results are in sufficient agreement with the experimental formulae of different investigators /I3 - I5/. (Some discrepancy with data of /I5/ is probably due to the fact that in the above work the Prandtl number relation is established based on the horizontal layer.) Based on a great amount of numerical information processed a dimensionless equation is built up in the form:

$$Nu = 0.22 (H/L)^{-0.25} \left( \frac{Pr}{0.2 + Pr} Ra \right)^{0.28} \qquad (I5)$$

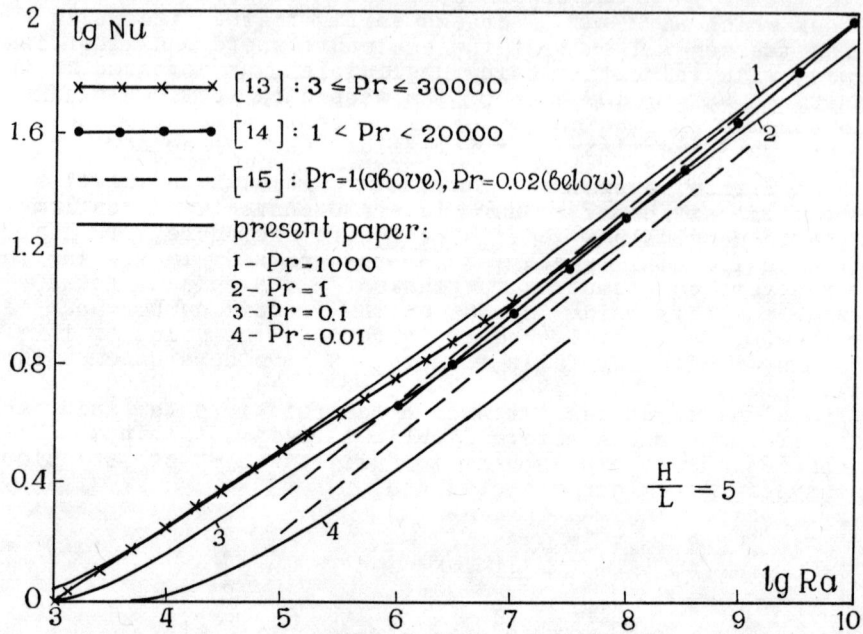

Fig.3. Effect of Rayleigh and Prandtl numbers on convective heat transfer in a vertical layer. H/L=5.

Within $2 \leq H/L \leq 10$, $10^{-3} \leq Pr \leq 10^5$, $10^3 \leq Ra \cdot Pr/(0.2+Pr)$ Eq.(15) approximates numerical relations obtained, with an accuracy of 10 %. At $1 \leq H/L \leq 2$ the enclosure geometry effect on heat transfer appeared to be negligibly small, so the Nusselt number is determined by

$$Nu = 0.18 \left( \frac{Pr}{0.2 + Pr} Ra \right)^{0.29} \qquad (16)$$

at $\quad 10^{-3} \leq Pr \leq 10^5, \quad 10^3 \leq Ra \cdot Pr/(0.2+Pr)$

APPENDIX

To elucidate the approximation and conservative properties of scheme (5) - (7), the latter was compared with two known monotonic conservative algorithms of the first and second orders. The problem for free convection in a square enclosure involving sine heating from above /4,5/ has been chosen as a test problem. In contrast to the heating from a side (12), specific for central symmetry of steady solutions, the above problem steady-state

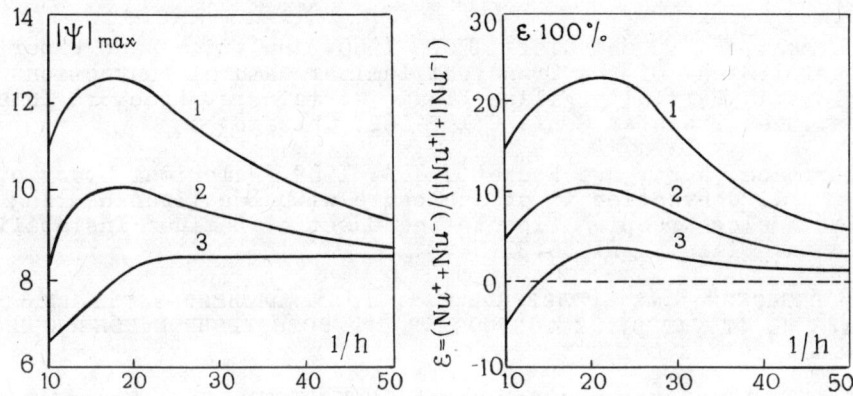

Fig.4. Comparison of difference schemes. Pr=I, Ra=$5 \cdot 10^6$.
I - first-order scheme [I6]; 2 - second-order scheme [3]; 3 - higher accuracy scheme.
$Nu^+$, total heat flux directed inside the enclosure;
$Nu^-$, heat flux from the enclosure.

solutions are symmetric relative to the vertical axial line. This leads to the unbalance of calculated heat fluxes through the region boundary. This unbalance is eliminated either by refining the grid or by increasing the order of approximation of the algorithm applied. The accuracy of the numerical results may be roughly estimated from the discrepancy value, since it is known beforehand from the problem formulation that the amount of heat supplied to the enclosure is equal to that withdrawn from it.

The calculations have shown that at Ra≤$10^5$ the results obtained on uniform grids at $h=h_1=h_2$ ≤ I/20 are essentially independent of the type of the realized scheme -- the discrepancy was no more than I5%. It is unreasonable to apply scheme (5)-(7) at such Ra, as due to the complicated formulation it takes too much computation time, not giving better accuracy.

However, as is seen from Fig.4, already at Ra~$5 \cdot 10^6$ maximum scatter in the results obtained on the grids 2I×2I and 4I×4I in size about 5% for scheme (5)-(7) and 20-30% for the schemes of the first and second orders. Using the second-order approximation, the solution found on the 2I×2I grid by scheme (5)-(7) is no less accurate than that on the 4I×4I grid. At high Rayleigh numbers (Ra≳$10^7$) high quality numerical data may be probably obtained only using the higher accuracy schemes, since successful realization of the difference methods of the first and second orders requires unreal capacities of the computer memory.

REFERENCES

1. Barakat, H.Z., and Clark, J.A. 1966. Analytical and Experimental Study of the Transient Laminar Natural Convection Flows in Partially Filled Liquid Containers. Proc.3rd Internat.Heat Transfer Conf. 2:152-162. Chicago.

2. Torrance, K.E., and Rockett, J.A. 1969. Numerical Study of Natural Convection in an Enclosure with Localized Heating from Below Creeping Flow to the Onset of Laminar Instability. Journ. Fluid Mech. 36:33-54.

3. Берковский Б.М., Полевиков В.К. 1973. Влияние числа Прандтля на структуру и теплообмен при естественной конвекции. ИФЖ . 24:842-849.

4. Berkovsky, B.M., and Polevikov, V.K. 1974. Heat Transfer at High-Rate Free Convection. Proc. 5th Internat.Heat Transfer Conf. 3:85-89. Tokyo.

5. Берковский Б.М., Полевиков В.К. 1975. Теплообмен в условиях высокоинтенсивной свободной конвекции. В сб. "Теплообмен, 1974. Советские Исследования" : 169-175, Москва, Наука.

6. Полевиков В.К. 1974. Схема повышенного порядка точности для задач высокоинтенсивного тепломассообмена. В сб. "Современные проблемы тепловой гравитационной конвекции" : 84-89. Минск, ИТМО АН БССР.

7. Самарский А.А. 1971. Введение в теорию разностных схем. Москва, Наука.

8. Самарский А.А. 1963. Схемы повышенного порядка точности для многомерного уравнения теплопроводности. ЖВМ и МФЗ : 812-840.

9. Дьяконов Е.Г. 1962. Разностные схемы с расщепляющимся оператором для многомерных нестационарных задач. ЖВМ и МФ2 : 549-568.

10. Самарский А.А., Гулин А.В. 1973. Устойчивость разностных схем. Москва, Наука.

11. Elder, J.W. 1965. Laminar Free Convection in a Vertical Slot. Journ. Fluid Mech. 23:77- 98.

12. Elder, J.W. 1965. Turbulent Free Convection in a Vertical Slot. Journ. Fluid Mech. 23:99-111.

13. Emery, A.F., and Chu, N.C. 1965. Heat Transfer across Vertical Layers. Journ. Heat Transfer, Trans. ASME 87:110-116.

14. Mac Gregor, R.K., and Emery, A.F. 1969. Free Convection through Vertical Plane Layers- Moderate and High Prandtl Number Fluids. Journ. Heat Transfer, Trans. ASME. 91:391-403.

15. Dropkin, D., and Somerscales, E. 1965. Heat Transfer by Natural Convection in Liquids Confined by Two Parallel Plates which are Inclined at Various Angles with Respect to the Horizontal . Journ. Heat Transfer, Trans. ASME 87: 79-84.

16. Gosman, A.D., Pun, W.M., Runchal, A.K., Spalding, D.B., and Wolfshtein, M. 1969. Heat and Mass Transfer in Recirculating flows. London and New York, Academic Press.

# BOUYANCY DRIVEN COUNTERCURRENT FLOWS GENERATED BY A FIRE SOURCE

**B. J. McCAFFREY and J. G. QUINTIERE**
*Center for Fire Research, National Bureau of Standards, U.S.A.*

ABSTRACT

The velocity and temperature fields were determined for fire induced flows in corridors. The effects of scale, fire size, and doorway openings are presented. Detailed measurements illustrate the complex recirculating three dimensional character of the flow field. Mass flow rates are determined for the doorway openings and to determine the extent of entrainment. A critical Richardson number criterion was used to identify the mixed and stably stratified regions of the flow field.

NOMENCLATURE

g      acceleration of gravity

H,z      height

l      corridor length

$\dot{m}$      mass flow rate

Q      energy release rate of fire

Re      Local Reynolds number, $\dfrac{\rho V H_c}{\mu}$

Ri      Richardson number, defined by equation (4)

T      temperature

$\Delta T$      temperature rise above ambient

V      velocity

W      width

$\rho$      density

$\mu$      viscosity

Subscripts

C     corridor or critical

D     room doorway

E     corridor exit doorway

jet    floor or ceiling jet

INTRODUCTION

A fire in a building compartment is a source of buoyancy which sets in motion the surrounding flow field. The fire induced gas flows in a corridor or room which connects a fire compartment to the ambient atmosphere is the subject of this study. The developed buoyancy and entrainment of the fire tends to draw in cool air along the bottom of the corridor and to force out hot combustion products along the ceiling. The character of the resulting flow field is important for several reasons. It will influence the spread of fire in the corridor by its effect on heat transfer and the distribution of air and combustion products. It will influence firefighters in attack and rescue operations by its effect on the height of the interface between the hot products and cold air. Additionally, the knowledge of the movement of smoke and toxic products can aid designers in establishing criteria necessary for smoke control and extraction systems.

In an earlier study [1], fire induced flows were examined in a scale model facility. Both visual observations using smoke tracer techniques and vertical velocity traverses demonstrated that the gases move in a complex recirculating manner exhibiting up to four distinct horizontal layers for any realistic corridor exit configuration. The purpose of the present work was to extend these measurements and to investigate the effect of additional parameters on the flow field. The factors that were considered included the effect of scale and the effect of internal obstructions on the flow phenomena. Also, reference 1 demonstrated that mass balances based on a single vertical velocity traverse could not be achieved. This suggested strong three dimensional effects. An understanding of the nature of this three dimensional flow was another objective of the present study. Finally, the nature and extent of mixing and turbulence were examined for these corridor flows.

EXPERIMENTAL APPARATUS AND PROCEDURE

A full-scale corridor burn room facility and a 1/7 scale model are shown in figure 1. Note the two configurations are not identical in every detail but for purposes of obtaining the gross features of the flow in the corridor they are sufficiently similar. The model is the previously described facility used in reference 1 with the 90° off-axis burn room blocked and replaced by a porous plate, diffusion-flame floor burner. A 0.35 m x 0.35 m room is shown aligned with the corridor.

After the burner was ignited a sufficient time was allowed for the temperature of the corridor walls and ceiling to come to steady state. After this time, temperature and velocity (T-V) traverses were made at various locations. The corridor velocities were determined using a bidirectional impact device [2] and a sensitive electric manometer, and temperatures were determined using chromel-alumel thermocouples. Additional probing in the cooler regimes was accomplished using a hot wire anemometer. A pitot-static probe was also used in measuring doorway velocities.

Throughout this work modifications were made in the nature and size of the room and corridor doorways shown for the model in figure 1. In all cases the height of the soffit (0.057 m) was kept constant and the width of the corridor doorway ($W_E$) was varied. The room doorway ($W_D$) was fixed when a doorway configuration was used, but the nature of this internal obstruction was varied. The fire size in both the full-scale and model systems was controlled by metering the gas flow to the burners. The energy release rate (Q) was varied to consider its effect on the flow, or to maintain a fixed temperature level in the room depending on the experiment.

RESULTS AND DISCUSSION

1. Effect of Scale

In reference 1, flow studies in the scale model revealed a multi-layer countercurrent flow structure in the corridor. The effect of corridor exit geometry on flow was dramatic. For a wide open exit (no soffit, no door frame) the combustion gases and ambient air exit and enter respectively through the corridor in a simple hot-cold, two layer fashion. By placing a soffit at the exit, a four layer flow pattern results. The size and strength of the intermediate, reversing layers increase as the ventilation is further decreased by adding a standard door frame. The associated temperature traces indicate a significant lowering of the cold-hot interface implying that hot gases return toward the burn area.

In order to determine if the phenomenon was scale related, similar experiments were conducted in the full-scale corridor shown in figure 1. Figure 2 is a plot of midwidth temperature and velocity at 1.8 m from the end of the full-scale corridor (0.8ℓ from the fire room) for three exit configurations and a fire of 210 kW (12,000 Btu/min). These results confirm that the general nature of the flow is not dependent on scale. Moreover, it can be shown that the velocity and temperature fields are in fair quantitative agreement between the model system and full-scale. This can be arrived at by assuming geometric similarity and assuming dynamic similarity holds if the Froude number is maintained constant. The Froude number criterion leads to equal corresponding temperatures between the model and full-scale systems. The relationship

$$\left(\frac{V}{\sqrt{H}}\right)_{F.S.} = \left(\frac{V}{\sqrt{H}}\right)_{MODEL}$$

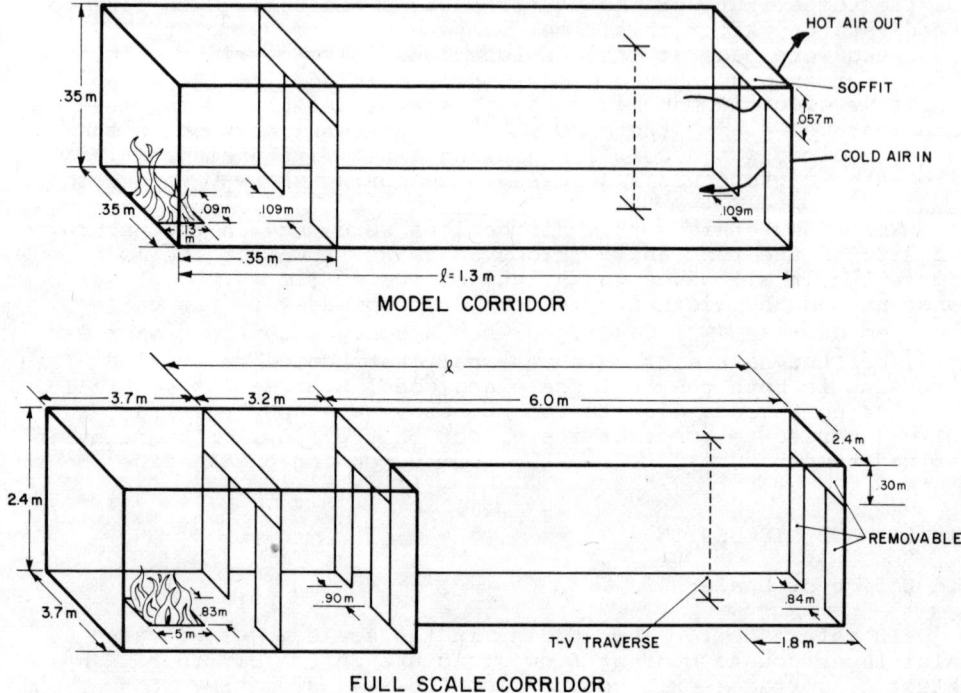

Figure 1. Experimental room and corridor configurations.

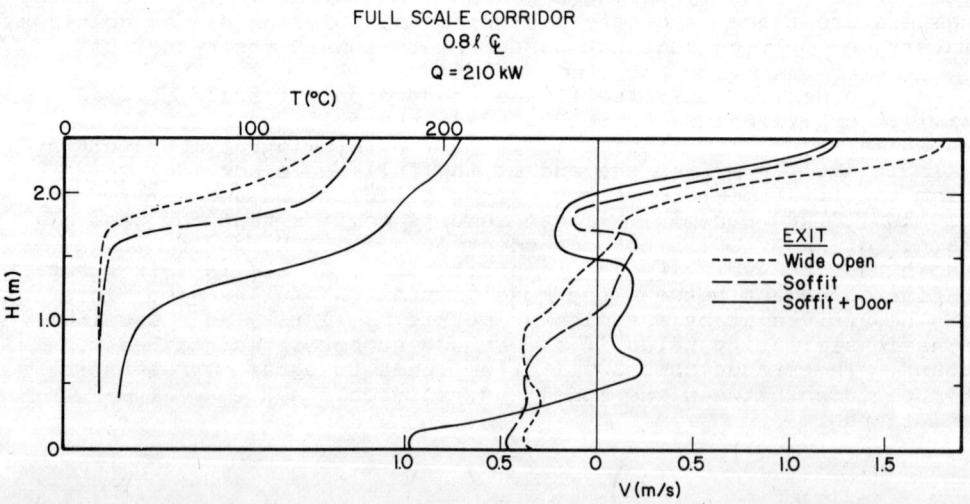

Figure 2. Full-scale velocity and temperature profiles as a function of corridor exit.

holds for velocity (V) where H is height, provided

$$\left(\frac{Q}{H^{5/2}}\right)_{F.S.} = \left(\frac{Q}{H^{5/2}}\right)_{MODEL}.$$

This criterion is approximately satisfied by the data shown in figure 3. The 1/7 scale model velocity data was scaled up in order to make a direct comparison with the full-scale velocities.

## 2. Effect of Fire Size

Figure 4 shows the relative insensitivity of the flow pattern to fire size for a fixed geometry. In particular, the recirculating layers are nearly identical in each case. The only detectable differences between the velocities are in the ceiling and floor jets, those being more intense for a larger fire. Of course the temperature traces reflect the increased fire size in this fuel-controlled burning regime.

The maximum temperature and velocity in the ceiling jet correlate as

$$\Delta T_{jet} = 5.1 \, Q^{2/3} \quad (°C) \tag{1}$$

and

$$V_{jet} = 0.0815 \, Q^{1/2} \quad (m/s) \tag{2}$$

for the data shown in figure 4 with Q in kW. Equation (1) is consistent with the theory given by Alpert [3] for an unobstructed ceiling jet generated by a symmetric line fire plume; however, that analysis yields $V_{jet} \propto Q^{1/3}$. The difference in form may be due to the presence of the room doorway.

## 3. Effect of Fire Room Configuration

It was initially shown for the model [1] and demonstrated in figure 2 for the full-scale corridor that the flow pattern is influenced significantly by the size and nature of the exit opening. The significance of internal obstructions had not been determined. This was examined by varying the nature and size of the opening between the room and corridor in the scale model.

A series of vertical, midwidth T-V traverses were made at 0.8ℓ in the model to determine the effect of varying the burn room geometry for a fixed corridor exit. Results indicate that altering the burn room doorway configuration does not have a dramatic effect on the corridor flow field compared to altering the corridor exit geometry.

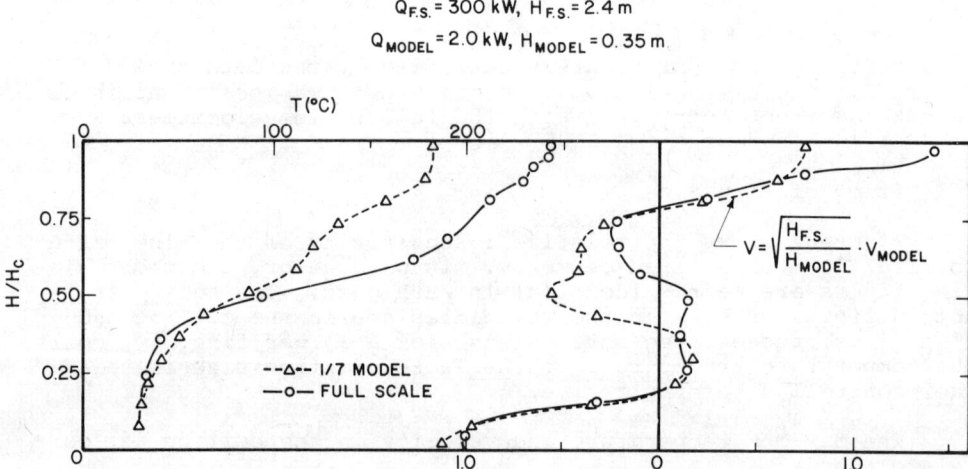

Figure 3. Comparison of full-scale and small scale results.

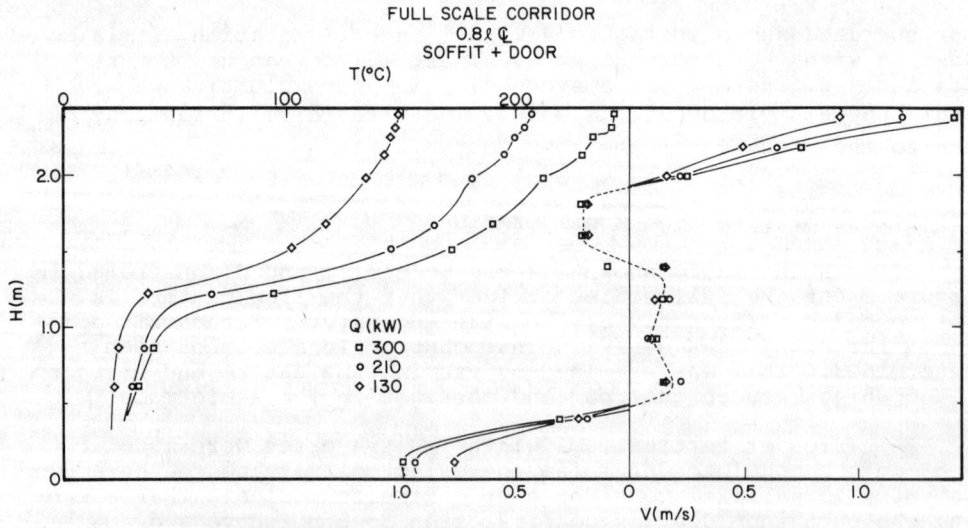

Figure 4. Effect of fire size (Q) on corridor temperature and velocity.

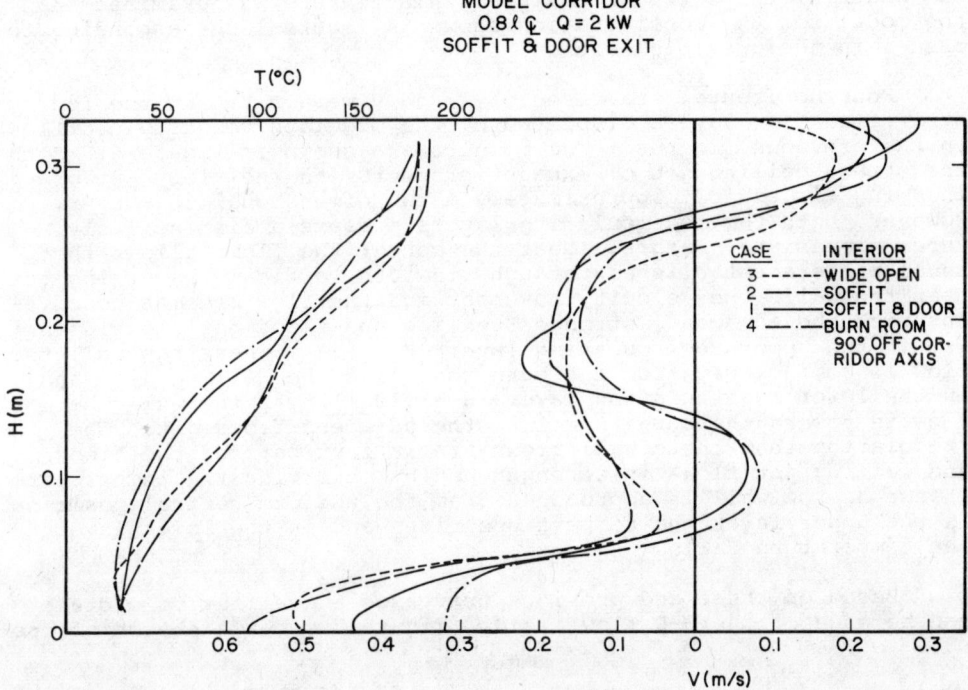

Figure 5. Effect of room doorway configuration on temperature and velocity.

Figure 5 compares T-V plots for four cases: (1) a standard soffit and door frame located one corridor width away from the burner end of the corridor, (2) a soffit on its own at the same position, (3) no obstruction, i.e. the corridor becomes one long room with a fire at one end, and (4) the 90° off-axis burn room with a standard doorway. All were for a single fire ($Q = 2$ kW) and a single corridor exit (soffit + door frame). The temperature profiles are not very different for all cases. Although the two least obstructed cases do not result in a four layer velocity regime the character of the three plots is similar. The second layer from the floor which travels away from the fire for the most obstructed configuration is replaced by a stagnant region for the least obstructed geometries. The insensitivity of the bulk corridor flow to burn room configuration is further illustrated by the dot-dash line in figure 5 which is the velocity (and temperature) profile for the case of the 90° off-axis burn room.

## 4. Three Dimensional Effects

Up to now the flow pattern has been discussed only for the midwidth plane of the corridor. This flow pattern remains relatively insensitive to position along the length of the corridor excluding end effects near the doorways. However, mass balances based on these midwidth velocity profiles have not been successful since the flow demonstrated a three dimensional character across the width of the corridor. Hence, this feature was examined in the model for the configuration shown in figure 1 corresponding to case 1 in the previous section.

Four horizontal traverses of T and V were taken in the four stratified flow layers displayed by the midwidth velocity profiles in the previous figures. The results are shown in figure 6. Both the strong ceiling jet of exhaust products and the floor jet of incoming air exhibit approximately a two dimensional character. However, both the recirculating layers possess a distinctively three dimensional nature. Near the center the flow follows the usual pattern established through midwidth vertical traversing but near the walls the velocity reverses and the flow becomes consistent with the adjacent, stronger ceiling and floor jet flows. That is, in the upper recirculating layer (H = .22 m) near the wall the flow is coming away from the burn room like the ceiling jet, and in the lower recirculating layer (H = .10 m) near the wall the flow is toward the burn room like the adjacent floor jet. The speculation that these wall flows are in fact part of the floor and ceiling jet flows is strengthened by observing the temperature change as the wall is approached from the uniform central position. In the upper layer the temperature rises and in the lower layer the temperature falls.

Based on these and previous traverses a qualitative picture can be made of the 3-D flow field. Figure 7 shows a sketch of one-

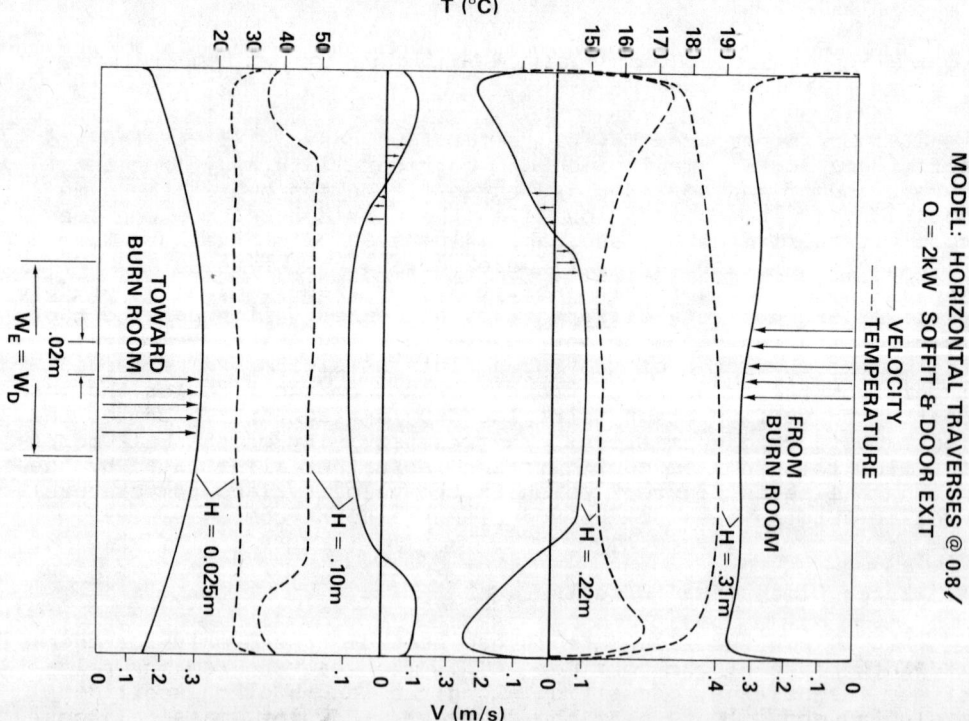

Figure 6. Corridor velocity and temperature profiles at several heights.

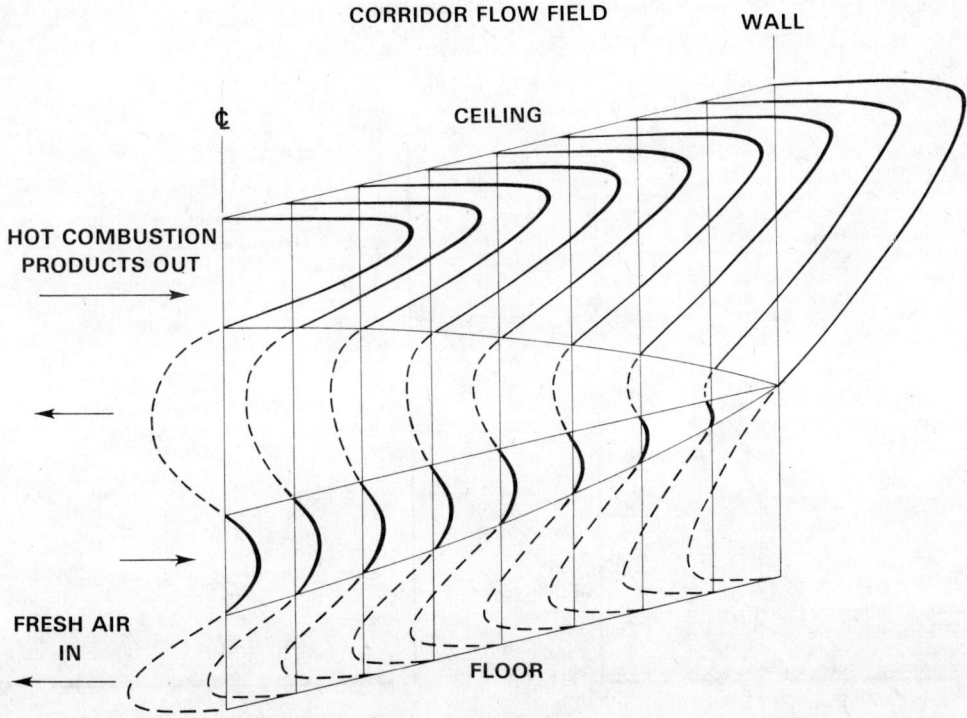

Figure 7. Three dimensional sketch of corridor flow field.

half the flow field in a typical vertical plane. Although a mass balance has not been performed based on the three-dimensional sketch in figure 7, an assessment of it suggests that it is consistent with mass conservation. For example, based on the mid-width profile and the assumption of two-dimensionality 10 g/s is flowing in and 4.5 g/s is flowing out [1]. Yet figures 6 and 7 indicate that the upper recirculating layer has about 1-1/2 times more mass flowing out than flowing in along the midwidth.

5. Factors Effecting Mixing and Recirculation

The complex flow patterns that have been shown to result in the corridor appear to be a function of two interactions. One is the mixing that occurs within the corridor where the effects of buoyancy and turbulence play a role. The other is the effect of doorway flows at both ends of the corridor. The quantity of mass flow through each doorway and the amount of entrained mass in the cold floor jet entering the corridor and hot jet leaving the fire room will surely influence the nature of recirculation within the corridor. Some of these features will be examined here.

5a. Nature of Corridor Entrance Jet

Figures 8 and 9 show various velocity measurements of the cold flow at the entrance of the corridor in the scale model. For all cases the model corridor is located above the floor of the

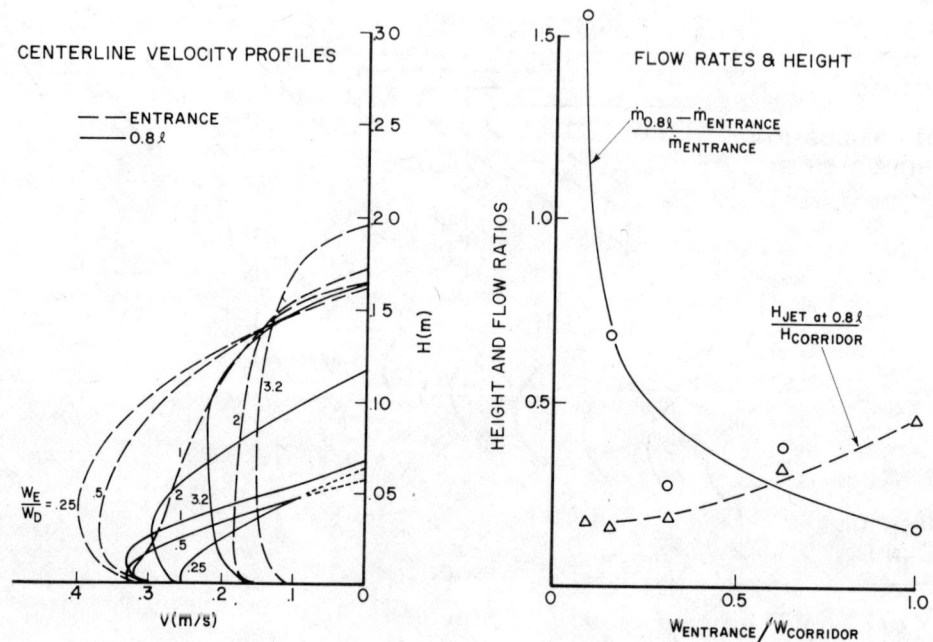

Figure 8. Characteristics of the cold flow entering the corridor as a function of exit width.

laboratory. There is no corridor floor upstream of the entrance which would induce the flow to become horizontal, the way it might do so in connecting corridors or the way, in fact, the previous full-scale measurements were obtained. No attempt has been made here to evaluate these effects.

Figure 8 shows a vertical, centerline velocity traverse of the inlet flow at the outside entrance plane (upstream of door frame) and at $0.8\ell$ within the corridor for various openings, $W_E/W_D$. $W_E$ is the width of the corridor exit door opening and $W_D$ is the width of the room door held fixed at 0.11 m. ($W_E/W_D = 3.2$ is for a soffit only exit.) The configuration is that of reference 1 where the burn room was located 90° off the corridor axis, and was maintained at a constant temperature. The dependence on door width of the cold flow is evident from the velocity profiles in figure 8. The falling behavior of the cold flow is quantified on the right-hand side of the figure. As the width ($W_E$) of the opening is decreased the height of the cold jet ($H_{jet}$ at $0.8\ell$) in the corridor is decreased. For a soffit in place and for $W_E = W_C$ cold flow in the corridor extends up to 45% of the corridor height and decreases to about 15% as the door is closed.

By using the velocity profiles on the left in figure 8, mass flow rates were computed to estimate the rate of entrainment. For this calculation, the flow was assumed to be two-dimensional which is a reasonable approximation for the entrance flow and floor jet. The ratio of estimated entrained flow to entrance flow

$$([\dot{m}_{0.8\ell} - \dot{m}_E] / \dot{m}_E)$$

is shown on the right in figure 8. For the wide open configuration there appears to be about 15% more flow at $0.8\ell$. As the opening is decreased to $W_E/W_D = 0.25$, flow increases to 1-1/2 times the entrance flow. The solid line indicates the trend of increased entrainment due to the entrance jet as the flow is squeezed through the smaller openings.

Figure 9 illustrates the jet from the plan view, showing horizontal traverses using a smaller probe (.005 m Dia.) at three locations, all .025 m above the floor. For this case the center

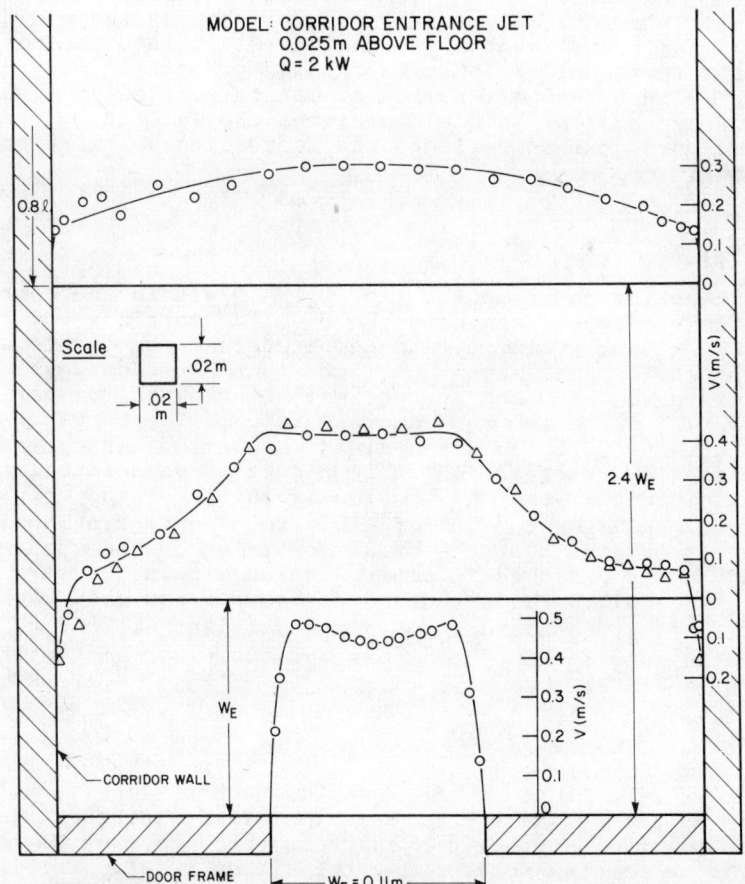

Figure 9. Transition of the corridor entrance jet near the floor.

burner was used and even in this symmetric configuration considerable difficulty was encountered in trying to obtain symmetric profiles. The flow field was very dependent upon external breezes caused by laboratory ventilation and baffling near the corridor entrance was required in order to minimize the asymmetry. Due to the flat nature of the profiles near the centerline this difficulty was not observed in the previous vertical traces.

The entrance traverse was taken at the inside plane of the door frame and shows the inviscid character of the contracted flow i.e. the outside edge velocities are in excess of the center or average value. The next traverse at one door width inside the corridor shows clearly the potential core and mixing region of the jet. Finally, at $0.8\ell$ ($2.4\ W_E$) the flow has appeared to reach its asymptotic character, and the total mixing appears to take place within about 2-1/2 door widths. The spread of the jet is considerably greater than an unconfined plane turbulent jet, the half angle of spread measuring about 40° versus 14° for the plane jet in the similarity regime. The demarkation of the mixing regime was also evident from the corresponding temperature traverses. The temperature at $0.8\ell$ is essentially flat or uniform across the width (see figure 6, lower curve) whereas at one doorwidth into the corridor the temperature is uniform only in the center portion for about 50% of the width. Thereafter, the temperature rises to a maximum of 6-7 °C above ambient as one approaches the wall and the excess temperature more or less reflects the decreasing velocity profile in the wings.

## 5b. Doorway Mass Flow Rates

In attempting to trace the flow pattern within the corridor, the mass flow rates through both the room and corridor exit doorways must be known. Velocity and temperature measurements were made in the scale model shown in figure 1 at both doorways. A vertical traverse was taken in the in-flowing fluid for various exit widths ($W_E$) with the room doorway fixed. The entrance profile of figure 9 justified a two-dimensional flow assumption. The derived mass flow rates are shown in figure 10 as a function of $W_E/W_D$. Also plotted were the previous results for the corridor doorway determined for the 90° off-axis room configuration [1]. It is clearly evident that the mass flow rates of each doorway differ and thus contribute to recirculation within the corridor. Moreover, for small $W_E/W_D$ hot products are recirculated back into the room; whereas, for large $W_E/W_D$, more ambient air enters the corridor than can flow into the room and just cool air enters the room.

## 5c. The Nature of the Flow

In flows with buoyancy the Reynolds number alone is not sufficient to define the nature of the flow, i.e. whether it is laminar or turbulent. Buoyancy can suppress turbulence at Reynolds numbers far beyond the non-buoyant transition point. For these situations the Richardson number is a measure of how efficiently the turbulent mixing is being suppressed by gravity. It is the ratio of the buoyancy production to stress production of turbulent kinetic energy. The gradient Richardson No. is made up of more easily measurable flow quantities:

$$Ri = g \frac{\partial T}{\partial Z} \bigg/ T \left(\frac{\partial V}{\partial Z}\right)^2 \qquad (3)$$

The sign of Ri is controlled by the temperature gradient. An increasing function of temperature with height, i.e. heat transfer downward, leads to a stable configuration or positive Ri. The magnitude of Ri will determine the amount of gravitational potential energy as compared to flow kinetic energy.

From the previous vertical traverses the quantities in equation (3) can be evaluated at each point in the vertical direction by measuring the local slopes of T and V versus H. Figure 11 is a T-V traverse of the full-scale corridor results alongside of which is plotted the gradient Richardson number, individual calculations being denoted by circles. Note near the floor and ceiling Ri becomes small due to a combination of both small vertical temperature gradients and large velocity gradients. In between these two areas Ri becomes large indicating that near the floor and ceiling the flow will be most turbulent and less so away from these shear regions. The question of a sufficiently small or large Ri to characterize a transition from turbulent to non-turbulent flow remains to be identified.

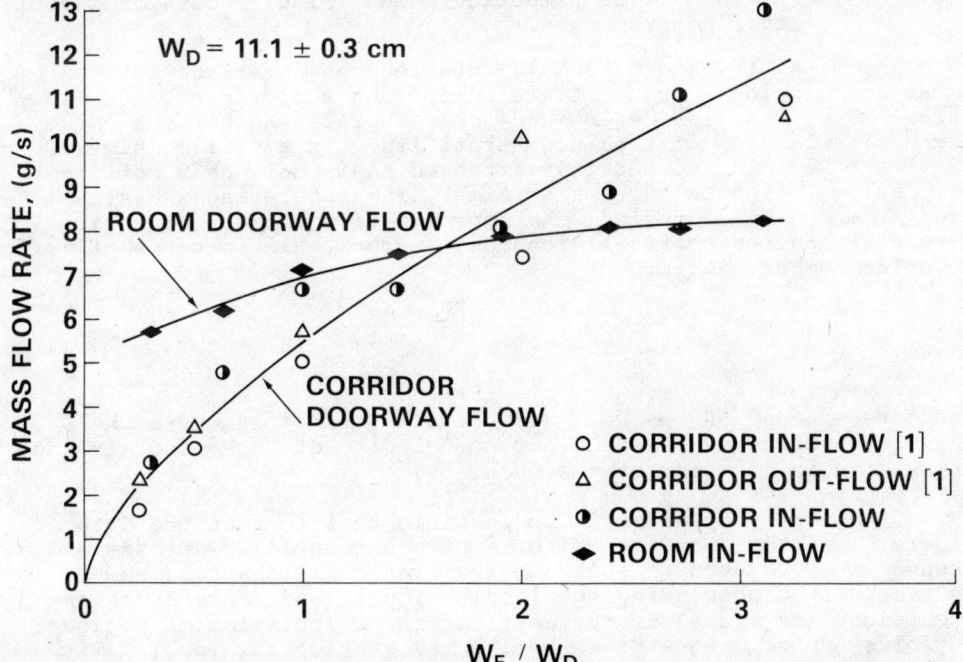

Figure 10. Doorway mass flow rates as a function of exit opening for in-line, and 90° off-axis room geometry taken from Ref. [1].

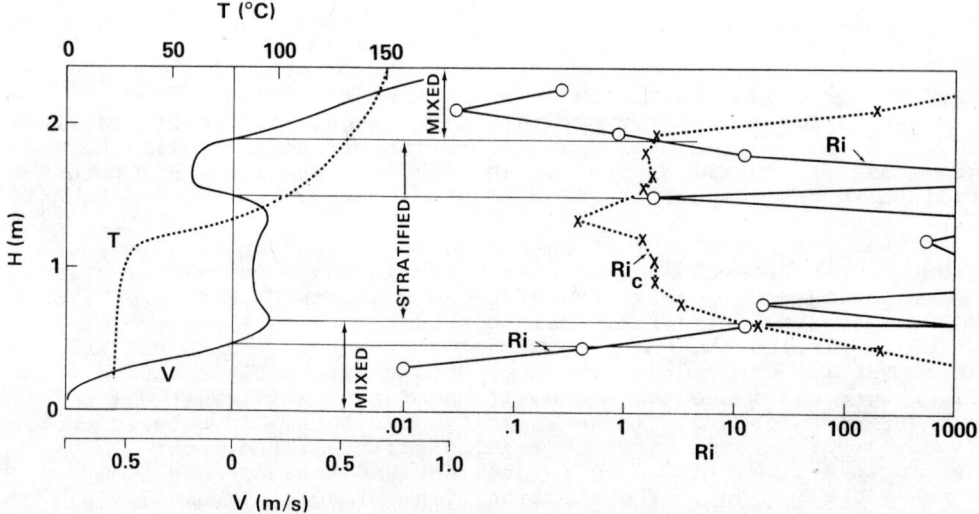

Figure 11. Application of Terai, et al. [4] criterion to identify stratified and mixed flow regimes in the corridor for one set of temperature and velocity data from full-scale measurements.

Terai, Nitta, and Matsuo [4] studied the interface between smoke and air in a long duct, for different temperatures and flow rates. They visually determined for each case whether the smoke mixed with the air or remained stratified. On a semilog plot of Ri versus Reynolds number they arranged their data as stratified, critical or mixed and drew a straight line separating the stratified from the mixed regime. An approximate equation for their line defining the critical Richardson number, $Ri_C$ in terms of Reynolds number, Re is:

$$Ri_C = 0.05\, e^{\frac{Re}{4200}} \qquad (4)$$

The data spanned $5000 < Re < 20000$ and $.1 < Ri < 10$ where the Reynolds number was based on a mean velocity of smoke and air and the hydraulic diameter of the duct.

Using their results, it is possible to interpret the data of figure 11 in terms of the criteria of equation (4). Crosses and a dashed line are used to indicate the critical Richardson number at the various heights using the local velocity and temperature and the height (or width) of the corridor for calculation of a 'local' Reynolds number. At a given horizontal station the local Ri is compared to $Ri_C$; for $Ri > Ri_C$ it would be expected that the flow would be stratified or tending toward stratification since the conditions of the two experiments are not identical. Similarly, for Ri < $Ri_c$ the flow should tend toward a mixed or turbulent state. Both the upper and lower intersections of the local Ri and $Ri_c$ are indicated in figure 12 and marked accordingly. Visual observation with smoke tracers indicate that the two recirculating layers appear laminar. Smoke injected in the midlength of the corridor at those two vertical levels travels away from the injection point with very little disturbance or mixing; the only visible motion being a low frequency periodic waviness of very weak surge-like movement. Smoke injected into the ceiling and floor jets on the other hand immediately breaks up in a random turbulent fashion.

The square root of the numerator of the gradient Richardson number is the Brunt-Väisälä frequency or the natural frequency of oscillation of buoyant disturbances, e.g. internal gravity or ocean waves [5]. To a first approximation it is the upper frequency cut off of an organized wave disturbance, waves of higher frequency will not be propagated. Calculation of the Brunt-Väisälä frequency in the stratified region of figure 11 including the three zero-velocity interfaces yields values of between 2 and 6 $s^{-1}$ or periods of 1 to 3 s, consistent with what is visually observed. The undulation noted in following the smoke traces in the recirculating layers are probably internal waves. These conclusions are consistent with preliminary turbulence measurements made in the model corridor where the higher frequency components of V' had virtually disappeared as the hot wire was raised out of the floor jet into the recirculating region.

CONCLUSIONS

These results have demonstrated some factors which influence the character of fire induced flows in corridors. The occurence of fire within a corridor or room adjoining a corridor will set in motion a stratified flow within the corridor with regions of turbulent and laminar flows and large recirculating eddies. It was found that these flow patterns were insensitive to scale size for geometrically similar configurations. Moreover the effect of fire size (Q) and the location and size of the room doorway had little effect on the overall corridor flow pattern. The nature and size of the corridor exit doorway had the most significant influence on the corridor temperature and velocity flow field.

It was demonstrated that the corridor flow field is three-dimensional primarily in the stably stratified layers between the floor and ceiling turbulent jets. Mixing occurs along these jets and at the doorway openings where fluid enters the corridor. It was shown that the ratio of entrained to entrance flow rate increases sharply as the exit door width decreases. Significant entrainment also occurs in the hot jet entering the corridor from the room. However, it was found that the mass flow rate through the room and corridor doorways were not identical. Finally, a criterion for predicting the regions of mixed and stably stratified flows developed by Terai et. al [4] yielded good results for one example representative of the corridor flows observed in the present study.

REFERENCES

1. McCaffrey, B. J. and Quintiere, J. G. 1975. Fire Induced Corridor Flow in a Scale Model Study. Proceedings Conseil International du Batiment (CIB) Symposium on the Control of Smoke Movement in Building Fires, Garston, U.K., Nov. 4-5. 34-47.

2. McCaffrey, B. J. and Heskestad, G. 1976. A Robust Bidirectional Low Velocity Probe for Flame and Fire Application. Combustion and Flame. 26(1): 125-127.

3. Alpert, R. L. Fire Induced Turbulent Ceiling-Jet. 1971. FMRC Ser. No. 19722-2, Factory Mutual Research Corporation, Norwood, Mass. 91 pp.

4. Terai, T., Mitta, K. and Matsuo, T. 1974. Richardson Number at the Interface of Smoke-Air Stratified Flow. Main Reports on Production, Movement, and Control of Smoke in Buildings, Occasional Report of Japanese Association of Fire Science and Engineering, No. 1. 189-190.

5. Turner, J. S. 1973. Buoyancy Effects in Fluids. Cambridge University Press, London.

# STABILITY OF FREE CONVECTION FLOW IN A SHALLOW CAVITY WITH AND WITHOUT ROTATION

GUENTER P. MERKER and ULRICH GRIGULL
*Institute A of Thermodynamics, D-8000 Munich, Postfach 202420, Germany*

ABSTRACT

The stability of the temperature and velocity field in a horizontal water layer contained in a rectangular container with aspect ratio $A = h/L = 1/15$ whose end walls were held at different but uniform temperatures was analytically investigated. The temperature difference between the isothermal end walls was adjustable in the range $0 \leq \Delta T < 50K$. The container could rotate around its vertical axis with angular velocities of $\Omega = 0$, 0.934, 1.321 and 1.868 1/s.

In the limit $A \to 0$, a linear stability analysis for the case $\Omega = 0$ shows that the resultant parallel flow structure in the core region of the containers is stable if the parameter $6rAK_1 < 1846$. This is in good agreement with the experimental results.

A linear stability analysis for the case $\Omega \neq 0$ with the restriction that the flow is isothermal indicates that the parallel flow structure at $\Omega = 0$ is stable in the limit $A \to 0$ if the Taylor number $Ta < 1330$.

INTRODUCTION

Convection due to buoyancy forces is often a dominant mode of heat and mass transport in environmental systems. For example, pollutants and heat waste in estuaries are transported significantly by the buoyancy-driven convective motion induced by gradients in salt concentration or temperature. Laminar flow in an enclosed rectangular cavity whose height to length ratio is small ($A \to 0$) with differentially but uniformly heated end walls has been studied extensively as an idealized and simple model of the estuary flow, see Figure 1.

Cormack, Leal and Imberger [1] have shown by use of matched asymptotic expansions that the flow consists of two distinct regimes in the limit $A \to 0$; a parallel flow in the core region and a secondary non-parallel flow near the ends of the cavity. They derived asymptotic expressions for the stream function and temperature field, valid in this limit and correct to $O(A)$. Cormack, Leal and Seinfeld [2] obtained numerically a solution of the full Navier-Stokes equations for the convection flow in a cavity with small aspect ratio. Their result show excellent agreement between the asymptotic and numerical solution provided that $A < 0.1$ and $Gr^2 Pr^2 A^3 < 10^5$. Imberger [3] studied experimentally the steady motion in an enclosed rectangular cavity with aspect ratios of $10^{-2}$ and $1.9 \times 10^{-2}$ and the result support the asymptotic solution. Merker and Grigull [4] studied the transition regime between the shallow-cavity limit (δr fixed, A→0) and the boundary-layer limit (A fixed, Gr→∞) for the two cases of a fixed container and a rotating container with slow angular velocity around its vertical axis, see Figure 2. The experimental results show that the parallel flow in the core region in the limit A → 0 is maintained if the angular velocity $\Omega$ is sufficiently small.

In the present paper, we use the standard linear stability theory to consider the parallel flow in the core region in the limit A → 0. We shall show that this flow structure is stable if GrA, (TaU), is sufficiently small in the case $\Omega = 0$, ($\Omega \neq 0$).

## MATHEMATICAL FORMULATION

We consider a closed rectangular three-dimensional cavity of length $\ell$, height $h$ and width $b$ which contains a Newtonian fluid and rotates around its vertical axis with angular velocity $\Omega$, and is shown schematically in Figure 1. The end walls are maintained at different but uniform temperatures $T_c$ and $T_h$, with $T_c < T_h$. The top, the bottom and the side walls are insulated, and all surfaces are rigid no-slip boundaries. The velocity distribution in a cavity with a free surface was experimentally found to be very similar to that where there is a lid at the surface; i.e. the measured value for the velocity at the surface was zero in both cases [3,5].

The appropriate governing differential equations in non-dimensional form for an incompressible Newtonian fluid with constant thermal properties, subject to the usual Boussinesq approximation, are

$$Gr^2 A^2 \left(\frac{\partial}{\partial t} + \vec{u} \cdot \nabla\right) u = -\frac{\partial P}{\partial x} + GrA \nabla^2 u + TaGrv \quad (1)$$

$$Gr^2 A^2 \left(\frac{\partial}{\partial t} + \vec{u} \cdot \nabla\right) v = -\frac{\partial P}{\partial y} + GrA \nabla^2 v - TaGru \quad (2)$$

$$Gr^2 A^2 \left(\frac{\partial}{\partial t} + \vec{u} \cdot \nabla\right) w = -\frac{\partial P}{\partial z} + GrA \nabla^2 w + Gr\theta \quad (3)$$

$$GrPrA \left(\frac{\partial}{\partial t} + \vec{u} \cdot \nabla\right) \theta = \nabla^2 \theta \quad (4)$$

$$\nabla \cdot \vec{u} = 0 \quad (5)$$

Following Cormack et al. [1], the characteristic scaling factors are, $\ell_{ref} = h$, $u_{ref} = g\beta h^3 (T_h - T_c)/(\nu \ell)$, $t_{ref} = \ell_{ref}/u_{ref}$ and $\theta = (T-T_c)/(T_h-T_c)$. The resultant dimensionless parameters are

$Gr \equiv g h^3 \beta (T_h - T_c)/\nu^2$, Grashof number

$Pr \equiv \nu/k$, Prandtl number

$Ta \equiv 2\Omega h^2/\nu$, Taylor number

$A \equiv h/\ell$, aspect ratio

The corresponding boundary conditions are

$u = v = w = 0$ on all solid boundaries
$\partial \theta / \partial z = 0$ on $z = 0, 1$
$\partial \theta / \partial y = 0$ on $y = \pm b$

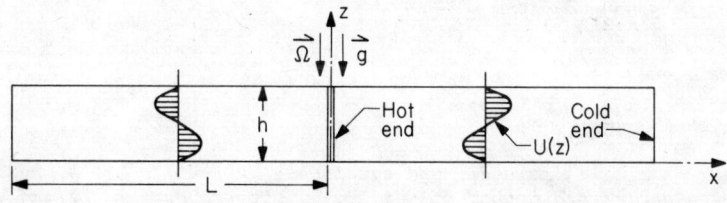

Figure 1. Schematic diagram of a shallow cavity system

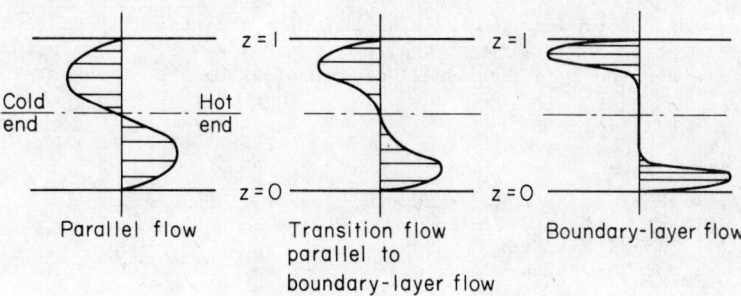

Figure 2. Different types of flow

$$\theta = 0,1 \qquad \text{on } x = O_1 A^{-1} \tag{6}$$

## ASYMPTOTIC SOLUTION FOR $\Omega = 0$

Eqns (1) - (5) can be solved for a two-dimensional cavity, i.e. side walls at $y = \pm \infty$ and $v \equiv 0$, $\partial/\partial y \equiv 0$, in the limit $A \to 0$. Introducing a stream function $\psi(x,z)$ and a vorticity function $\omega(x,z)$ with

$$U = \frac{\partial \psi}{\partial z} \text{ and } W = -\frac{\partial \psi}{\partial x}, \quad \omega = \frac{\partial U}{\partial z} - \frac{\partial W}{\partial x}$$

reduces eqns. (1)-(5) to the following set of eqns.

$$GrA^2 \frac{\partial(\omega,\psi)}{\partial(x,z)} = A\nabla^2 \omega - \frac{\partial \theta}{\partial x} \tag{7}$$

$$\nabla^2 \psi = \omega \tag{8}$$

$$GrPrA \frac{\partial(\theta,\psi)}{\partial(x,z)} = \nabla^2 \theta \tag{9}$$

Cormack et al. [1] solved these equations, subject to the boundary conditions (6), by use of the method of matched asymptotic expansions and derived the following solution for the core flow, valid in the limit $A \to 0$ and correct to $O(A^3)$,

$$W \equiv 0$$

$$U(z) = K_1 F'''(z) \tag{10}$$

$$\theta(x,z) = K_1 Ax + K_1^2 GrPrA^2 F(z) + K_2$$

where

$$F(z) = \frac{z^5}{120} - \frac{z^4}{48} + \frac{z^3}{72}$$

$$K_1 = 1 - 3.48 \times 10^{-6} \times Gr^2 Pr^2 A^3 + O(A^5)$$

$$K_2 = (1-K_1)/2 - K_1^2 GrPrA^2/1440$$

## PERTURBATION EQUATIONS

The aim of the present paper is to consider the stability of the above asymptotic solution for the two cases, $\Omega = 0$ and $\Omega \neq 0$. The perturbation equations are derived by adding small disturbances to the steady state solution.

$$\begin{aligned} u &= U(z) + \varepsilon \cdot \tilde{u} \\ v &= \varepsilon \cdot \tilde{v} \\ w &= \varepsilon \cdot \tilde{w} \\ p &= P(x,z) + \varepsilon \cdot \tilde{p} \\ \theta &= \Theta(x,z) + \varepsilon \cdot \tilde{\Theta} \\ \Omega &= \varepsilon \tilde{\Omega} \end{aligned} \tag{11}$$

Substituting (11) into (1)-(5) and equating terms of like order in $\varepsilon$ gives at $O(\varepsilon)$,

$$Gr^2 A^2 \left( \frac{\partial \tilde{u}}{\partial t} + U \frac{\partial \tilde{u}}{\partial x} + \tilde{w} \frac{\partial U}{\partial z} \right) = -\frac{\partial \tilde{p}}{\partial x} + GrA\nabla^2 \tilde{u} \tag{12}$$

$$Gr^2A^2 \left(\frac{\partial \tilde{v}}{\partial t} + U \frac{\partial \tilde{v}}{\partial x}\right) = -\frac{\partial \tilde{p}}{\partial y} + GrA\nabla^2\tilde{v} - TaGrAU \qquad (13)$$

$$Gr^2A^2 \left(\frac{\partial \tilde{w}}{\partial t} + U \frac{\partial \tilde{w}}{\partial x}\right) = -\frac{\partial \tilde{p}}{\partial z} + GrA\nabla^2\tilde{w} + Gr\tilde{\theta} \qquad (14)$$

$$GrPrA \left(\frac{\partial \tilde{\theta}}{\partial t} + U \frac{\partial \tilde{\theta}}{\partial x} + \tilde{u}\frac{\partial \Theta}{\partial x} + \tilde{w}\frac{\partial \Theta}{\partial z}\right) = \nabla^2\tilde{\theta} \qquad (15)$$

$$\frac{\partial \tilde{u}}{\partial x} + \frac{\partial \tilde{v}}{\partial y} + \frac{\partial \tilde{w}}{\partial t} = 0 \qquad (16)$$

Squire's theorem for parallel flows shows that for every three-dimensional disturbance which grows in time there is a corresponding two-dimensional disturbance which also grows, and the two-dimensional disturbance becomes unstable at a lower Reynolds (Grashof) number than the three-dimensional disturbance. One can consider disturbances in the x-z and x-y plane.

THE CASE $\Omega = 0$.

Equation (12)-(16) are selfconsistent for disturbances in the x-z plane only if $Ta \equiv 0$. Setting $\tilde{v}$ and $\partial/\partial y$ equal to zero, introducing a stream function $\tilde{\psi}$, same definition as for $\psi$, and eliminating the pressure terms, gives

$$GrA^2 \left[\left(\frac{\partial}{\partial t} + U \frac{\partial}{\partial x}\right)\nabla^2\tilde{\psi} - U'' \frac{\partial \tilde{\psi}}{\partial x}\right] = A\nabla^2\tilde{\psi} - \frac{\partial \tilde{\theta}}{\partial x} \qquad (17)$$

$$GrPrA \left[\left(\frac{\partial}{\partial t} + U \frac{\partial}{\partial x}\right)\tilde{\theta} + \frac{\partial(\Theta,\psi)}{\partial(x,z)}\right] = \nabla^2\tilde{\theta} \qquad (18)$$

If one seeks a solution by separation of variables, one finds that the stream function $\tilde{\psi}$ and the temperature field $\tilde{\theta}$ must be a sum or integral of terms in the form

$$\{\tilde{\psi},\tilde{\theta}\} = \exp[ik(x-ct)]\cdot\{\phi(z), \nu(z)\} \qquad (19)$$

where k denotes a Fourier wave number in x direction and c is a complex dimensionless velocity,

$$c = c_R + ic_i$$

A disturbance which grows corresponds to $c_i > 0$, one which decays to $c_i < 0$, and neutral stability to $c_i = 0$. More details can be found by Yih [6] and Denn [7]. Substituting (19) into (17) and (18) reduces the partial differential equations to a fourth and second order ordinary differential equation for the unknown stream function $\psi$ and temperature field $\nu$,

$$A(\phi^{IV}-2k^2\phi''+k^4\phi) - ikGrA^2[(U-c)(\phi''-k^2\phi)] - U''\phi = k^2\nu \qquad (20)$$

$$\nu''-k^2\nu-ikGrPrA\ (U-c)\nu = GrPrA(\Theta_x\phi'+ik\Theta_z\phi) \qquad (21)$$

will be boundary conditions

$$\begin{aligned}\phi &= \phi^* = \phi' = \phi^{*\prime} = 0\\ \nu &= \nu^* = \nu' + \nu^{*\prime} = 0\end{aligned} \quad \text{at } z = 0,1 \qquad (22)$$

where $\phi^*, \nu^*$ denotes the conjugate complex function of $\phi,\nu$.

We assume the solution of (20) and (21) may be obtained as a regular expansion in the small parameter $A$,

$$\phi(z) = \phi_o(z) + A\phi_1(z) + A^2\phi_2(z) + \ldots \qquad (23)$$

$$\nu(z) = \nu_o(z) + A\nu_1(z) + A^2\nu_2(z) + \ldots$$

Substituting equations (23) into (20) and (21) and equating terms of like orders in $A$ gives at $O(1)$

$$\nu_o = 0 \qquad (24a)$$

$$\nu_o - k^2\nu_o = 0 \qquad (24b)$$

Since the only solution which satisfies equation (24b) and the boundary conditions (22) is $\nu_o \equiv 0$ the general solution for the temperature function is $\nu(z) \equiv 0$ in the limit $A \to 0$, which can be seen by equating terms of higher order in equation (21). With this result, equation (20) reduces in the limit $A \to 0$ to the well known Orr-Sommerfeld equation

$$\phi^{IV} - 2k^2\phi'' + k\phi - ikGrA[(U-c)(\phi'' - k^2\phi) - U''\phi] = 0 \qquad (25)$$

The result $\nu(z) = 0$ in the limit $A \to 0$ is consistent with the physical picture of an isothermal fluid which is maintained not by buoyancy forces but by pumping fluid in and out at the hot and cold end walls. This approximation seems to be reasonable because the first correction in the velocity profile $U(z)$ due to convection is of $O(A^3)$, see equation (10).

## SUFFICIENT CONDITION FOR STABILITY IF $\Omega = 0$

Since it is not an easy matter to find a general solution for equation (25) one is interested in establishing a sufficient condition for stability which allows an estimate of the parameters $GrA$. Such a sufficient condition can be derived as follows [8]. Equation (25) is multiplied by $\phi^*$, the conjugate complex of $\phi$, and integrated between $z = 0$ and $z = 1$. Adding to the resulting equation its own complex conjugate and integrating by parts leads to the relation

$$c_i kGrA(J_1^2 + k^2 J_o^2) \leq kGrA|U'_{max}|J_o J_1 - (J_2^2 + 2k^2 J_1^2 + k^4 J_o^2) \qquad (26)$$

where

$$J_o^2 \equiv \int_0^1 \phi\phi^* \, dz$$

$$J_1^2 \equiv \int_0^1 \phi''\phi^* \, dz = -\int_0^1 \phi'\phi^{*'} \, dz$$

$$J_2^2 \equiv \int_0^1 \phi^{IV}\phi^* \, dz = \int_0^1 \phi''\phi^{*''} \, dz$$

$|U'_{max}|$ denotes the maximum of the velocity gradient $U' \equiv dU/dz$ which appears at $z = 0.5$.

The relation (26) gives an upper bound for $c_i$. If $kGrA$ is small enough so that the upper bound is negative for all functions $\phi(z)$ satisfying the boundary conditions (22), the flow is stable. Thus, a sufficient condition for stability is

$$GrA \, |U'_{max}| < \min \frac{J_2^2 + 2k^2 J_1^2 + k^4 J_o^2}{k J_o J_1} \tag{27}$$

To find the minimum of the right-hand side one has to calculate the corresponding eigenfunctions. However, if the eigenfunctions are not known, relation (21) can be still used to estimate the parameter GrA by choosing a reasonable trial function $\phi$ which satisfies the boundary conditions.

RESULTS FOR THE CASE $\Omega = 0$

Choosing

$$\phi(z) = 1 - \cos(2\pi n z), \qquad n = 1,2,3,\ldots$$

leads to the result that the parallel flow structure is stable if

$$GrAK_1 \leq 1846 \tag{28}$$

where $K_1$ can be calculated either with the asymptotic solution, equation (10), or with the numerical solution calculated by Cormack et al. [2].

The stability diagram, Figure 3, shows relation (28) where $K_1$ is determined from the analytical (dotted lines) and numerical (solid line) solution. A comparison with the experimental data shows that most of Imberger's [3] experiments were done in the stable region whereas the data obtained by Merker [5] are clearly in the transition from the parallel flow to the boundary layer flow regime.

From equation (10) follows the definition of the Reynolds number

$$Re \equiv \frac{U_{max} \cdot h}{\nu} = 8 \times 10^{-3} \, GrAK_1$$

This gives the simple stability criteria

$$Re \leq 14.78 \tag{29}$$

The function $Re = Re(k,n)$ is shown in Figure 4. It is interesting to compare the critical value (29) for the Reynolds number with a simpler approximation [4]. They assumed that the unknown velocity profile for parallel flow can be approximated with the function $\sin(2\pi n z)$, and obtained (after correction an error) for the critical Reynolds number the value of 12.24 which is in excellent agreement with the result of the present analysis, particularly in view of the fact that the exact velocity profile was not known.

THE CASE $\Omega \neq 0$

By setting $\tilde{w}$ and $\partial/\partial z$ equal to zero in equation (12)-(16) it can be seen that the resultant set of equations are selfconsistent only for an isothermal flow. As already pointed out for the case $\Omega = 0$ the assumption of an isothermal flow seems to be a reasonable approximation in the limit $A \to 0$ because the first correction term due to convection is of $O(A^3)$. Hence, one obtains from equation (12)-(16)

$$Gr^2 A^2 \left( \frac{\partial \tilde{u}}{\partial t} + U \frac{\partial \tilde{u}}{\partial x} \right) = -\frac{\partial \tilde{p}}{\partial x} + GrA \nabla^2 \tilde{u} \tag{30}$$

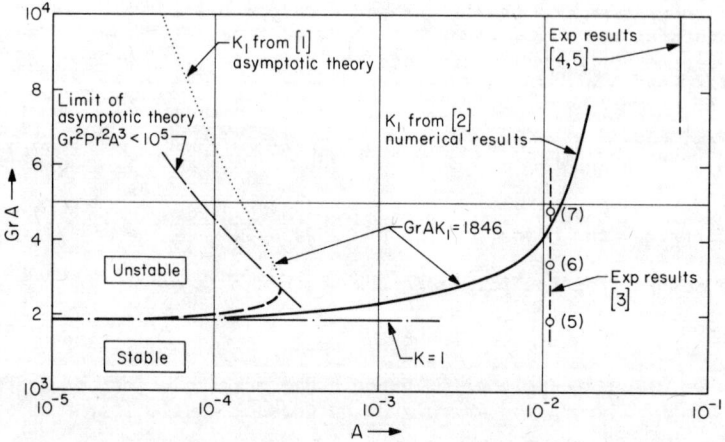

Figure 3. Stability diagram, $GrA = f(A, K_1)$ for $Ta \equiv 0$

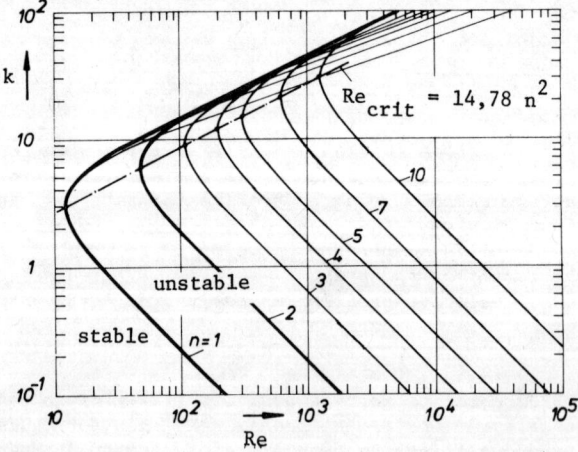

Figure 4. Stability diagram, $Re = f(k, n)$ for $Ta \equiv 0$

$$Gr^2 A^2 \left( \frac{\partial \tilde{v}}{\partial t} + U \frac{\partial \tilde{v}}{\partial x} \right) = - \frac{\partial \tilde{p}}{\partial y} + GrA\nabla^2 \tilde{v} - TaGrAU \tag{31}$$

$$\frac{\partial \tilde{u}}{\partial x} + \frac{\partial \tilde{v}}{\partial y} = 0 \tag{32}$$

Introducing the stream function $\tilde{\psi}$, separating the variables $x$ and $y$ with

$$\{\tilde{\psi}, \tilde{p}\} = exp[ik(x-ct)]\{\Lambda(y), q(y)\} \tag{33}$$

and eliminating the pressure terms reduces equation (30)-(32) to a single fourth order ordinary differential equation for the function $\Lambda(y)$, namely

$$GrA(\Lambda^{IV} - 2k^2 \Lambda'' + k^4 \Lambda) - ikGr^2 A^2 (U-c)(\Lambda'' - k^2 \Lambda)$$

$$= ikTaGrAU \; exp[-ik(x-ct)] \tag{34}$$

The corresponding boundary conditions are

$$\Lambda = \Lambda^* = \Lambda' = \Lambda^{*'} = 0 \quad \text{at } y = \pm b \tag{35}$$

A sufficient condition for stability can be derived by applying the same procedure on equation (34) as it was done on equation (25) for the case $\Omega = 0$. Finally, one obtains

$$c_i kGr^2 A^2 (I_1^2 + k^2 I_o^2) \leq kGrATaUI_o - GrA(I_2^2 + 2k^2 I_1^2 + k^4 I_o^2) \tag{36}$$

where the integrals $I_o^2$, $I_1^2$, and $I_2^2$ are defined as in equation (25) if one replaces the functions $\phi$ and $\phi^*$ by $\Lambda$ and $\Lambda^*$ and integrates from $-b$ to $+b$.

From equation (36) follows the sufficient condition for neutral stability, $c_i \equiv 0$,

$$TaU \leq \min \frac{I_2^2 + 2k^2 I_1 + k^4 I_o^2}{kI_o} \tag{37}$$

To compare the result of the present theory with the experimental data of Merker and Grigull [4] we take $b = 2$. Choosing

$$\Lambda(y) = 1 + \cos\left(\frac{\pi}{2} ny\right) \quad , \quad n = 1,3,5,\ldots \tag{38}$$

as trial function one obtains the limit for the Taylor number

$$Ta \leq 1330 \quad \text{in the limit } A \to 0 \tag{39}$$

The stability diagram, Figure 5, compares the result of the present theory with our previous experimental investigations. Since the critical Reynolds number is independent of the parameter $K_1$ and the aspect ratio $A$ the Reynolds number is used instead of the Grashof number in Figure 5. It should be noted that the experimental values are obtained in the transition from the parallel flow to the boundary-layer regime. To the author's knowledge no investigation, to date, has dealt with the rotational instability of the parallel flow structure in a cavity of small aspect ratio.

Part of this work was done while G. P. Merker was on a NATO Fellowship as Research Fellow at the Department of Chemical Engineering, California Institute

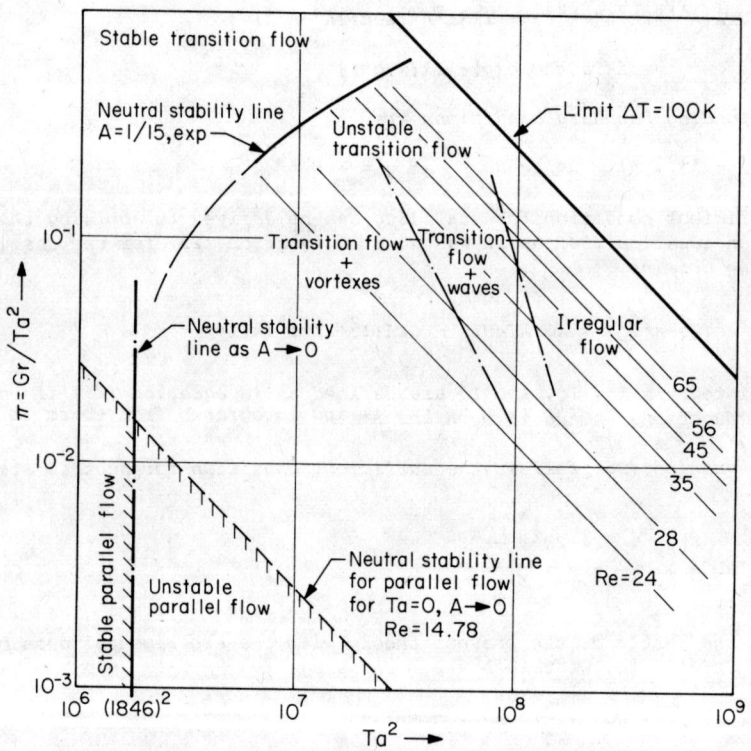

Figure 5. Stability diagram, $\pi = f(Ta^2, Re)$

of Technology, Pasadena, CA   91125, U.S.A.

REFERENCES

1. Cormack, D. E., L. G. Leal and J. Imberger, 1974. Natural Convection in a Shallow Cavity with Differentially Heated End Walls. Part 1. Asymptotic Theory. J. Fluid Mech. 65, pp. 209-229

2. Cormack, D. E., L. G. Leal and J. H. Seinfeld. 1974. Natural Convection in a Shallow Cavity with Differentially Heated End Walls. Part 2. Numerical Solutions. J. Fluid Mech. 65, pp. 231-246

3. Imberger, J., 1974. Natural Convection in a Shallow Cavity with Differentially Heated End Walls. Part 3. Experimental Results. J. Fluid Mech. 65, pp. 247-260.

4. Merker, G. P., and Grigull, U. 1975. Freie Konvektion in einem flachen Behalter mit und ohne Rotation. Waerme-und Stoffuebertragung 8, pp. 101-112

5. Merker, G. P., 1975. Freie Konvektion in einem rechtwinkligen Trog mit und ohne Rotation, Dissertation, Technische Universitaet Muenchen

6. Yih, C. S., 1969. Fluid Mechanics, McGraw Hill, Inc.

7. Denn, M. M., 1975. Stability of Reaction and Transport Processes, Prentice-Hall, Inc.

8. Lin, C. C., 1967. The Theory of Hydrodynamic Stability, Cambridge University Press.

# FREE CONVECTION
# IN ENGINEERING EQUIPMENT

# THE MODEL OF TURBULENT FREE CONVECTION NEAR A VERTICAL HEAT TRANSFER SURFACE

S. S. KUTATELADZE
*Siberian Branch of the USSR Academy of Sciences*

The problem of heat convection is one of the classical those in the heat transfer theory. However, if well developed as to laminar flows, this problem is so far poorly understood for turbulent flows.

The first systematical experiments on a free convection near vertical tubes were carried out by M.V.Kirpichev and A.A.Gukhman [1] in the twentieth years in the State Physicotechnical Laboratory in Leningrad. One of these installations is shown in Fig.I.

The shadow pictures of flows were investigated and the temperature distribution was measured along the height of a heater with a constant density of heat flow. From these experiments, the peculiar flow pattern with a free convection was distinctly observed. However, the notion "curl-like" flow introduced by those investigators did not lead to any concrete physicomathematical models. Somewhat later [2] I proposed the model of free heat turbulent flow along a vertical wall. Its basic elements were: a viscous sublayer and an outer flow out of free turbulent jets. This publication was not given attention although a hypothesis of the occurrence of the Reynolds eigennumber for a viscous sublayer appeared to be first suggested in it.

The current splash of interest to a free turbulent convection arose in the fiftieth years. It is succeed both by the occurrence of some physical and mathematical models and, it is very important, by the progress in a new sensitive experimental technique.

One of the groups of investigators actively working at free heat convection was organized at the Institute of Thermophysics of the Siberian Branch of the USSR Academy of Sciences. Some results of the work of this group underlay the present report.

Let us consider the available models of turbulent heat gravitational convection. The main geometries are: a)the heat transfer surface normal to a gravitational acceleration vector (a horizontal plane), b)the heat transfer surface parallel to a gravity vector ( a vertical surface).

For a horizontal layer heated from below, mean velocities are equal to zero and hypotheses connecting the turbulent viscosity and mean velocities are unsuitable in this case.

Nowadays, one can distinguish the following medels of a turbulent transfer mechanism in a horizontal layer.

Fig.I. Photograph of the 8m vertical tube for investigation of free thermal convection.
State Phys.-Techn. Lab., Leningrad, 1928.

The model of Pristley [3] is based on the assumption that the viscosity and heat transfer far from the heat exchange surface play no role in forming a mean velocity profile. It follows from the analysis of dimensionalities that a mean temperature far from the heat transfer surface changes by the law $y^{-1/3}$.

The hypothesis of Malkus [4] on a maximum transfer consists in the following. Practically realized solutions are those conforming to the maximum heat flow through a layer at the set temperature drop. This hypothesis essentially simplifies a real problem of turbulent transfer in a horizontal layer at heat gravitational convection. The heat transfer can be expected to be maximum with large numbers of Ra in common with cases where the velocity and temperature fields are not assumed to satisfy the Boussinesq's equations and to do only boundary and homogeneity conditions and two the simplest integral consequences of a disturbed flow being known as energy integrals.

Further developing the hypothesis of Malkus, Howard [5] and

Busse [6] proved a maximum heat transfer range with $Ra \to \infty$ to be determined by the relation $Nu \sim Re^{1/2}$.

Experiments indicate in practice this law to hold for the very heated horizontal plate, which is out of keeping with the data of theoretical investigations obtained by using the hypothesis of Malkus.

According to Chu and others [7], $T \sim y^{-2}$ far from the transfer surface which falls out either the results of Prisley ($T \sim y^{-1/3}$) or the result of Malkus ($T \sim y^{-1}$).

In calculating the heat and mass transfer in convective zones of both stars and the sun "a mixing way" and dimensions of convective elements are supposed to be commensurable with the homogeneous atmosphere height.

Some difficulties in creating complete models of the heat transfer in a turbulent free convection horizontal layer arise for lack of experimental data on the structure of velocity fields.

Under conditions covered one-point hydrodynamic moments are equal to zero. Thus, in developing semiempirical theories the main information should be given by one-point second moments as well as two-point moments (for determination of dimensional scales).

The study of the flow pattern at the heat transfer surface in order to establish a mechanism for transfer in a wall boundary layer and an extension of the "viscous" field is worth special notice. In this case the essential differences will be observed if a surface is rigid or free.

In the case of a free heat transfer surface such as basin surfaces of different scales the heat gravitational convection may be followed by the thermocapillary convection due to the temperature variation of a surface tension.

In our laboratory it was established that the intensive cellular flows near a free heat transfer surface in a turbulent regime of flow originate from the joint effects of thermogravitational and thermocapillary convections. Scales of these flows are of the order of some millimetres but the surface friction and velocity due to thermocapillary forces are sometimes appreciable at temperature drops less than one degree between the surface and volume of fluid. Thus along with centimetre scale effects for consideration in oceanology, it is essential to study microscale flows due to the joint actions of thermocapillary and thermogravitational convections adjacent to a cooled free surface which control all transfer processes in the vital upper layers of different scale basins [II].

There are average temperature and velocity fields adjacent to a vertical heat transfer surface. Consequently, in this case, some methods for investigation of a turbulent boundary layer in a forced flow of heat transfer surfaces may be used. The principal problem is to establish the relationship among the turbulent viscosity, turbulent transfer and time-average velocities and temperatures.

As above mentioned, as early as 1935 [2] I hypothesized the occurrence of a viscous sublayer with the invariable value of the Reynolds proper number in a turbulent boundary layer on vertical heat transfer surfaces at free convection. Due to experimental difficulties in investigating a low velocity nonisothermal stream until recently no this model has been tested.

Subsequently, methods for calculation of a heat flow were developed according to Eckert and Jackson [8].

The principal assumptions of the theory of Eckert and Jackson are the following: a) laws of friction and a heat transfer are the same in a forced flow and at conditions of free convection; b) Reynolds analogy is true for free-convective flows.

The correlation of experimental and calculated values of heat transfer coefficients by the theory of Eckert shows that the close agreement is quantitatively observed in the change range of the Rayleigh number $Ra_{cr} < Ra < 10^{12}$. However, the theoretical exponent $n = 2/5$ and not $1/3$ as can be seen from the experiment.

The correlations of experimental and theoretical values of mean velocities indicated discrepancies in maximun velocity values ($u_m$) of over two fold as in thicknesses of a dynamic layer. Discrepancies in local velocity values proved to be even greater.

In our laboratory some investigations were carried out in order to examine structural features of a free convective turbulent boundary layer.

Measurements of the temperature field in a boundary layer at the isothermal vertical plate were taken by means of microthermocouples made of non-chromium and constantan wires of 0,06mm in diameters. Thermocouples were tared with an accuracy of $0,08°C$. The average values of the electromotive force of the thermocouple were obtained with a potentiometer in recording a non-compensated part of a signal on the automatic potentiometer. The mean values were calculated by records with the interval of about 5 min.

The velocity of a turbulent boundary layer was measured by the stroboscopic visualization method with using an electronic stroboscope developed by V.V.Orlov in our Institute. The method is to photograph the injected particles of $5 \div 15 \mu$ size in a fixed film with the pulse side illumination at the set pulse separations. The measurement error of the instantaneous value of longitudinal velocity component is $\pm 0,27$ mm/sec.

The measured temperature fields in laminar, transient and turbulent flow regime permitted the heat flows to be calculated by the mean temperature gradient adjacent to a wall.

All experimental data indicated the theoretical result of Ostrach to describe the real fluid flow: $Nu = 0,66(Ra/4)^{0,25}$ in the laminar flow regime ($Ra < 3 \cdot 10^9$). For transient conditions, the dimensionless coefficient of heat transfer $Nu$ is not a function of the only criteria $Ra$, $Pr$ and for the turbulent flow field ($Ra > 2 \cdot 10^{10}$), the coefficient of heat transfer is independent of the longitudinal coordinate: $Nu = 0,106(Ra)^{1/3}$. In the turbulent flow regime the velocity and temperature profiles were measured at the same conditions; at heights $x = 275$; 400; 500, 600 mm at three thermal pressures $\Delta T = 7, 10, 12°C$. This has made it possible to estimate a wall friction using the integral pulse relation.

The boundary layer has been found to have three representative regions: 1) the wall region $y < \delta_1$ where $\delta_1$ is the bed depth defined by the linear law of temperature variations against lateral coordinates; 2) the transition region $\delta_1 < y < y_m$; 3) the outer region $y > y_m$, where $m$ is the parameter index at the maximum velocity.

Let us consider the wall flow region where we may observe the linear law of a mean temperature variation against lateral coordinates. We shall let the coordinate $y = \delta_1$ to be the outer boundary of this layer defined by deviation of the temperature field dependence from the linear law. In this flow region it will

# MODEL OF TURBULENT FREE CONVECTION

be noted that: **a)** the temperature gradient normal to a wall $\partial T/\partial y$ is independent of the longitudinal coordinates with a fixed value of $\Delta T$ ; b) the flow resembles flat-parallel that: $V \approx 0$ and the convective transfer may be neglected.

Let us analyse the effect of the Prandtl's number ($Pr_T = \nu_T/a_T$) on the laws of temperature and velocity variations with $y$ adjacent to a wall with Prandtl's different physical numbers. If convective terms in equations of pulses and energy are neglected, thermal flows and the wall friction are calculated by the known formulas:

$$\left. \begin{array}{l} q = (a + a_T)\rho C_p \dfrac{dT}{dy} \\[6pt] \tau = (\nu + \nu_T) \dfrac{du}{dy} \\[6pt] \tau = \tau_o - \displaystyle\int_0^y \rho \beta g (T - T_\infty) dy \end{array} \right\} \quad (I)$$

Following the standard analysis of forced flows, we let $Pr_T$=const and transform the equation (I) as follows:

$$\left. \begin{array}{l} \dfrac{dT}{dy} = \dfrac{q_o}{\rho C_p \left( \dfrac{\nu}{Pr} + \dfrac{\nu_T}{Pr_T} \right)} \\[10pt] \dfrac{du}{dy} = \dfrac{\dfrac{\tau_o}{\rho} - \displaystyle\int_0^y \beta g (T - T_\infty) dy}{\nu + \nu_T} \end{array} \right\} \quad (2)$$

As seen from the equation (2), the occurrence of turbulent viscosity at the certain distance from the wall with numbers $Pr > 1$ leads to the irregularity of the linear law of temperature variation with $y$. This irregularity is the greater, the larger the numeral value of $Pr$.

With large numbers of $Pr$ the relative change of the turbulent transfer effect $\nu_T$ increases. So, for $Pr > 1$ the viscous sublayer thickness is to be estimated by the deviation of mean temperature profile from the linear law, i.e. the most sensitive to the occurrence of turbulent transfer.

For $Pr \ll 1$ the non-linearities of the $T(y)$ variation are slightly affected by the turbulent transfer mechanism. In this case, the velocity profile $U(y)$ is more sensitive to the occurrence of turbulent viscosity whereby it is necessary to estimate a viscous sublayer thickness.

The wall region $y < \delta_1$ is noteworthy. Here, there is a linear variation of the mean temperature against lateral coordinates; the mean temperature gradient $(\partial T/\partial y)_o$ is independent of longitudinal coordinates and the average flow may be regarded as one-dimensional.

Neglecting the turbulent friction, we have the conditions:

$$\left. \begin{array}{l} y < \delta_1 \; ; \quad \dfrac{\partial^2 T}{\partial y^2} = 0 \; ; \quad \nu \dfrac{\partial^2 u}{\partial y^2} = -\beta g (T - T_\infty) \\[8pt] u(0) = 0, \quad u(\delta_1) = u_1, \quad T(0) = T_o, \quad T(\delta_1) = T_1 \end{array} \right\} \quad (3)$$

The occurrence of this heat transfer field is borne out by experimental data shown in Fig.2, i.e. there is a postulated linear temperature distribution:

$$\frac{T_o - T}{\Delta T} = \frac{T_o - T_1}{\Delta T} \qquad (4)$$

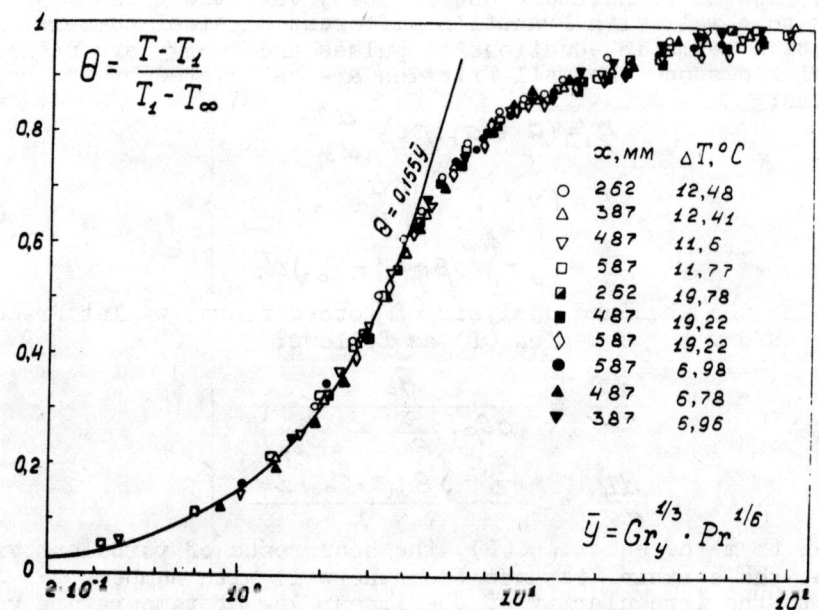

Fig.2. Temperature distribution in the vicinity of a vertical wall. Pr = 16.

For the velocity profile we obtain

$$\bar{u} = \bar{y}\left[1 + \frac{Gr_1}{Re_1}\left(\frac{1}{2} - \frac{T_o - T_1}{6\Delta T}\right)\right] - \frac{Gr_1}{Re_1}\left(\frac{\bar{y}^2}{2} - \frac{T_o - T_1}{6\Delta T}\bar{y}^3\right) \qquad (5)$$

where $\bar{u} = u/u_1$, $\bar{y} = y/\delta$, $Gr_1 = \beta g \Delta T \delta_1^3/\nu^2$, $Re_1 = u_1 \delta_1/\nu$, $\Delta T = T_o - T_\infty$.

With changing the Prandtl's number of over 20 folds (0,7 < Pr < 17) the value $\left[1/2 - (T_o - T_1)/6\Delta T\right]$ varies less than by 5 per cent.

As shown by experiments in Fig.3, the relation $Gr_1/Re_1$ is independent of either longitudinal coordinates or the thermal pressure and it is equal to 1,1 ÷ 1,15 but $Re_1$ and $Gr_1$ are constant with Pr = Const. If $Re_1$ and $Gr_1$ are assumed to be power function of the Prandtl's number ($Re_1 = B/Pr^n$, $Gr_1 = A/Pr^n$) then

$$\delta_1 = \left(A\nu^2/(\beta g \Delta T Pr^n)\right)^{1/3}; \qquad (6)$$

$$\frac{\tau_o Pr^{2/3}}{\rho(\nu \beta g \Delta T)^{2/3}} = \frac{B}{A^{2/3}}\left[1 + \frac{A}{B}\left(0,5 - \frac{T_o - T_1}{6\Delta T}\right)\right] \qquad (7)$$

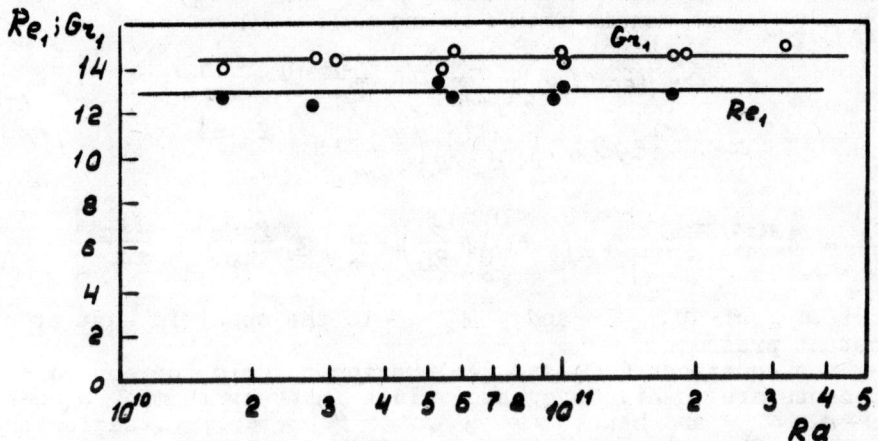

Fig.3. Values of the numbers $Gr_1$ and $Re_1$ characterizing a thickness of viscous sublayer at free convection near a vertical wall. $Pr = 16$.

For $Pr = 16; 0,7$ $Gr_1$ are 15 and 78 respectively, hence $A = 60$, $n = 0,5$. According to our experiments ($Pr = 16$), $\tau_o/\rho(\beta g \Delta T \nu)^{2/3} = 3,2$. From the relation (7), the values $Re_1 = 13,2$ and $Re = 55$ are determined. The relation obtained for $Re_1$ is in good agreement with the data [9].

The flow at the outer part of the turbulent boundary layer ($y > y_m$) has characteristics common to free jets. A linear dependence of a thickness of the outer part of the boundary layer on $x$ is observed, i.e. isotaches of the dimensionless velocity $U/U_m$ are the rays which intercross at the pole $x_o$. From the presented experiments, the angle factor of the ray $U/U_m = 0,5$ is $\delta_{1/2}/(x - x_o) = 0,02$ where $\delta_{1/2} = y_{1/2} - y_m$, $y_{1/2}$ - is the velocity coordinate, $U = 0,5 U_m$; $x = x_{cr}(1 - \alpha_\lambda/\alpha_T)$; $x_{cr}$ is the turbulent boundary layer origin defined by the height distribution of the heat transfer coefficient $\alpha$ and at $Pr = 16$ and $0,7$ $Ra_{cr} = 1,7 \cdot 10^{10}$ and $4,3 \cdot 10^9$; $\alpha_\lambda$ and $\alpha_T$ are mean values of heat transfer coefficients respectively for laminar and turbulent flows. The velocity and temperature profiles $u/u_m = f_1((y-y_m)/\delta_{1/2})$ and $\theta/\theta_m = f_2((y-y_m)/\delta_{1/2})$ respectively are affine-like (Figs 2, 3) where $\theta = T - T_\infty$, $\theta_m = T_m - T_\infty$.

For $y > y_m$ we have the equations

$$\int_{y_m}^{\infty} \frac{du^2}{dx} dy - \int_{y_m}^{\infty} \beta g \theta dy + \frac{\tau_m}{\rho} - (uv)_m = 0 \qquad (8)$$

$$\frac{q_m}{\rho C_p} = \int_{y_m}^{\infty} \frac{du\theta}{dx} dy - (v\theta)_m \qquad (9)$$

Let the following values to be $\theta_m = \bar{\theta}_m \bar{x}_m$, $u_m = \bar{u}_m \bar{x}^n$, $\tau_m = \bar{\tau}_m \bar{x}^\kappa$. Neglecting the terms $(\theta v)_m, (uv)_m$, using the parameter differentiation rule and allowing for that $\partial y_m / \partial x = \text{Const.}$, $q_o = q_m$, we obtain from (8) and (9):

$$\bar{x}^{2n}\bar{u}_m^2 \left[ (2n+1)C \int_0^\infty \left(\frac{u}{u_m}\right)^2 d\left(\frac{y-y_m}{\delta_{1/2}}\right) + \frac{dy_m}{dx} \right] = \bar{x}^{(m+1)} \beta g \bar{\theta}_m C \int_0^\infty \frac{\theta}{\theta_m} d\left(\frac{y-y_m}{\delta_{1/2}}\right) - \frac{\tau_m x^\kappa}{\rho} \quad (10)$$

$$\frac{q_o}{\rho C_p} = \bar{x}^{(m+n)} \bar{\theta}_m \bar{u}_m \left[ C(m+n+1) \int_0^\infty \left(\frac{\theta}{\theta_m}\right)\left(\frac{u}{u_m}\right) d\left(\frac{y-y_m}{\delta_{1/2}}\right) + \frac{dy_m}{dx} \right] \quad (11)$$

where $\bar{x} = x - x_o$ and $C_p$ is the specific heat at a constant pressure.

The equations (10) and (11) having to hold for any $x$ then exponents are to have the same values in every term $2n = (m+1) = \kappa$, $m + n = 0$ and hence $n = 1/3$, $m = -1/3$, $\kappa = 2/3$. Neglecting the term $\tau_m/\rho$ owing to its minimum we obtain

$$\frac{u_m}{(q_o x \beta g/(\rho C_p))^{1/3}} = C_1 , \quad \frac{\theta_m (\beta g \bar{x})^{1/3}}{(q_o/(\rho C_p))^{2/3}} = C_2$$

As found from our experiments, $C_1 = 2,1$; $C_2 = 16,4$ for Pr= 16. According to [9] $C_1 = 2,2$; $C_2 = 16,5$ for Pr = 0,7.

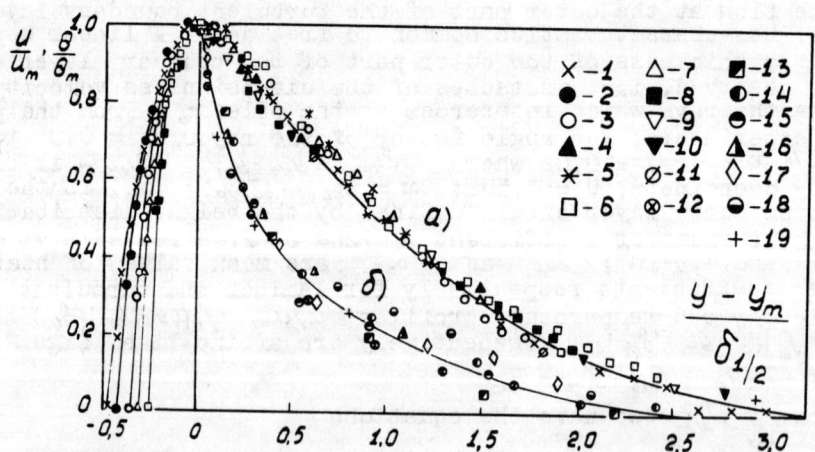

Fig.4a. Velocity distribution in the wall and outer regions of free convection at a vertical wall. The processing of data by the "jet" model.

The figure 4a shows the processing of velocity profiles by mean of the "jet" model. One can see that there is a differentiation of experimental data in the wall region while the excellent generalization of all the data is observed in the outer region of flow. These results are in full accord with our model of a free thermal convection at a vertical wall [2].

The appropriate generalization of experimental data on temperatures in the outer jet region is given in Fig.4b.

| Parameters | 1 | 2 | 3 | 4 | 5 | 6 | 7 | 8 |
|---|---|---|---|---|---|---|---|---|
| $\Delta t$, °C | 6,8 | | | 11,6 | | | | 19,5 |
| $x \cdot 10$ m | 4,0 | 5,0 | 6,0 | 2,75 | 3,63 | 5,0 | 6,0 | 6,0 |
| $Ra \cdot 10^{-10}$ | 3,0 | 5,8 | 10,3 | 1,9 | 5,0 | 11,7 | 20 | 37,1 |
| $u_m$ mm/sec | 25,0 | 26,6 | 27,6 | 30,2 | 33,6 | 35,5 | 39,2 | 47,0 |
| $\delta_{1/2}$ mm | 5,8 | 7,2 | 9,6 | 4,1 | 5,9 | 8,0 | 10,5 | 10,3 |

| Parameters | 9 | 10 | 11 | 12 | 13 | 14 | 15 | 16 | 17 | 18 | 19 |
|---|---|---|---|---|---|---|---|---|---|---|---|
| $\Delta t$, °C | 35,0 | 30,3 | 56,16 | 7,0 | 12,5 | 12,4 | 11,6 | 11,8 | 19,2 | 30,3 | 57,9 |
| $x \cdot 10$ m | 20,0 | 26,0 | 26,0 | 3,87 | 2,62 | 3,87 | 4,87 | 5,87 | 5,87 | 26,0 | 26,0 |
| $Ra \cdot 10^{-10}$ | 2,12 | 4,0 | 6,05 | 2,9 | 1,74 | 5,45 | 10,0 | 18,4 | 35 | 4,0 | 6,04 |

Fig.4b. Temperature distribution in the outer region of a boundary layer. Numbers 9 - 11, 18, 19 are given from the data [9]

## REFERENCES

1. Physical and technical laboratory in Leningrad. Thermotechnics Department. STO, Fedorov Printing-house, Leningrad,1928.
2. Kutateladze S.S.   Zh.Techn.Physics, v.5, N10, p.1706, 1935.
3. Pristley C.H.B.   Austral. J.Physics, v.7, p.176, 1954.
4. Malkus W.V.R.   Proc. Soc., A, v.225, p.196, 1954.
5. Howard L.N.   J.Fluid Mech., v.17, p.405, 1963.
6. Busse F.H.   J.Fluid Mech., v.37, p.457, 1969.
7. Chu T.Y., Goldstein R.I.   J.Fluid Mech., v.60, pt.1, pp. 141-159, 1973.
8. Eckert E.R.G., Jackson T.W.   NACA Rep. No.1015, 1951.

9. Cheesewright R.J. Heat Transfer, Trans. ASME, v.90, p.1-8, 1968.
10. Kutateladze S.S., Kirdyashkin A.G., Ivakin V.P. The turbulent natural convection at a vertical plate. Dokl. Akad. Nauk SSSR, v.217, N6, 1974.
11. Berdnikov V.S., Kirdyashkin A.G. The turbulent convection i in a horizontal fluid layer with a free upper boundary. Parts 1 and 2. Proceedings of the XVIII Siberian Thermophysical Seminar. Institute of Thermophysics, Siberian Branch of the USSR Academy of Sciences, 1975.

# HEAT AND MASS TRANSFER ACROSS GAS FILLED ENCLOSED SPACES BETWEEN A HOT LIQUID SURFACE AND A COOLED ROOF

J. C. RALPH and A. W. BENNETT

UKAEA Harwell, Oxfordshire, England

## ABSTRACT

A detailed knowledge is required of the amounts of sodium vapour which may be transported from the hot surface of a fast reactor coolant pool through the cover gas to cooler regions of the structure.

Evaporation from the unbounded liquid surfaces of lakes and seas has been studied extensively but the heat and mass transfer mechanisms in gas-vapour mixtures which occur in enclosed spaces have received less attention. Recent work at Harwell has provided a theoretical model from which the heat and mass transfer in idealised plane cavities can be calculated. An experimental study is reported in this paper which seeks to verify the theoretical prediction. Heat and mass transfer measurements have been made on a system in which a heated water pool transfers heat and mass across a gas-filled space to a cooled horizontal cover plate. Several cover gases were used in the experiments and the results show that, provided the partial density of the vapour is low compared with that of the gas, the heat transfer mechanism is that of combined convection and radiation. The enhancement in heat transfer due to the presence of the vapour is broadly consistent with assumption of a direct analogy between heat and mass transfer neglecting condensation in the interspace. The mass transfer measurements, in which water condensing on the cooled roof was measured directly, showed for low roof temperatures an imbalance between the mass and heat transfer. This observation is consistent with the theoretical predictions that heat transfer in the convecting system should be independent of the amount of condensation and 'rain-back' within the cavity.

The results of tests with helium showed that convection was entirely suppressed by the presence of the water vapour. This confirms the behaviour predicted for gas-vapour mixtures in which the vapour density is of the same order as the gas density.

## INTRODUCTION

A knowledge of the behaviour of sodium vapour in the cover gas of sodium-cooled fast breeder reactors is important for design purposes and in determining the operating procedures for such systems.

The topic has been the subject of a recent paper by Clement and Hawtin {1} who presented an idealised model for the transport of sodium from a hot pool through a gas space to a cooled horizontal roof.

The theory presented by Clement and Hawtin for an argon cover gas was based on the previously published work of Hills and Szekely {2} who considered the evaporation of hot molten metals into cold surroundings.

It was shown in {1} that significant vapour condensation can be expected in the cover gas when the temperature of the roof is significantly lower than the sodium pool. It was not possible to calculate the amount of the vapour condensing in the interspace which was subsequently transported to the cooled roof. However, it was suggested by Clement and Hawtin that because the evaporation from the liquid pool was enhanced above that which occurs if condensation in the gas is absent, then in practice the amount of fluid transported to the roof may be similar to that calculated without allowing for condensation.

## THEORETICAL

Broadly, the theory {2} applied to the closed cavity {1} introduces a modification to the heat and mass transfer analogy, commonly used for mass transport calculations at low rates, to allow for condensation. Condensation steepens the concentration gradient above the evaporating liquid surface leading to increased mass transfer from the surface. In general, however, the convection flow patterns in the system are not expected to be radically changed from those without the presence of vapour, condensing or otherwise.

### Suppression of convection

Before presenting the key equations developed in {1} which can be applied to the convection experiments described in this paper, an important observation also made in {1} can be stated which shows that radically different behaviour can be expected if a saturated vapour is added to some dry gases. It is postulated by Clement and Hawtin that for a system where the mean vapour density is similar to that of the cover gas the mixed mean density as a function of temperature can cause an inversion preventing convection, even in large cavities. Sodium vapour in helium, for example, should at a fixed total pressure around 1 bar and at 500°C prevent convection. The phenomenon should also be demonstrable in mixtures of water vapour and helium.

### Free convection with mass transfer

In free convection the heat transfer is increased above that of pure conduction by the velocity field in the system. The velocity field also augments the diffusive mass transfer in an analogous manner when phase changes are present in a gas-vapour mixture.

In general terms

$$Nu = f(Gr\ Pr). \tag{1}$$

$$Sh = f(Gr\ Sc). \tag{2}$$

Provided that the respective Grashof number functions are the same, and there is ample evidence of this for low vapour concentrations, then clearly the Sherwood number will equal the Nusselt number provided

$$\frac{Sc}{Pr} = 1 = \frac{k}{\rho C_p D} \equiv \text{Lewis No.} \qquad (3)$$

Since $\quad Nu = \frac{q\ell}{k\Delta T} \qquad (4) \quad$ and $\quad Sh = \frac{i\ell}{D\Delta\rho} \qquad (5)$

then $\quad i = \frac{q}{C_p} \frac{\Delta m}{\Delta T}. \qquad (6)$

Equation (6) enables the mass transfer rate to be calculated from a knowledge of the dry heat transfer, q, provided the concentration of vapour is small and condensation does not occur in the bulk fluid and provided Le = 1. The latter is approximately satisfied for a number of gas-vapour systems although there is often difficulty in calculating the value of Le accurately due to lack of sufficient data for the diffusion coefficient.

If Le ≠ 1 then the generalisation of (6) may be used:

$$i = \frac{q}{C_p} \frac{\Delta m}{\Delta T} \frac{1}{(Le)^{1-n}} \qquad (7)$$

when n is the index of the Grashof number function in (1).

The total heat transfer rate due to the combined convection and simultaneous mass transfer is:

$$q_T = q + i \left[ C_{p\ell} \Delta T + L \right] \qquad (8)$$

which substituting from (6) yields

$$q_T = q \left[ 1 + \Delta m \left( \frac{C_{p\ell}}{C_p} + \frac{L}{C_p \Delta T} \right) \right] \equiv q\,'F'. \qquad (9)$$

To reiterate, the above equations applied to convecting gas plus vapour in a close cavity between a hot pool and a cooled roof are for no condensation in the bulk fluid. The vapour evaporating at the hot surface is assumed to cross the interspace in a supersaturated condition and condense on the cooled roof. When the amount of supersaturation is large then clearly the probability that condensation will occur in the interspace is large.

Because the condensation will steepen the vapour phase concentration gradient above the evaporating pool a higher mass flux than calculated from (6) should result. However, this enhancement should not be reflected in the amount of mass transferred to the cooled roof since the 'fog' droplets formed are most unlikely to be all carried onto the roof.

To calculate the mass transfer in the case of condensation Hills and Szekely took the saturated vapour equilibrium form of 'm', the concentration of the evaporating species, to be given by the equilibrium partial pressure so that

$$m = \exp(B - C/T) \qquad (10)$$

where B and C are constants. The same procedure was followed in {1} and resulted in the expressions summarised below.

Evaporation rate from pool surface:

$$i_E = \frac{m'_{(T_o)}}{1 + \frac{L}{C_p} m'_{(T_o)}} \left[ \frac{\Delta T + \frac{L}{C_p} \Delta m}{\Delta m} \; i \right] \tag{11}$$

where $\quad m'_{(T_o)} = \frac{C}{T_o^2} m_o .$ \hfill (12)

Condensation rate on cooled roof:

$$i_C = \frac{m'_{(T_1)}}{1 + \frac{L}{C_p} m'_{(T_1)}} \left[ \frac{\Delta T + \frac{L}{C_p} \Delta m}{\Delta m} \; i \right] \tag{13}$$

where $\quad m'_{(T_1)} = \frac{C}{T_1} m_1 .$ \hfill (14)

In (11) and (13), i is calculated from (6) or (7).

It is shown in {1} that the total non-radiative heat transfer should be independent of the presence or amount of condensation and also independent of the form $m_{(T)}$ within the cavity. This is a somewhat surprising result, a simplified physical picture being that the latent heat of any condensing vapour in the interspace must be given up locally to the gas which subsequently transfers this heat to the cooled roof by convection.

For temperatures typical of sodium-cooled-pool-type fast reactors it was shown in {1} that the dominant heat transfer mechanism from the pool surface to the surroundings is radiation whatever the cover gas.

The object of the work described in this paper was to verify if possible, using simple water vapour analogue experiments, the theories suggested in {1} and {2} and summarised above.

EXPERIMENTAL

The experiments consisted essentially of measuring the total heat and water transfer rate from a hot water pool to a cooled roof in an enclosed volume with insulated vertical boundary walls. The work was carried out in several pieces of apparatus. That shown in Fig 1 provided a large width-to-height ratio to minimise edge effects but permitted heat transfer measurements only. The apparatus shown in Fig 2 however also enabled condensation on the cooled roof to be measured directly. To facilitate this a convex glass roof, roughened slightly to promote film-wise condensation, was used. A glass roof of this type is not ideal as a constant temperature boundary surface and only approximates to a flat surface. Thus a further rig was also used in which the convex glass roof was replaced with a flat copper one and the heat transfer results compared with both of the other rigs.

# HEAT AND MASS TRANSFER ACROSS GAS FILLED ENCLOSED SPACES

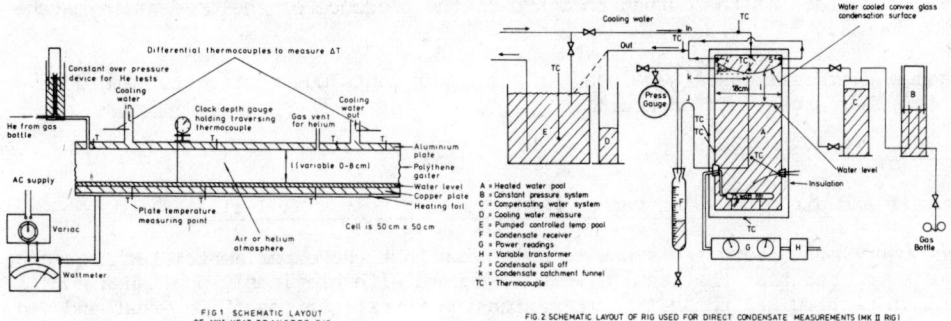

FIG 1 SCHEMATIC LAYOUT OF MKI HEAT TRANSFER RIG

FIG 2 SCHEMATIC LAYOUT OF RIG USED FOR DIRECT CONDENSATE MEASUREMENTS (MK II RIG)

The initial tests were carried out on the Mk I rig with air in the apparatus in the absence of water to establish that the heat transfer coefficients for the dry gas were in agreement with published data. In this way the reliability of the apparatus could be established. Plate gaps varying from a few mm up to several cm were used. The bottom plate temperatures were measured for a range of input powers with a fixed top plate temperature. The results for the narrow gaps were used to establish the critical spacing below which only conduction and radiation across the gap occurred. This in turn enabled an emissivity to be established for the dry surfaces (painted matt black) thus permitting the radiation correction to be made for all of the dry gas cases. The procedure was repeated with a water layer about 0.5 cm deep on the bottom plate. For the tests with water present a total emissivity value of 0.95 for both surfaces was taken from the literature. Fortunately, in the wet tests the radiation correction is relatively small and is not therefore as sensitive to the assumed value of e as in the dry tests. Account was also taken of the radiation absorption by the water vapour but the effect is small even at the largest combination of spacing and vapour pressure.

The procedure for tests using the apparatus shown in Fig 2 was as follows. Test vessel 'A' was filled via vessel 'C' to the level of the condensate collection funnel and a gas purging sequence carried out throughout the system. The purpose of vessel 'B' was to enable a fixed small overpressure to be applied to the gas space in the test vessel. The test vessel immersion heater was switched on and a running temperature set on the pool thermostat control. The constant temperature cooling supply to the top plate was then adjusted to the required temperature and flow and the apparatus allowed to come to equilibrium. During this time the height of vessel 'F', with its attached spill-off leg 'J', was such that condensate simply spilled back to the water pool. At the start of a run the vessel 'F' was lowered so that condensate could enter via 'J'. The level in the test vessel was then noted and the condensation rate measured by timing the spill-off of 10 ml quantities into 'F'. Since the removal of 10 ml quantities from the test vessel is equivalent to only about 0.4 mm level change, the gas space height remained approximately constant during a single condensate measurement.

The procedure was repeated for a range of pool-to-cooling-plate distances. During each run the heat removed via the cooling plate water was calculated from the measured mean temperature rise and the mass flow.

As mentioned above, the heat transfer results were also checked with the curved glass top plate replaced by a flat copper one and the condensate catchment funnel was also removed. This provided a measure of the effect of the unavoidable but non-ideal shape of the curved roof and of the presence of the condensate catchment funnel.

Several different gases were used in the experiments but most tests were done with either argon, air or helium.

RESULTS AND DISCUSSION

Dry air and air plus water vapour heat transfer tests (Mk I rig)

The object of the dry tests was to establish that the experimental techniques were sound and that the results were in accord with previously published work and, in particular, those from the extensive investigations of De Graaf and Van der Held {3}. Only a limited number of dry runs were done. Since agreement with the De Graaf correlations was reasonable, subsequent comparisons with dry gas convection were made using the latter. Fig 3 shows the calculated dry air convection heat transfer and the significantly increased measured heat transfer with water present for several gas gaps as a function of water pool temperature. Theoretical estimates of the heat transfer in the 'wet' case, calculated from (9), are also given in Fig 3. The apparent anomaly in the theoretical line for the 2 cm depth is caused by the fact that for this depth the 'dry' convection flux q used in (9) comes from the laminar region of the De Graaf correlation (see Fig 4). The other theoretical lines display the relative insensitivity of the heat flux to cell depth which is characteristic of 'dry' convection in the turbulent regime.

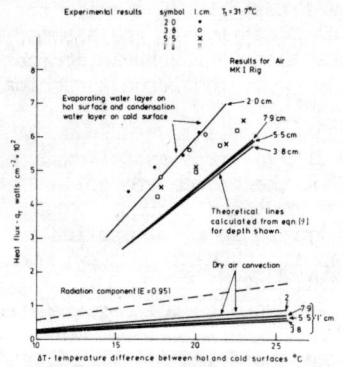

FIG.3 COMPARISON BETWEEN THEORETICAL ESTIMATES BASED ON NO CONDENSATION IN THE GAP AND DIRECT HEAT TRANSFER - MASS TRANSFER ANALOGY EQN [9] AND TEST RESULTS

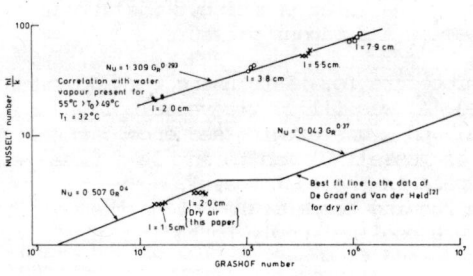

FIG 4 RELATIONSHIP BETWEEN NUSSELT NUMBER AND GRASHOF NUMBER FOR FREE CONVECTION BETWEEN HORIZONTAL SURFACES

It can be seen that there is reasonable agreement between the theoretical and observed results for the 2 cm gap but that for the larger gaps the theoretical results underestimate the observed heat flux somewhat. The dependence of the observed heat flux on the gap is also slightly different from that predicted. The results for the 2.0 cm gap are close to the results for the other gaps in practice, suggesting that the transition region observed in the dry gas may occur at lower Grashof number than that in the 'wet' case.

If the correction for Lewis number (see Table I) is made by using (7) with $n = \frac{1}{3}$ this would increase the theoretical predictions by about 10% giving quite good agreement with the experiments.

As will be noted from (9), the heat transfer increase caused by the water vapour is sensitive to its concentration difference between the pool and the cooled roof. The effective Nusselt number in the wet case should be increased from that given by (4) by the multiplying factor for q in (9). This is provided the vapour has an insignificant effect on the thermal conductivity of the gas. Thus a Nusselt number vs Grashof correlation should only strictly be attempted for fixed temperature levels and differences. Recognising this fact it is however still instructive to present the results for the 'wet' tests as effective Nusselt numbers as a function of Grashof number based on dry air properties. This has been done in Fig 4 together with the original dry air correlation of De Graaf and Van der Held.

The best fit line through the 15 'wet' data points yields the following correlation:

$$Nu = 1.309 \, Gr^{0.293} \tag{15}$$

where $1.3 \times 10^4 < Gr < 1.2 \times 10^6$, $49°C < T_o < 55°C$ and $T_1 = 32°C$.

In the relatively narrow range of Grashof number covered by changing the value of the pool temperature ($T_o$) at a fixed roof temperature ($T_1$) and gas gap the Nusselt number increases much more rapidly than the overall correlation given by (15). This reflects the observed and theoretically predicted behaviour shown in Fig 3. The correlation coefficient for (15) is 0.993 indicating that in spite of the 'local' changes of Nusselt number (due to rapid vapour concentration increase with temperature) the overall correlation based on dry air properties is still meaningful. However it would be less so if the correlation was extended over much wider ranges of $T_o$ and $\Delta T$.

It will be noted that the index of the Grashof number in (15) indicates a turbulent behaviour for the mixed air-water vapour system.

### Argon-water vapour mass and heat transfer measurements (Mk II rig)

A number of tests were carried out on this apparatus in which both heat and mass transfer measurements were made for a range of gas space heights and temperatures. The results showed a similar weak dependence on cavity height to that found on the Mk I rig, indicating turbulent free convection. However, the most revealing result was that obtained for a fixed cavity height and water pool temperature in which the cooled roof temperature was progressively reduced.

#### Heat transfer

The results in terms of heat transfer are shown in Fig 5. The one set of experimental points shown is for the non-radiative heat flux onto the cooled roof measured from a heat balance around the roof and corrected for radiation as on the Mk I rig.

The second set is from the measured condensation rate using equation (8). It can be seen that as the roof temperature is reduced the latter points fall significantly below the directly measured heat flux values. The divergence between the sets of points seems to start at pool to roof temperature differences of around 30°C and is particularly marked at the maximum $\Delta T$ used of 45°C.

Our observations are consistent with the expected behaviour of condensation in the interspace at high $\Delta T$s. The theoretical heat flux calculated from (9), in which condensation is ignored, follows the observed results. This indicates that even in conditions of known condensation the heat transfer may be calculated from the simple no-condensation theory. More evidence is clearly needed of the effect of the vapour and droplet presence on the flow patterns and velocities which have been assumed to be relatively unaffected.

## Mass transfer

The measured mass transfer rate of water onto the cooled roof is shown in Fig 6 together with the predicted rates for the 'no-condensation in the gap' theory from (7). Also shown are the increased evaporation rate from the water pool surface and the reduced condensation arriving on the cooled roof when condensation in the gap is allowed for using equations (7), (11), (12) and (13). At low $\Delta T$s where little condensation is expected in the gap all three calculated mass transfer rates are in close agreement with each other and with the experimental observations. At high $\Delta T$s the observed mass transfer rate lies somewhat above that calculated from (13), which assumes that all water droplets formed in the interspace rain back into the water pool. This was the sort of behaviour expected; Clement and Hawtin {1} conjectured that perhaps about half of the interspace condensation would reach the roof while the other half was returned to the pool. Our observations certainly indicate that water droplets did form in the interspace and it appears that rather more than half returned to the pool.

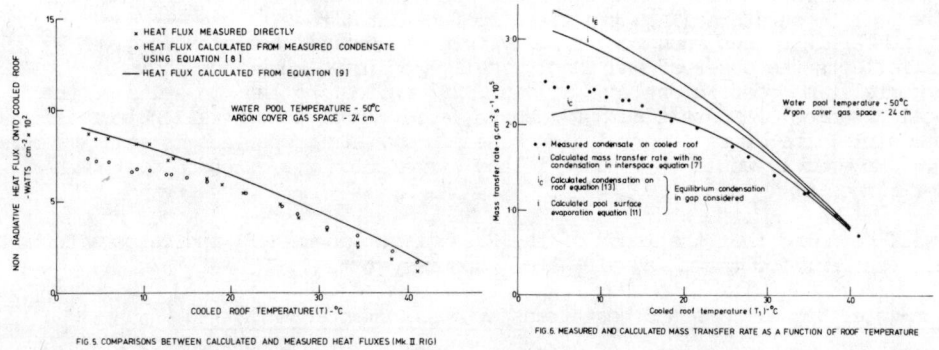

FIG 5 COMPARISONS BETWEEN CALCULATED AND MEASURED HEAT FLUXES (Mk II RIG)

FIG 6 MEASURED AND CALCULATED MASS TRANSFER RATE AS A FUNCTION OF ROOF TEMPERATURE

## Heat transfer measurements with water vapour in argon, air neon and xenon (Mk III rig)

As was pointed out above the unavoidable shape of the roof on the Mk II rig makes the application of the De Graaf and Van der Held correlation for horizontal enclosed surfaces of doubtful accuracy. A direct comparison was therefore made using a flat metal water-cooled roof in which heat transfer only was measured. Several gases of widely differing density and thermal properties were also used in this rig as a further test of the theories.

Measured heat fluxes as a function of gas space for a fixed pool temperature and several roof temperatures are given for argon, air neon and xenon in Fig 7. The results for air and argon are consistent with those obtained on the Mk I and Mk II rigs. This indicates that the curved roof of the Mk II rig and the rather large height-to diameter ratio on both the Mk II and Mk III rigs did not radically change the convection from the situation covered by the De Graaf and Van der Held correlation.

The relative differences exhibited by the different cover gases are best illustrated on an effective Nusselt number vs Grashof number plot. This is done in Fig 8 where it can be seen that the results for each cover gas are displaced from one another but lie on similar gradients. For dry gas, of course, provided the Prandtl numbers are about the same, no such displacement should be seen. A possibility is that the physical properties of the dry gases, particularly the thermal conductivity, are changed by mixing with the water vapour. Calculations

# HEAT AND MASS TRANSFER ACROSS GAS FILLED ENCLOSED SPACES

show that only relatively small changes would result from the latter.

The displacement of the lines is in fact consistent with the calculated enhancement of heat transfer with the Lewis number taken into account, ie, equation (9) with 'i' calculated from (7). The enhancement factor 'F', which multiplies the dry Nusselt number, is given for the various gases at the relevant temperature in Table I.

It was necessary to use calculated diffusion coefficients for water vapour in neon and xenon since measured values could not be found. However, with that proviso, it can be seen that since the above theories imply that

$$Nu_{WET} = 'F_M' \, Nu_{DRY} \tag{16}$$

then from Table I it can be seen that the experimental lines on Fig 8 are displaced to about the expected positions.

TABLE I

| Physical properties at $40^0C$ & 1 bar for → | Air | Argon | Neon | Xenon | Helium |
|---|---|---|---|---|---|
| $\rho$ kg.m$^{-1}$ | 1.1 | 1.5 | 0.76 | 4.9 | 0.16 |
| $C_p$ kJ.kg$^{-1}$ $^0C^{-1}$ | 1.07 | 0.52 | 1.03 | 0.16 | 5.2 |
| k W.M$^{-1}$ $^0C^{-1}$ | 0.0275 | 0.018 | 0.051 | 0.0056 | 0.015 |
| $\mu$ N.S.M$^{-2}$ x $10^5$ | 1.77 | 2.35 | 3.28 | 2.36 | 2.2 |
| D water vapour M$^2$.s$^{-1}$ x $10^4$ | 0.27 | 0.27 | 0.72$^+$ | 0.4$^+$ | 1.01 |
| Le | 0.84 | 0.86 | 0.91 | 0.18 | 1.97 |
| 'F' $T_0 = 50^0C$, $T_1 = 30^0C$ (eqn (9)) | 7.87 | 10.44 | 10.46 (7.8 with $T_1 = 10^0C$) | 10.37 | |
| $'F_M' \equiv 'F'/Le^{0.66}$ | 8.7 | 11.6 | 11.06 (8.3 with $T_1 = 10^0C$) | 30 | |

+ calculated values

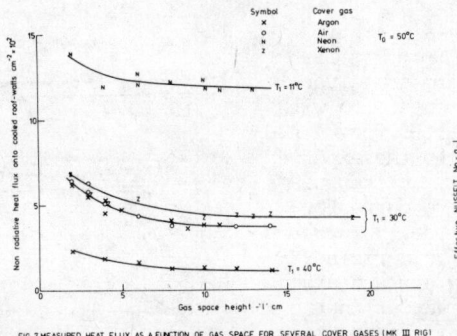

FIG. 7 MEASURED HEAT FLUX AS A FUNCTION OF GAS SPACE FOR SEVERAL COVER GASES (MK III RIG)

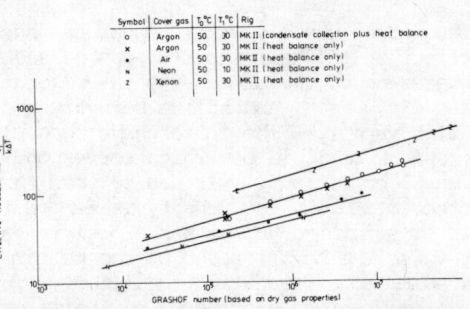

FIG 8 RELATIONSHIP BETWEEN NUSSELT AND GRASHOF NUMBERS FOR WATER SATURATED GASES

## Heat and mass transfer in helium plus water vapour

The condensation rates on the roof of the Mk II rig, both measured directly and calculated from (8) (with q=o), are given as a function of effective gas height in Fig 9. The effective gas space ($\ell$+2.5 cm, see Fig 2) is an approximate allowance for the convex shape of the cooled roof.

It can be seen that the behaviour of the system is markedly different from the other gases studied and shows clearly that diffusion is the controlling factor. It may be noted that the lowest Grashof numbers are all above the critical one and thus convection would be expected in the dry gas. The experiment was repeated on the flat roof rig (Mk III) and excellent agreement obtained. The results of the latter tests are given in Fig 10.

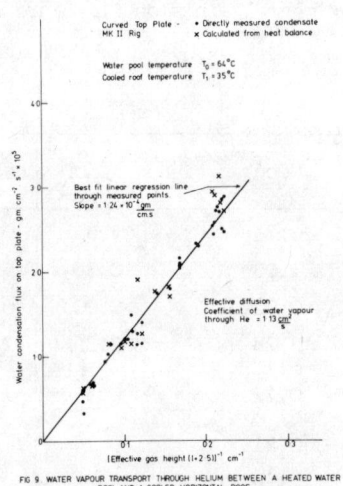
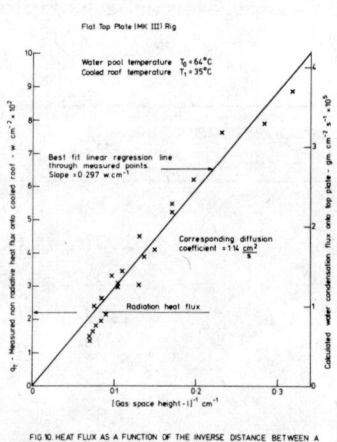

FIG 9 WATER VAPOUR TRANSPORT THROUGH HELIUM BETWEEN A HEATED WATER POOL AND A COOLED HORIZONTAL ROOF

FIG 10 HEAT FLUX AS A FUNCTION OF THE INVERSE DISTANCE BETWEEN A HEATED WATER POOL AND A COOLED HORIZONTAL ROOF WITH A HELIUM COVER GAS

The calculated diffusion coefficients for the tests in the Mk II and Mk III rigs were 1.13 and 1.14 $cm^2.s^{-1}$ respectively. A rather more limited set of tests was done on the Mk I rig {4} and this yielded a diffusion coefficient of 1.03 $cm^2.s^{-1}$.

The values of the diffusion coefficients obtained in our tests are in good agreement with those of Monchick{5} who quotes a measured value of 1.01 $cm^2/s$ for water vapour in helium at $55^{0}C$.

The reason why the water vapour has such a dramatic effect on the helium system can be seen from Fig 11, where the curves of density vs temperature are given for air, argon, and helium both dry and saturated with water vapour. The curves have been plotted for a total pressure of 1.013 bar which corresponds to the experimental conditions. It can be seen that, unlike the other mixtures, the density vs temperature curve for water-saturated helium has a positive slope. This will reverse the normal trend of the hotter fluid rising, leading to a stagnation condition. The air and argon are only slightly affected by the water vapour in the range of temperature shown and thus retain the normal convection behaviour as seen in the tests.

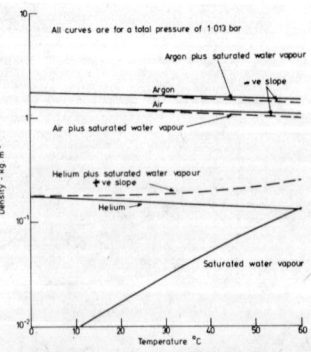

FIG 11 VARIATION OF DENSITY WITH TEMPERATURE FOR AIR AND HELIUM, DRY AND SATURATED WITH WATER VAPOUR

## CONCLUSION

The heat and mass transfer between a hot water pool and a cooled horizontal roof has been studied experimentally for several cover gases at atmospheric pressure. It is shown that with the exception of helium the heat transfer observations were consistent with a system in which combined radiation and convection was occurring. Even in cases of known water droplet formation and 'rain back' into the pool the measured total non-radiative heat transfer was consistent with the assumption of a direct heat and mass transfer analogy ignoring condensation. This confirms the behaviour postulated by Clement and Hawtin {1}.

The mass transfer measurements showed that in cases of significant temperature differences between the liquid pool and the cooled roof, condensation and rain back into the pool do occur and must be allowed for. The theory of Hills and Szekely {2} which was applied by Clement and Hawtin to the closed system under study gives a close estimate to the observed results.

When helium was used as a cover gas the presence of water vapour suppressed convection. Heat and mass transfer then took place by radiation conduction and diffusion. This confirms the theory of Clement and Hawtin, who suggested the possibility during their analysis of sodium vapour transport rhrough fast reactor cover gas.

## ACKNOWLEDGEMENT

The able assistance of Mr P Cannon, who constructed the apparatus and carried out most of the experiments, is gratefully acknowledged.

## NOTE ADDED IN PRESS

A recently published paper by John F van de Vate and A Plomp {6} has also confirmed the suppression of convection in certain gas-vapour mixtures including the water vapour helium system.

## REFERENCES

1. Clement C.F. and Hawtin P.   "Transport of sodium through the cover gas of a sodium cooled fast reactor".   Paper to International Conference on Liquid Metal Technology in Energy Production (Pennsylvania, May 1976).

2. Hills A.W.D. and Szekely J.   Chem Eng. Sc. 19, 79 (1964).

3. De Graaf J.G.A. and Van der Held.   "The relation between the heat transfer and the convection phenomena in enclosed plane air layers".   Applied Science and Research Secta Vol 3 (1953).

4. Ralph J.C. and Sugarman P.   "Free convection heat transfer across gas-filled enclosed spaces between a hot liquid surface and a cooled roof". UKAEA Harwell Report R8244 (1976).

5. Mason E.A. and Monchick L.   J. Chem. Phys. 36, 2746 (1962).

6. Van de Vate John F. and Plomp A.   "Atmospheric stability inside containments with a heated layer of liquid on the floor".   Nuc. Sci. & Eng. 56 196 (1975).

## NOMENCLATURE

| | |
|---|---|
| $C_p$ | Specific heat of gas (J/kg °K) |
| $C_{p\ell}$ | Specific heat of liquid condensate (J/kg °K) |
| D | Diffusion coefficient of vapour in gas (m²/s) |
| e | Total emissivity of surface |
| g | Acceleration due to gravity (m/s²) |
| h | Heat transfer coefficient (W/m² °C) |
| i | Mass transfer rate with no condensation (kg/s m²) |
| $i_C$ | Condensation rate on roof (kg/s m²) |
| $i_E$ | Evaporation rate from pool surface (kg/s m²) |
| k | Gas thermal conductivity (W/m² (°C/m)) |
| $\ell$ | Height of cavity (m) |
| L | Latent heat of vaporisation at vapour source (J/kg) |
| m | Concentration ratio of vapour ($\rho_v/\rho$) |
| $\Delta m$ | Concentration ratio difference between liquid pool and roof ($M_o - M_1$) |
| q | Non-radiative heat flux in dry gas (W/m²) |
| $q_T$ | Total (non-radiative) heat flux in vapour and gas (W/m²) |
| T | Temperature (°K) |
| $\Delta T$ | Temperature difference between hot pool and cooled roof (°K) |
| Gr | Grashof number $\frac{\beta g \rho^2 \ell^3 \Delta T}{\mu^2}$ |
| Le | Lewis number $\frac{k}{\rho C_p D} = \frac{Sc}{Pr}$ |
| Nu | Nusselt number $\frac{h\ell}{k}$ |
| Pr | Prandtl number $\frac{C_p \mu}{k}$ |
| Sc | Schmidt number $\frac{\mu}{\rho D}$ |
| Sh | Sherwood number $\frac{i\ell}{D \Delta \rho}$ |
| $\beta$ | Gas thermal expansion coefficient ($\frac{1}{T}$) (°K⁻¹) |
| $\mu$ | Gas dynamic viscosity (N s/m²) |
| $\rho$ | Gas density (kg/m³) |
| $\rho_v$ | Saturated vapour density (kg/m³) |

## Suffices

| | |
|---|---|
| o | Value at liquid surface |
| 1 | Value on cooled roof |
| $\ell$ | Liquid |
| g | Gas |
| v | Vapour |

# FLOW AND HEAT TRANSFER IN A RECTANGULAR CAVITY AND THEIR DISTURBANCE BY BUOYANCY

DOMINIQUE GRAND

D.T.C.E., Centre d'Etudes Nucléaires, Grenoble, France

ABSTRACT

A two dimensional flow is induced in a cavity by forced flow in a channel adjacent to it. Buoyancy forces can act upon the flow since a positive temperature difference is maintained between one of the walls of the cavity and the fluid in the channel. Quantitative results on the dynamic field include mean velocity and fluctuating velocity profiles. Obtained in isothermal flow, they are only valid for the forced convection limiting case. The departure from this limiting case (with increasing buoyancy effects) is studied for two different thermal boundary conditions. Experimental results include temperature profiles, flow visualisations and overall heat transfer across the cavity.

NOMENCLATURE

| | |
|---|---|
| $g$ | acceleration of gravity (cm/sec$^2$) |
| $k$ | thermal conductibility (cal/cm sec K) |
| $L$ | length of the cavity (cm) |
| $P$ | power transferred across the cavity (W) |
| $T$ | temperature (K) |
| $\bar{u}\ (\bar{v})$ | horizontal (vertical) component of the mean velocity (cm/sec) |
| $u'\ (v')$ | horizontal (vertical) component of the fluctuating velocity (cm/sec) |
| $U$ | reference velocity (cm/sec) |
| $\beta$ | coefficient of thermal expansion (K$^{-1}$) |
| $\Delta T$ | temperature difference across the cavity (K) |
| $\nu$ | kinematic viscosity (cm$^2$/sec) |
| $\psi$ | streamfunction (cm$^2$/sec) |
| $\omega$ | component of the vorticity (sec$^{-1}$) |

Subscripts

| | |
|---|---|
| e | entry of the test section |
| o | central region |

## INTRODUCTION

The french fast breeder reactors are pool-type reactors. Consequently, huge volumes of liquid sodium undergo forced convection (due to the circulation caused by the pumps) and buoyancy effects (due to temperature differences present). The knowledge of the resulting temperature field is important since it contributes to the stresses supported by the vessels. Recirculating flows, which are characterised by closed streamlines, appear in many parts of the reactor.

Recirculating flows appear in many other practical situations, and thus have been the object of numerous studies. But the main theoretical results concerning these flows are for the laminar regime and the two dimensional case. When the Reynolds number is high, the flow can be divided into two parts : an <u>inviscid central region</u> and a <u>viscous wall region</u>. Furthermore, these flow models assume that the viscous wall region is thin so that the inviscid central region tends to occupy the whole flow area. This set of hypothesis is confirmed by experience for flow domains where the lengths in two directions are roughly the same. Also it should be noted that experimentaly the flow is laminar for high Reynolds numbers when the flow boundaries are solid walls /̄1̲/.

For the inviscid central region, the pioneering work of Prandtl /̄2̲/ showed that the vorticity is uniform, i-e the flow is in solid body rotation. Batchelor /̄3̲/ gave a better mathematical basis of this result for isothermal flows. Burgraf /̄4̲/, Grimshaw /̄5̲/ showed that the temperature is, like the vorticity, uniform in non isothermal flows, even when buoyancy effects are present. However, when buoyancy effects are present, the assumption that the viscous region is thin is not true in many cases. So the following discussion is restricted to the isothermal case.

At this stage streamlines can be found by solving the Poisson's equation :

$$\Delta \psi = -\omega_o \tag{1}$$

over the flow domain.

$\omega_o$ is the constant unknown value of the vorticity in the inviscid central region. For complete knowledge of the flow we must match $\omega_o$ to the boundary conditions throughout the wall region. Batchelor /̄3̲/ gave an exact solution to this problem for a circular cavity. Burggraf /̄4̲/ applied it to a rectangular cavity, thus neglecting the influence of the pressure gradient which distinguishes the rectangular cavity from the circular one. Grand /̄6̲/ gave an approximate solution for the rectangular cavity, taking into account the pressure gradient. This method gives good agreement with numerical calculations of Navier-Stokes equations on results such as the length of corner eddies and the strength of the recirculating flow.

For the turbulent regime, most results present in the literature are experimental. They concern a rectangular cavity open to an external flow and are restricted to the forced convection case. Velocity measurements are few and are restricted to the wall region /̄7̲/, or to the mixing layer /̄8̲/, /̄9̲/, /̄10̲/.

The heat transfer was studied by Fox /̄8̲/ and Seban /̄11̲/.

## EXPERIMENTAL PROCEDURE

<u>Test section</u>. Fig. 1 gives a description of the test section. Water is the fluid used in these experiments. A closed loop imposes a forced flow in the channel at the bottom of the test section. A heat exchanger in the loop insures a constant temperature at the entrance of the test section. The flow in the channel is fully-developed upstream of the test section.

The skeleton of the test section is a cube whose framework is made of stainless steel. The elements which form the sides of the cavity are mounted on the framework. Optical windows are used for the lateral sides to allow for

Fig. 1 - The test section

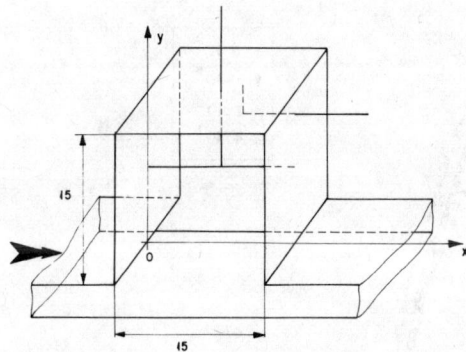

visualisation studies. The other sides can be heating plates or insulated walls and can be arranged in such a way as to explore the following thermal conditions :
- the top wall is heated and the two vertical sides insulated
- one of the vertical walls is heated (the upstream or downstream one in reference to the flow in the channel), and the two other sides are insulated.

Two values of the aspect ratio are also explored : 1 to 1 and 2 to 1. The aspect ratio is the ratio of the height to the width of the cavity.

The heating elements are conceived in such a way as to insure a constant temperature condition. A heating wire is uniformly distributed on a thin steel plate and covered with a layer of silver. Its great thermal diffusivity ensures good longitudinal conduction and a quasi uniform temperature.

For the estimation of the thermal losses, we use the following procedure : the test section is emptied, i-e the water flows only in the channel. Then the heat transfer across the cavity is negligible, and the heat necessary to maintain a temperature difference between the heated plate and the channel balances the thermal losses of the test section. Most of the experiments are done with a temperature difference of less than 40°C. Under these conditions, the heat losses are 25 W when the top wall is heated and 50 W when the vertical walls are heated. In many cases the heat losses are a small fraction of the heat input which varies from 100 to 800 W.

Measurements. Temperature measurements are made with 0.5 mm thermocouples. In the fluid, the vertical translation of a small horizontal rod with 8 thermocouples permits the covering of the vertical mid-plane of the cavity. Under the top wall, the temperature field is measured by 4 thermocouples mounted on a vertical rod which can be translated horizontaly (see Fig. 1). In the heating elements, 3 thermocouples give the wall temperature. $\delta T$ is the maximum temperature difference between the thermocouples of the same heating element and $\Delta T$ the temperature difference across the cavity (more precisely, the temperature difference between the maximum temperature in the heating element and the temperature of the fluid at the entrance of the test section). The ratio $\delta T/\Delta T$ gives an idea of how precisely an isothermal condition is realized on the

Fig. 2 - Horizontal components of mean and fluctuating velocities
        (vertical axis)

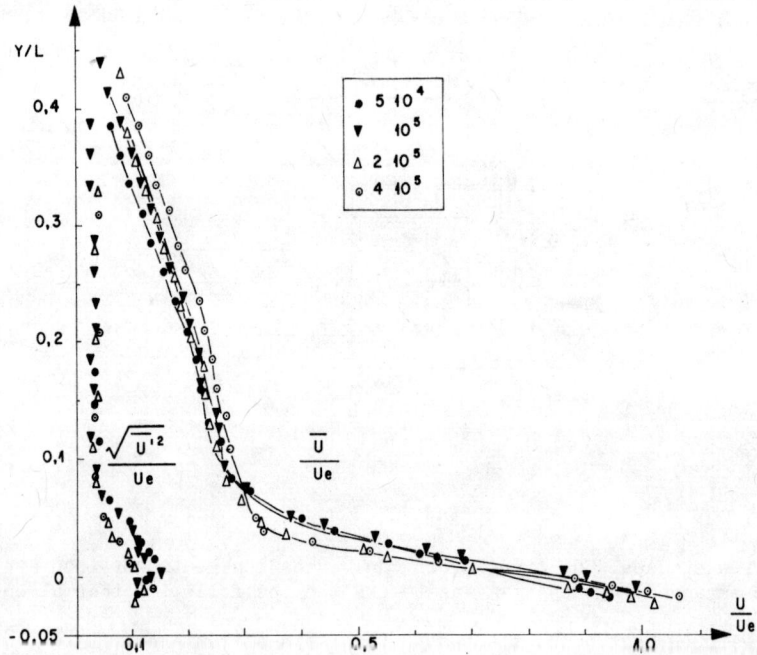

heating element. This ratio is less than :
    10 % for 40 % of the experimental runs
    20 % for 75 %    "      "       "           "
    30 % for 95 %    "      "       "           "
This is quite satisfactory when the difficulty of insuring isothermal conditions is taken into account. The runs with the downstream vertical wall heated are mostly responsible for the departure from constant temperature. In anisothermal studies, the temperature measurements are complemented by visualisations using the colored Schlieren method described in $/ \overline{12} /$. This gives a qualitative description of both the dynamic and the temperature fields which is very instructive.

Quantitative results concerning the dynamic field are restricted to the isothermal case. Velocity is measured with a Doppler laser anemometer developped by the Département Essais of the Electricité de France $/ \overline{13} /$. The device used in these experiments does not have frequency shifting. This prohibits measurements near the center of the recirculating region where the instantaneous velocity may change sign.

DYNAMIC FIELD IN FORCED CONVECTION

All the results presented in this section and in the following concern only the square cavity. Lengths and velocities are made non dimensional using the length L of the cavity and the mean velocity $U_e$ in the channel. Thus the Reynolds number is defined by :

$$Re = \frac{U_e L}{\nu} \qquad (2)$$

Fig. 3 - Vertical components of mean and fluctuating velocities (horizontal axis)

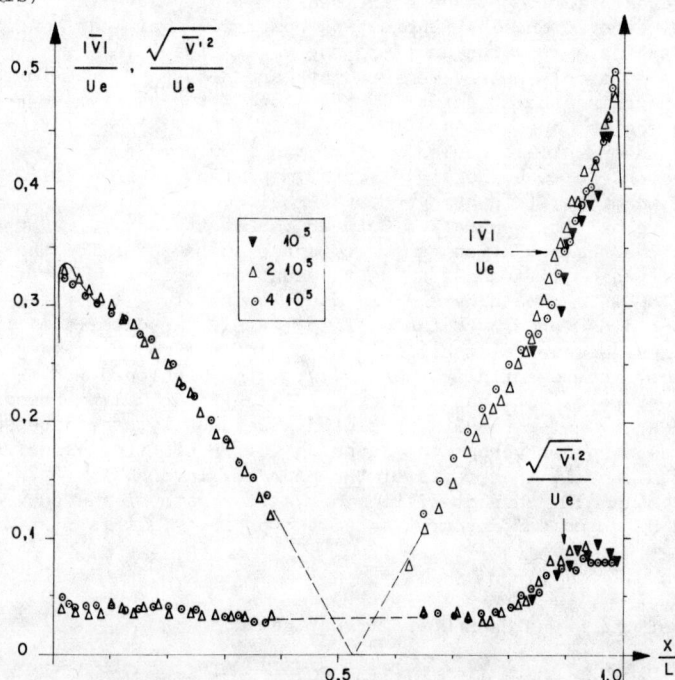

Results are given for values of Re in the range 50,000 to 400,000.

The velocities are measured in the vertical mid plane. Preliminary measurements have shown that the flow in the cavity is two-dimensional (except very close to the lateral sides).

Fig. 2 gives the horizontal components of the mean velocity and of the fluctuating velocity along the lower half of the vertical axis of the cavity. The origin of the vertical y-axis lies in the interface between the channel and the cavity.

With the non-dimensional velocities, the results for different Reynolds numbers are very similar. We can distinguish two regions from these velocity profiles :
- near the interface between the channel and the cavity, the mean velocity profile has an important gradient. The fluctuating velocity is at a maximum in this region of strong turbulence production. Both results are characteristic of a mixing layer.
- inside the cavity, the mean velocity tends to be linear. The constant velocity gradient is approximatively 20 times less than the maximum one in the mixing layer. In the same region, the fluctuating velocity has a uniform value, independent of the Reynolds number.

Fig. 3 gives similar quantities for the vertical components along the horizontal axis of the cavity. The origin of the horizontal x-axis is at the upstream corner of the cavity. Because of limitation on the graphical presentation, only the absolute value of the mean velocity is given. Again two regions are apparent in this profile :
- the wall region where the mean velocity has its extrema. The fluctuating velocity is also at a maximum at the same place. Note the non-symmetric character of these profiles. In the ascending flow, both mean and fluctuating velocities are greater than in the descending flow. This in good agreement with measurements of Roskko /_7_/, represented in this figure by the solid lines

- the central region where the mean velocity profile again is quasi linear. This result is characteristic of a constant mean vorticity, i-e the fluid is in a solid body-like rotation. The fluctuating component again has a uniform distribution with the same value as along the vertical axis.

Although some complementary measurements are necessary, the following assumptions concerning the structure of the flow in the cavity can reasonnably be made. Three regions appear :
- the central region which occupies most of the flow domain. Here, where the viscous effects tend to be negligible, the mean-flow is in a solid body-like rotation and submitted to a homogeneous turbulence.
- the mixing layer at the separation between the cavity and the channel flow which drives the recirculation zone. It is the main center for the production of turbulence which diffuses towards the central region.
- the wall-region which contains the boundary layers along the solid walls. We could not come close enough to the wall to measure the details of these boundary layers. However, a stability analysis of the boundary layers /̲ 6 /̲ shows that they are essentially laminar. Although developed for a laminar central region, we expect that this result is qualitatively correct. Anticipating the heat transfer results presented subsequently, the dependance of the mean Nusselt number versus the Reynolds number confirms experimentally the predominance of the laminar parts of the boundary layers.

An important overall characteristic is the flow rate of the recirculating eddy. Obtained by graphical integration of the profiles, it is :

$$\frac{q}{U_e L} = 0{,}105 \pm 0{,}003 \qquad (3)$$

for the range of the Reynolds numbers investigated.

PERTURBATION OF FORCED CONVECTION BY BUOYANCY

Buoyancy effects introduce the following governing parameter :

$$Ge = Gr/Re^2 = \frac{g\ \beta \Delta\ TL}{U_e^2} \qquad (4)$$

where $\Delta T$ is the difference between the maximum temperature of the heated wall and the temperature of the fluid at the entry of the test section. The other quantities are defined in the nomenclature.

With the Reynolds number given above and the Prandtl number, $Pr = \nu/\alpha$, we have the usual set of three non dimensional parameters necessary for mixed convection studies. The fluid properties are evaluated at the entry temperature which is 17 C for all runs.

The temperature field. Fig. 4, gives some mean temperature profiles along the vertical axis of the cavity. The non dimensionalized temperature plotted along the horizontal axis is :

$$\tilde{T} = \frac{T - T_e}{\Delta T} \qquad (5)$$

In Fig. 4a, where the top wall is heated, the different profiles present a constant temperature region which meets the maximum temperature on the top through a constant gradient region. Visualisations by the Schlieren method showed that the recirculating region is limited to the area where the temperature is uniform. Convective processes are damped in the stratified region. For completeness, we must say that the two runs shown here are representative of the different experimental runs. Also, both visualisations and thermocouples showed that the constant value of the temperature prevails in all of the recirculating region. What distinguishes different experimental runs is the extent of the constant temperature region and the value of this constant temperature. This

Fig. 4 - Temperature profiles along the vertical axis.
   a. top wall heated          b. downstream wall heated

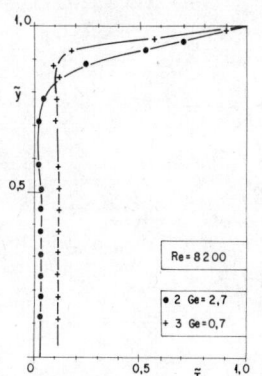 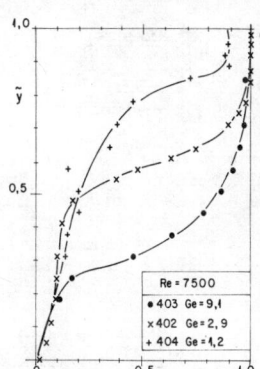

gives experimental evidence of a uniform distribution of the temperature in a region of closed streamlines, even when the flow is turbulent and submitted to buoyancy. It is also important to note that the value of the temperature in the recirculating region is close to the entry temperature. This shows that most of the resistance to the heat transfer through the cavity is caused by the boundary layer under the heated wall.

The two temperature profiles given in Fig. 4a are for the same value of the Reynolds number and two different values of $Gr/Re^2$. The increase of $Gr/Re^2$ -i-e of the buoyancy effects- diminishes both the extent of the constant temperature region and the value of the temperature. We have seen above that, in the limiting case of forced convection, we can model the flow under the top wall as a boundary layer submitted to a pressure gradient imposed by the inviscid central region. When buoyancy effects appear, they cause a stable stratification in the boundary layer. It can be shown that this stable stratification acts like an adverse pressure gradient $/\!\!\_14\_/$ increasing the boundary layer thickness and moving the separation point upstream. Thus the extent of the wall region and the increase of its resistance to heat transfer explain the behavior of the central region ; a smaller area and a lower temperature with increasing $Gr/Re^2$.

In Fig. 4b results are shown concerning the case where the downstream vertical side is heated. By comparison with Fig. 4a, a first and obvious result is that the temperature profile across the cavity depends strongly on the way the stratification is produced. However we still have two regions :
- a region where the temperature is quasi uniform. And visualisations show that it corresponds to the recirculation zone itself.
- a region where visualisations show very little convection and where the temperature profile presents a steep temperature gradient above the recirculation zone which is surmounted by quasi uniform temperature . In a large part of this region, the temperature profile corresponds to conduction.

Comparison of the profiles for different values of $Gr/Re^2$ shows that the central region has a smaller area with increasing values of $Gr/Re^2$. This is a-priori a surprising result since the natural convection along the downstream vertical wall causes an aiding-flow. However, because of the thermal insulation of the other sides of the cavity, the fluid heated along the vertical wall is trapped at the top of the cavity and causes the development of this stagnant region.

The heat exchange. The mean Nusselt number is defined by :

$$Nu = \frac{P}{kL\Delta T} \tag{6}$$

Fig. 5 - Effect of the buoyancy on the heat exchange across the cavity

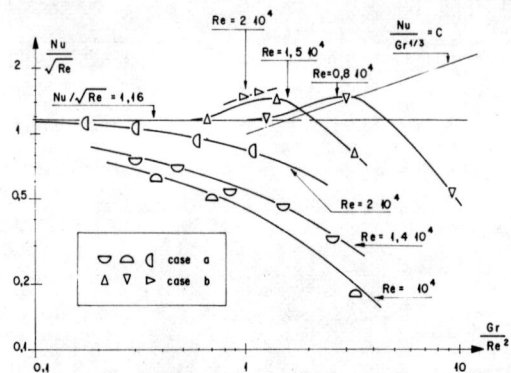

for the square cavity. P is obtained by substracting the thermal losses to the power input.

In the limiting case of forced convection, the correlation for the mean Nusselt number is :

$$Nu = 1,16 \ Re^{0,5} \tag{7}$$

with a standard coefficient of dispersion of $10^{-2}$, and for a range of Reynolds numbers from $3 \ 10^4$ to $3 \ 10^5$. In the case of the square cavity, this holds for both thermal conditions investigated : top wall or downstream vertical wall heated. The value of the exponent confirms the predominance of the laminar regime in the boundary layers along the walls.

How the buoyancy affects the heat transfer is given quantitavely in Fig. 5. The ratio $Nu/\sqrt{Re}$ is plotted against $Gr/Re^2$.

When the top wall is heated, the curves for different values of the Reynolds number exhibit a monotonous decrease with increasing values of $Gr/Re^2$. All the points are below the horizontal line which corresponds to the limiting case of forced convection. For a given value of $Gr/Re^2$, they tend towards this line for increasing values of the Reynolds number. All these results are consistant with the two facts mentionned above : the overall heat transfer across the cavity is governed by the boundary layer and the stable stratification acts like an adverse pressure gradient for the boundary layer.

When the downstream vertical wall is heated, the Nusselt number first increases with increasing values of $Gr/Re^2$, due to the positive influence of natural convection. This is confirmed by the tendency of the experimental points to follow :

$$\frac{Nu}{Gr^{1/3}} = \text{const.} \tag{8}$$

which is characteristic of turbulent natural convection along a vertical wall. But for another increase of $Gr/Re^2$, the hot fluid carried to the top of the cavity cannot be washed out by the recirculating flow. A stable layer develops at the top of the cavity and diminishes the length of the plate influenced by convection. This explains the subsequent decrease of the heat exchange with increasing values of $Gr/Re^2$.

CONCLUSIONS

The turbulent flow in a cavity can be divided into three parts :
- a central region where mean vorticity, mean temperature and turbulent intensity have uniform distributions ;
- a mixing layer which drives the motion inside the cavity ;
- a wall region which controls the heat exchange across the cavity. Of the three regions it is the most sensitive to buoyancy effects.

The changes in the flow configuration and heat transfer which occur with different values of the parameters are associated with the development of the wall region.

REFERENCES

1. Pan, F. and Acrivos, A. 1967. Steady Flows in Rectangular Cavities. J. of Fluid Mech. 28 : 643-656.

2. Prandtl, L. 1904. in : Gesammelte Abhandlungen, 2 : 576.

3. Batchelor, G.K. 1956. On Steady Laminar Flow with Closed Streamlines at Large Reynolds Numbers. J. of Fluid Mech. 1 : 177 - 190.

4. Burggraf, O.R. 1965. A Model of Steady Separated Flow in Rectangular Cavities at High Reynolds Numbers. Proc. of the 1965 Heat Transfer and Fluid Mechanics Institute. Standford Univ. Press.

5. Grimshaw, R. 1969. On Steady Recirculating Flows. J. of Fluid Mech. 39 : 695-703

6. Grand, D. 1975. Contribution à l'Etude des Courants de Recirculation. Thèse de Doctorat d'Etat. Grenoble.

7. Roshko, A. 1955. Some Measurements of Flow in a Rectangular Cutout. NACA TN 3488.

8. Fox, J. 1965. Heat Transfer and Air Flow in a Transverse Rectangular Notch. Int. J. Heat Mass Transfer. 8 : 269-279.

9. Haugen, R.L. and Dhanak, A.M. 1966. Momentum Transfer in a Turbulent Separated Flow Past a Rectangular Cavity. J. of Applied Mechanics. 641-646.

10. Kistler A.L. and Than, F.C. 1967. Some Properties of Turbulent Separated Flows. The Phys. of Fluids Supp. : S 165-S 173.

11. Seban R.A. 1965. Heat Transfer and Flow in a Shallow Rectangular Cavity with Subsonic Turbulent Air Flow. Int. J. Heat Mass Transfer. 8 : 1353 - 1367.

12. Hauff, H. and Grigull, U. 1970. Optical Methods in Heat Transfer. Adv. in Heat Transfer. 6 : 133-366.

13. Ayache, C., Benque, J.P., Dessus, B. and Pernecker, L. 1973. Vélocimètre à Laser pour Liquides. CEA-EDF. Cycle de Conférences sur les Techniques de Mesure dans les Ecoulements. Ermenonville. 24-28 Sept. 1973. Eyrolles, Editeur, Paris, 26 p.

14. Sparrow, E.M. and Minkowycz, W.J. 1962. Buoyancy Effects on Horizontal-Boundary Layer Flow and Heat Transfer. Int. J. Heat Mass Transfer. 5 : 505-511.

# THE STRUCTURE OF TURBULENT FREE-CONVECTIVE FLOWS FROM HEAT TRANSFER SURFACES OF VARIOUS ORIENTATION

A. G. KIRDYASHKIN

*Siberian Branch of the USSR Academy of Sciences, Novosibirsk*

The determination of diagonal scales of recurrent flows and their patterns is one of the complicated problems of the cellular convection in a horizontal layer. The linear theory makes it possible only to compare the velocity-increasing disturbances and to distinguish the highest velocity disturbances. As to the supercritical region, a disturbance spectrum increases together with a maximum disturbance.

All experiments were run in the installation consisting of two flat horizontal heat-exchangers, one of which being transparent. The temperature constancy of heat transfer surfaces was achieved by using a thermostatic water circulating in a cavity of heat-exchangers [1].

Ethyl alcohol was used as a tested liquid ($Pr \approx 16$). Our investigations were performed in the installations having such sizes of heat-exchange surfaces as $400 \times 360$ mm$^2$, $280 \times 210$ mm$^2$, $130 \times 130$ mm$^2$ and layer depths of $2 \div 6$ mm. Aluminum particles of sizes $5 \div 15 \mu$ being in the layer were photographed with a side lighting so that a fixed film showed a representation of flow in a horizontal plane.

The wavelength ($L'$) measurements were made with 5-fold enlarging the photograph showing the flow in a horizontal plane in the interval along the roller commensurable with the roller size. The measurements made are illustrated in probability distribution curves of wavelengths wherefrom it is seen that (Fig.I): a) the more probable wavelength increases with the number Ra; b) the wavelength spectrum expands with increasing Ra snifting into a long wave region; c) the distribution curve is asymmetric against the more probable value of wavelength and the more sloping branch does in the long wave region. The mean wavelength value for a rolling flow pattern increases with the number Ra. The mean relative wavelength is seen to depend on a layer height.

For a liquid possessing physical properties being weakly dependent of temperature (such as ethyl alcohol), the convective flow has a rolling flow pattern with $Ra_{cr} \leqslant Ra \leqslant 10 \cdot Ra_{cr}$. The choice of wave number ($K$) for a stable rolling structure is accidental.

Comparison of experimental boundaries of steady rollers and theoretical those obtained by Busse showed that the long wave branch of Busse corresponds with the most appropriate value of wavelength rather than the stability limit of the rolling flow.

The short wave branch of Busse agrees better with the experimental limit of a rolling regime (Fig.I).

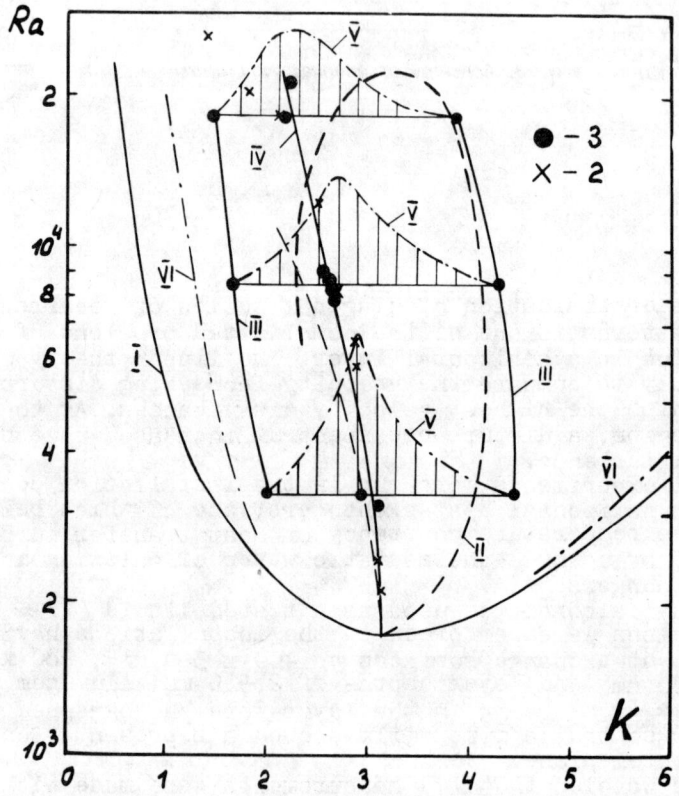

Fig.I.   Stability limit of convective rollers:
I - neutral stability curve; II - stability region curve of Busse's rollers; VI - stability curve of Vasin's and Vlasyuk's rollers.

2 - mean experimental value of the wave Krishnamurty number. Our experiment for Pr = 16:
III - stability limit of rollers; 3 and IV - mean value of a wavelength; V - probability distribution curve.

In cases of large quantities of particles in the layer which gets non-transparent three-dimensional patterns were steady, i.e. polygonal flow patterns occur up to $Ra \approx 6 \times 10^4$ and then they are replaced with unsteady rolling flows involving a short wave overlaped flow along the axis of the latters. The relative dimensions of cells were ($\ell'/\ell \approx 2$) larger than those of rolling patterns ($\ell'/\ell \approx 1,2$).

Instantaneous and average values of temperatures [2] were

recorded at the turbulent free convection conditions in a horizontal layer of the free heat-transfer surface liquid. Such problems as the study of flow patterns at the free heat-transfer surface, a mechanism for transfer processes and the establishment of the flow conditions at a free surface were particularly considered.

Experimental investigations were carried out in the installation consisting of a copper lower heat-exchanger ($397 \times 412 mm^2$) and a transparent upper heat-exchanger whose temperatures required were kept by pumping a thermostatic water through the heat exchanger cavity.

The temperature fluctuations were measured with using thermocouples of 0,3mm in diameter. Mean temperature profiles were measured with a platinum resistance thermometer.

The thermocouple transmitter controling a liquid level measured temperature fluctuations just on the free surface. A thermocouple signal was taken with a potentiometer where the constant signal component was compensated. A variable component passed to the input of a direct current amplifier.

The amplified signal passed through a low-frequency filter onto the loop oscillograph. The section of free liquid surface involving a thermojunction was photographed in step with recording the variable temperature component. The driving of the picture camera was performed by the motor with a speed providing the filming speed of 4,5 frame per second, and also by the own motor of camera (10 $\div$ 24 frame/sec). The eccentric closing contacts of a frame marker is mounted on a shaft of the camera obturator.

The calibration signal of 10 $\mu V$ in the photofilm was about 35 mm. The line thickness due to remained noises was 2 mm. The error in determining the line centre is no more than 0,5 $\div$ 1 mm. So, the resolution of the oscillograph band was 0,01 $^\circ$C.

The estimated value of a thermal inertia constant for the 0,03 mm dia thermocouple under steady conditions is $\varepsilon = 0,0096$ sec. The operating frequency band is 0 $\div$ 10 c.p.s.

The second thermocouple transmitter coupled with a platinum wire composed the probe used for measuring temperature fluctuations and a mean temperature in a liquid volume. The 16 mm dia platinum wire operated as a resistance thermometer.

The wire length of probe (125 mm) was taken so that to cross at least three large-scale cells. The recording time was determined experimentally for every regime and chosen as minimum-required for the reliable averaging of records by planimetry.

The system resolution amounted to 0,01$^\circ$C. The absolute error in a horizontal position of the wire and in determining coordinates is no more than 0,01 mm. $\ell = 45$ mm, i.e. $B/\ell > 10$ where $B$ is the lateral layer dimensions. The fluid layer (ethyl alcohol) at the upper free heat-transfer surface was investigated. The free heat-transfer surface cooling was effected through the air layer by the upper transparent heat exchanger. The layer height-average temperature profiles with one free heat-transfer surface are given in Fig.2. The temperature profiles adjacent to the heat-transfer surface are shown scaled up. The average temperature profiles have reverse temperature gradients ($\partial T/\partial y$). This circumstance favours the occurrence of the large scale cellular motion commensurable with the layer height.

The temperature gradient (accurate to $\pm 3\%$) and the heat flow were determined over the linear range of temperature measurements

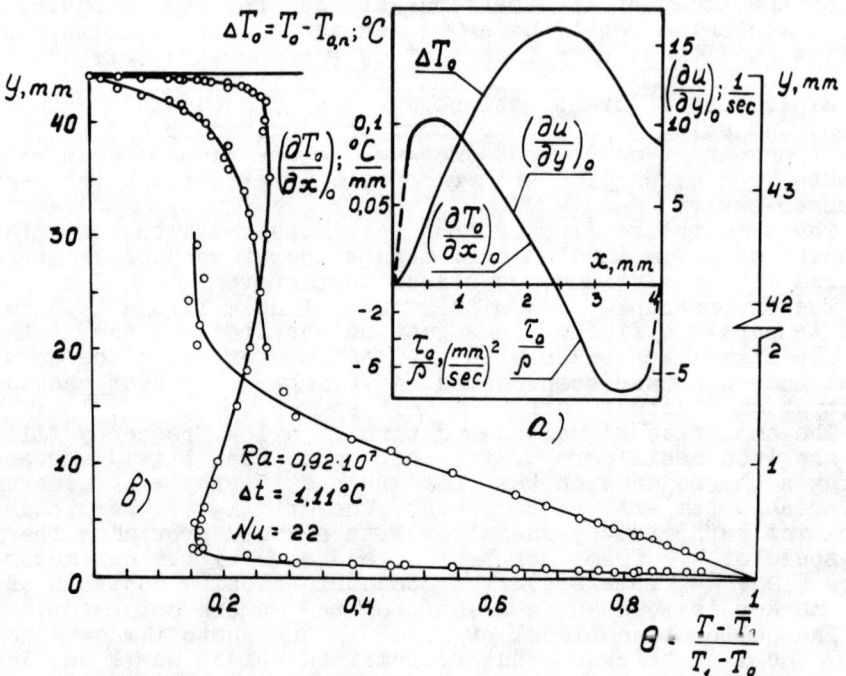

Fig.2. a) Mean temperature profiles in a horizontal layer with free surface in a turbulent regime.
b) Temperature and friction distributions on a surface of a microscale cell.

from the coordinate " $y$ " at a rigid heat transfer surface.

The measurement results are given in Table I. As can be seen from the table the essential differences in the transfer rate are observed near the heat transfer surface. The values of $Nu_x/Nu_1$ for different regimes are summarised in Table I, point 5 where $Nu_x = \alpha_x \ell/\lambda$, $Nu_1 = \alpha_1 \ell/\lambda$, $\alpha_x = q/\Delta T_x$, $\alpha_1 = q/\Delta T_1$; $\Delta T_x$, $\Delta T_1$ are the temperature drops in thermal layers at the free and rigid heat transfer surfaces respectively.

Visual investigations indicated the small-scale cellular flows to occur adjacent to a free heat-transfer surface against the large-scale flows commensurable with the layer thickness (Fig.3). Small-scale cells "generate" in the region of upward flows of large scale motions, "drift" at the average rate of large scale motions ( $U_\varrho$ ), and have a certain "life-time" ( $\mathcal{V}$ ). From the film we determined: a)the wave length spectrum ( $L'$ ) of a small-scale motion; b)the cell "drift" ( $U_\varrho$ ).

Measurements showed that: a)the more probable wave-length of a small scale motion diminishes with increasing $\Delta T$ ; b)the wave length spectrum narrows with increasing $\Delta T$ . The cell sizes account for some millimetres and are much less than a layer depth:

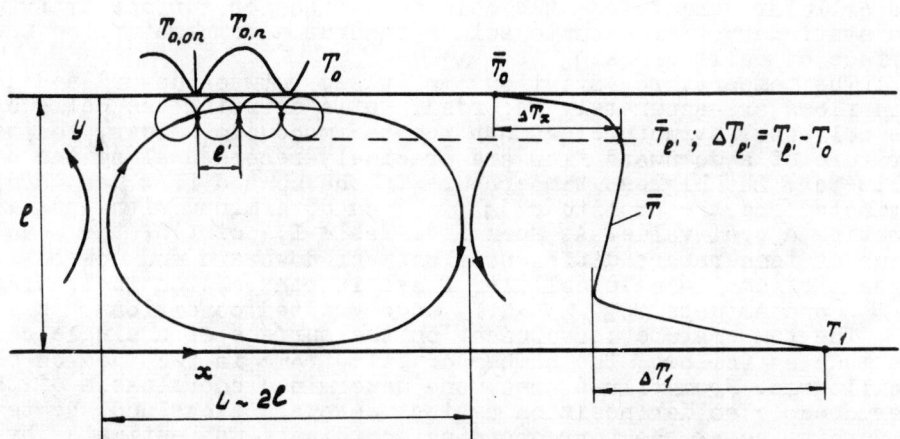

Fig.3. Flow circuit in a turbulent horizontal liquid layer with a free upper boundary.

$\ell'/\ell = 0,1; 0,03 \div 0,06$.

The filming with using a camera was made at the certain rate and on a graphic scale. Over the certain number of frames the shift of single cells on the free surface was determined and in this manner the cell drift velocity was estimated as one integral unit at the heat-exchanger surface. The cell drift velocity is also that of a liquid flow in a large scale cell.

With increasing the temperature drop and layer height the drift rate increases, a spectrum of the latter extending, shifting into a zone of higher velocities (Table I, point 7).

As a result of the film examination, it was found that a single cell occured in the frame visibility field for all the drift-time. Thus, the life-time may be evaluated by the relation $\mathcal{V} = \ell/u_o$.

The synchronous filming of flow patterns and the temperature record made it possible to place the position of a thermocouple solder and the temperature in the given point of cell, i.e. to determine the surface temperature of certain cells.

The correlation of oscillogram and films showed that in the downward flow region the sharp change of the temperature gradient sign is observed on the oscillogram. In the region of the upward flow the surface temperature varies slightly. So, the temperature difference between the downward and upward flows in the small scale cells is determined by the mode of temperature variations on the oscillogram and by correlating cynchronous measurements of film frame temperatures.

The probability distribution curves of temperature differences between the downward and upward flows on the heat transfer surface are obtained $\Delta \overline{T}_o = (\overline{T_{o,on} - T_{o,n}})$, where $T_{o,on}$, $T_{o,n}$ are the temperatures in the downward and upward flows on a free surface of the unit cell.

The measurement of the value $\Delta T_o$ was made with the oscillogram of temperature fluctuations on the surface at one point. The cellular structures at a cold heat-transfer surface drifting the stationary thermocouple solder records a temperature on the surface of cells crossing it.

The temperature variation amplitudes between upward and downward flows are accurately recorded. While drifting the cells of the polyhonal structure against the thermocouple solder, the temperature of a downward flow are precisely recorded along the cell perimeter. The highest temperature in the upward flow was hardly estimated and the amplitude $\Delta T_o$ was determined with underestimating a real value. As seen from Table I, point 8, the mean value of temperature differences between downward and upward streams of small-scale cellular flows increases with increasing $\Delta T$ and amounts to $\Delta T_o / \Delta T \sim 0,1$. under tested conditions.

The temperature measurement on the surface of a single cell was made as follows. The number of film-frame is recorded on the oscillogram. From film-frames, one determined coordinates of the thermocouple solder position against downward flows and the temperature value at the corresponding coordinate was estimated by the oscillogram.

In Fig. 2b are set out the temperature profile on the single microcell surface; the temperature gradient along the surface $(\partial T/\partial x)_o$; the horizontal velocity gradient at the surface $(\partial u/\partial y)_o$, the surface friction $\tau_o = \mu (\partial u/\partial y)_o = \partial \sigma / \partial T (\partial T/\partial x)_o$, where $\sigma$ is the surface tension, $\beta = \partial \sigma / \partial T$.

The mean values for $(\partial u / \partial y)_o$, $\bar{u}_o$ on the surface are given in Table I (points 9, 10).

Thus on the free heat transfer surface together with the thermal gravitational convection, the other mechanism, namely, thermocapillary convection, is included which acts in one direction with the first one.

The values of gravitational and thermocapillary forces in the surface cellular layer are estimated. The cells near the surface are supposed to resemble rollers having a height and a length $\ell' = L'/2$.

The values of the Marangony number defined from the average temperature difference in the rising and falling flows at the small-scale cell and the average size of one roller are summarized in Table I, point 11. The values of Rayleigh and Marangony numbers defined by the layer thickness $\bar{\ell}'$, if a roller is regarded as symmetrical, and by the temperature difference at the depth of the velocity layer $\bar{\ell}'$: $\Delta T_{\bar{\ell}'} = \bar{T}_{\bar{\ell}'} - \bar{T}_o$ are given in Table I, points 13 and 14. The Marangony and Rayleigh numbers being proportional to $\Delta T$, the relation between them is linear; when the thickness of layer and the physical parameters of liquid are fixed: $Ra = \dfrac{\rho \beta g (\bar{\ell}')^2}{\beta} \cdot Ma$.

The value of dimensionless complex $\rho \beta g (\bar{\ell}')^2 / \beta$ for a small-scale flow amounts to $0,1 \div 1$. That is to say, the action of the thermogravitational forces will be similar to that of the thermocapillary forces when the number Ra will be commensurable or by one order larger than values Ma. As can be seen from Table I, the values $Ma_{\bar{\ell}'}$ and $Ra_{\bar{\ell}'}$ are of the same order of magnitude.

So, in the horizontal layer of liquid with one free and one rigid heat transfer surfaces in the turbulent flow regime there are large scale flows in which a transfer is effected by the thermogravitational convection. There are cellular flows near the free heat transfer surface wherein the transport processes are determined by the joint action of thermocapillary and thermoconvective forces.

The differences in the heat-transfer coefficients on the free (cold) heat-transfer surface and as well as on rigid (heated) one are accounted for that two transfer mechanisms, i.e. thermocappilary and thermoconvective those, operate jointly at a free surface; the free convection transfer taking place at the rigid heat transfer surface.

3. The general structural features of the free convection boundary layer on the isothermal vertical plate are stated in the lecture of S.S.Kutateladze.

The experimental profiles of the mean velocity and temperature as well as the friction distribution along the boundary layer thickness ( [4] , Fig.4) permitted three characteristic regions to be distinguished: 1) near the wall $y < \delta_1$ where $\delta_1$ is the thickness of layer determined by the linear law of temperature variations versus the transversal coordinate; 2) the transition one $\delta_1 < y < y_m$ ; 3) the outer region $y > y_m$ where $m$ is the index of parameters at maximum velocity.

The characteristic feature of the turbulent boundary layer in the natural convection is its large alternation [3] .

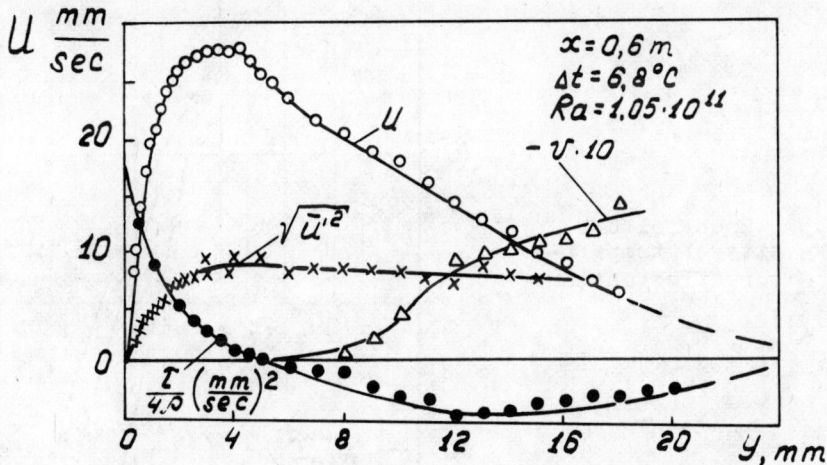

Fig.4. Distribution of longitudinal ( $u$ ), transversal ( $v$ ) components of velocity, friction ( $\tau/4\rho$ ) and $(\overline{u'^2})^{1/2}$ along the boundary layer thickness on a vertical isothermal plate.

The alternation factor $\varphi$ is the magnitude determined as a ratio of the life-time of turbulent fluctuations to the total record time. The calculations of alternation factors from the records of instantaneous temperature values indicated the alternation factor to be equal to 1 in the viscous sublayer and transient region of a boundary layer. The temperature fluctuations in the viscous sublayer and transient region where the alternation factor is equal I can be studied by statistical methods with calculating the spectral distribution of temperature fluctuations [3].

The results of investigations showed that the profiles of mean square values of the longitudinal velocity component fluctuations are automodeled at coordinates $(\overline{u'^2})^{1/2}/u_m$, $(y-y_m)/\delta_{1/2}$. The maximum mean square value of the velocity fluctuations is displaced into the outer region of the boundary layer and equal to $0,3\, u_m$. The profiles of the absolute mean values of the longitudinal velocity component fluctuations are automodeled at coordinates $|\overline{u'}|/u_m$, $(y-y_m)/\delta_{1/2}$.

The mean values of absolute quantities of the velocity fluctuation for the outer region of a boundary layer depend weakly on $y$. Expressing $|\overline{u'}|$ in the form of $|\overline{u'}|=c\ell|\partial u/\partial y|$ and considering $\partial u/\partial y \sim u_m/\delta$ it follows that $\ell \sim x$ in the outer region of the boundary layer.

The good correlation of temperature fluctuations along the coordinate ($R_y \geqslant 0,7$) perdendicular to the wall is observed in the viscous sublayer whereas the temperature fluctuation dispersion decreases of over 30 fold.

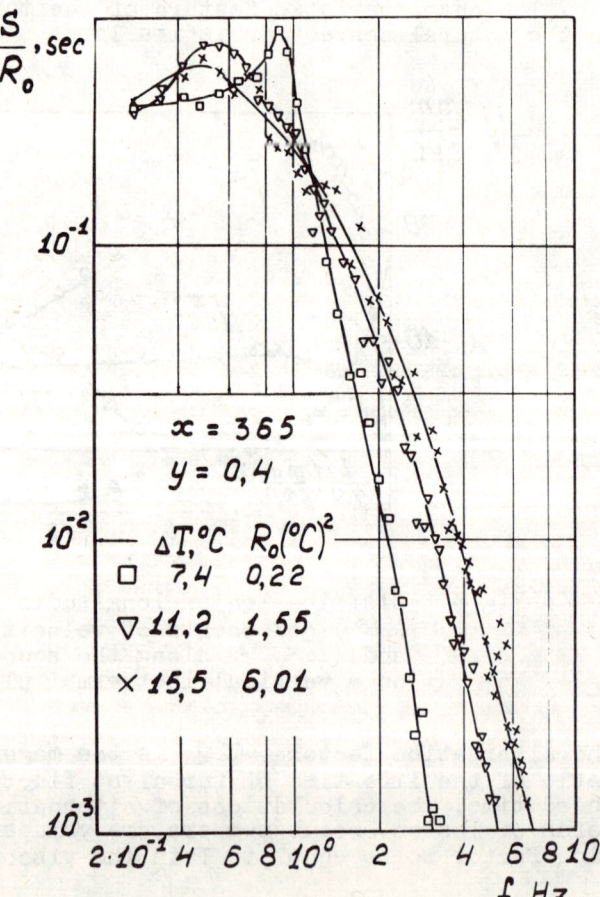

Fig.5. Spectral density of temperature fluctuations

The distribution of spectral density of temperature fluctuations was calculated by experimental autocorrelated functions.

The temperature fluctuation spectra are low-frequency and with increasing the temperature pressure the spectrum shifts into the higher frequency region. With maximum (in our case) temperature pressure $\Delta T = 15,5°C$, the spectrum values are in the frequency region up to 8 c.p.s. (Fig.5) [3]. The temperature fluctuation spectrum was found to shift into the low-frequency region as it approaches to the wall.

Table I.

| № | Regime | | I | II | III | IV |
|---|--------|---|------|------|------|------|
| 1 | $\ell$ | mm | 28,68 | 44,45 | 44,0 | 42,9 |
| 2 | $\Delta T = \overline{T}_o - T_1$ | °C | 0,53 | 0,76 | 1,11 | 2,0 |
| 3 | $10^{-7} Ra = 10^{-7} \frac{\beta g \Delta T \ell^3}{a \nu}$ | — | 0,13 | 0,69 | 0,92 | 2,0 |
| 4 | $Nu = \alpha \ell / \lambda$ | — | 13,0 | 18 | 21,9 | 28 |
| 5 | $Nu_x / Nu_1 = \Delta T_1 / \Delta T_x$ | — | 2 | 2,9 | 3,1 | 4 |
| 6 | $\overline{\ell}' = \overline{L}'/2$ | mm | 3,05 | 2,45 | 1,75 | 1,3 |
| 7 | $\overline{u}_g$ | mm/sec | 1,9 | 2,6 | 4,7 | 6,3 |
| 8 | $\Delta \overline{T}_o = \overline{(T_{o,n} - T_{o,on})}$ | °C | 0,094 | 0,088 | 0,096 | 0,19 |
| 9 | $\overline{(\partial u / \partial y)_o}$ | 1/sec | 2,7 | 3,2 | 6,9 | 18 |
| 10 | $\overline{u}_o = \frac{\overline{\ell}'}{2} \left(\overline{\frac{\partial u}{\partial y}}\right)_o$ | mm/sec | 4,5 | 3,9 | 6 | 11,5 |
| 11 | $Ma_o = \beta \Delta \overline{T}_o \overline{\ell}' / \mu a$ | — | 290 | 280 | 170 | 357 |
| 12 | $\Delta \overline{T}_{\overline{\ell}'} = \overline{T}_{\overline{\ell}'} - T_o$ | °C | 0,18 | 0,19 | 0,28 | 0,32 |
| 13 | $Ra_{\overline{\ell}'} = \beta g \Delta T_{\overline{\ell}'} (\overline{\ell}')^3 / a \nu$ | — | 480 | 250 | 140 | 120 |
| 14 | $Ma_{\overline{\ell}'} = \beta \overline{\ell}' \Delta \overline{T}_{\overline{\ell}'} / \mu a$ | — | 540 | 470 | 490 | 600 |

REFERENCES

1. Kutateladze S.S., Kirdyashkin A.G., Berdnikov V.S. The velocity field in a convective cell of a horizontal liquid lay-

er at thermal gravitational convection. Izv. Akad. Nauk SSSR, Physics of Atmosphere and Ocean, v.I0, N2, 1974.
2. Berdnikov V.S., Kirdyashkin A.G. Turbulent convection in a horizontal liquid layer with a free upper boundary. Sbornik "Wall turbulent flow", part 2. Proceedings of the XVIII Siberian Seminar on Thermophysics. Novosibirsk, Institute of Thermophysics of the Siberian Branch of the USSR Academy of Sciences, 1975.
3. Ivakin V.P., Kirdyashkin A.G., Chernyarsky L.I. The investigation of a turbulent boundary layer structure with natural convection at a vertical plate. Sbornik "Wall turbulent flow", part 2. Institute of Thermophysics of the Siberian Branch of the USSR Academy of Sciences, Novosibirsk, 1975.
4. Kutateladze S.S., Kirdyashkin A.G., Ivakin V.P. Turbulent natural convection at a vertical plate. Dokl. Akad. Nauk SSSR, v.217, N6, 1974.
5. Leontiev A.G., Kirdyashkin A.G. Experimental study of flow patterns and temperature fields in horizontal free convection liquid layers. Int. J.Heat Mass Transfer, v.II, pp. 1461-1466, 1968.

# A THEORETICAL INVESTIGATION OF BUOYANCY-INDUCED FLOW STRATIFICATION IN THE CYLINDRICAL OUTLET PLENUM OF A LIQUID-METAL-COOLED FAST BREEDER REACTOR

NICHOLAS C. G. MARKATOS

*Imperial College of Science and Technology, London, SW7*

## ABSTRACT

The method and results of a numerical procedure for solving the elliptic differential equations governing the transient turbulent flow and heat transfer of a heavy fluid jet issuing vertically into a volume of relatively light fluid are presented. This situation arises in the outlet plenum of a Liquid-Metal-Cooled Fast Breeder Reactor (LMFBR) during reactor scram transients. The time-averaged conservation equations for momentum and heat transfer along with a two-equation model of turbulence and proper modelling of the buoyancy terms, were solved on a CDC 6600 digital computer for various inlet transients and fluids and the results are presented in the form of vector plots and temperature contours. All results are in agreement with expectations, invariably establishing the short-circuiting of the plenum.

## NOMENCLATURE

| | |
|---|---|
| $C_1$, $C_2$, $C_\mu$, $C_D$ | constants in the turbulence model |
| $\overline{C}_p$ | specific heat at constant volume (mean value, J/Kg K) |
| h | stagnation enthaly (J/Kg) |
| k | turbulence kinetic energy ($m^2/sec^2$) |
| p | static pressure ($N/m^2$) |
| t | time (sec) |
| T | temperature (K) |
| u, v | axial and radial velocity components (m/sec) |

## GREEK SYMBOLS

| | |
|---|---|
| $\beta$ | coefficient of volumetric expansion ($K^{-1}$) |
| $\Delta \rho$ | change in density |

| | |
|---|---|
| ε | dissipation rate of turbulence energy ($m^2/sec^3$) |
| μ | molecular viscosity (Kg/ms) |
| ρ | density ($Kg/m^3$) |

SUBSCRIPTS

| | |
|---|---|
| eff | effective |
| l | laminar |
| m | mean value |
| t | turbulent |

INTRODUCTION

A heavy fluid jet issuing vertically into a volume of relatively light fluid will not rise indefinitely but will reach some maximum height and subsequently flow downward.

At the time this work was finished (April 1975) no previous work had been done on dense jets entering a finite vessel. This situation arises in the outlet plenum of a Liquid-Metal-Cooled Fast Breeder Reactor (LMFBR) during reactor scram transients. In the primary coolant loop of the Reactor, sodium coolant enters the pressure vessel through three inlet nozzles and flows upward from an inlet plenum, through the reactor core and other components and instrument trees requiring heat removal, into an outlet plenum. Figure 1 presents a simplified sketch of the desired outlet plenum, in the idealised form adopted for the present computations.

Under steady state isothermal operation, sodium enters the plenum through an inlet structure, and after mixing within the plenum exits through three outlet nozzles. The present model includes simulated core outlet (the actual open area being less than the total cross-sectional area illustrated in the above figure), simulated plenum outlet and simulated pool cover interface [1].

During a reactor scram transient, the plenum experiences an abrupt decrease in the entering temperature and consequently an increase in the entering fluid density. This temperature decrease is accompanied in a normal reactor scram by an exponential flow coastdown to about 10% of the initial flowrate. The design of the outlet plenum and components in a LMFBR reactor, is largely governed by their response to thermal transients. The large fluid volume in the outlet plenum provides a region in which thermal transients at the inlet can be significantly mitigated, due to flow mixing in the plenum.

If, however, the cooler, denser sodium has insufficient inertia upon entering the plenum to overcome the negative buoyancy forces, the incoming fluid will immediately be forced downward and outward toward the exit nozzles, thus short-circuiting the plenum and creating a stratified flow pattern. Potential areas of stagnation could result in temperature differentials in the outlet region.

THE MATHEMATICAL FORMULATION

The path of the coolant flow is vertically upward in the

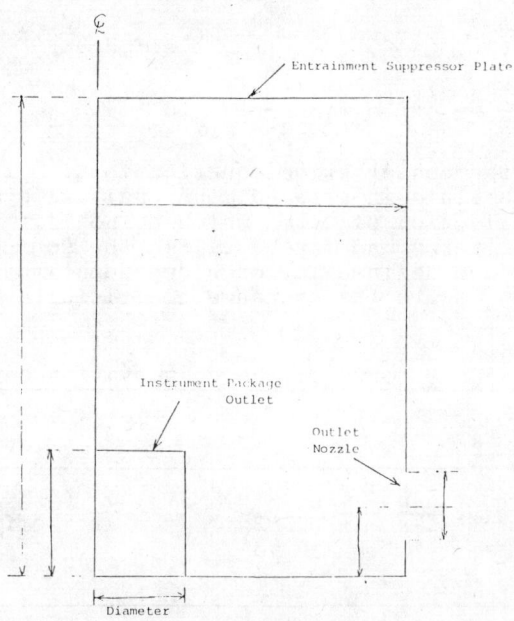

Fig. 1  LMFBR Outlet Plenum Elevation Schematic

core zone. The flow then deflects horizontally at, or near the pool cover-gas interface, passes outward at the top and then passes downward near the vessel wall toward the outlet nozzles.

For the present problem, the plenum domain will be modelled in cylindrical co-ordinates as a two-dimensional axi-symmetric field. Accordingly, the three outlet nozzles will be modelled as a continuous annular slot of area equivalent to the three outlet nozzles, and with its centre at the proper elevation. The upper surface of the vessel is assumed to be a solid surface in contact with the plenum fluid and located at the same elevation as the entrainment suppressor plate.

This flow depends on two space variables; but, there is no dominant direction of flow; such a flow can be described mathematically by partial differential equations of the elliptic type.

Under most circumstances the flow field within the plenum will be turbulent. To account for turbulence, a two-equation model of turbulence is included. This model which is described in (2, 3, 4) uses as dependent variables the kinetic energy of turbulence (k) and its dissipation rate ($\varepsilon$).

Within the above framework, the independent variables are the axial (x) and radial (r) components of a cylindrical-polar co-ordinate system and the time (t); the dependent variables (time-averaged values) are u,v,p,k,$\varepsilon$ and h.

The differential equations for the above dependent variables can be expressed in the single general form:

$$\frac{\partial}{\partial x}(\rho u\phi) + \frac{1}{r}\frac{\partial}{\partial r}(r\rho v\phi) = \frac{\partial}{\partial x}(\Gamma_{eff,\phi}\frac{\partial \phi}{\partial x}) +$$

$$+ \frac{1}{r}\frac{\partial}{\partial r}(r\,\Gamma_{eff,\phi}\frac{\partial \phi}{\partial r}) + S_\phi \quad (1)$$

This is the conservation equation for the transport of a property $\phi$ of the fluid in a two-dimensional axi-symmetric domain, where $S_\phi$ is a collection of terms which do not fit in the framework of the other terms and may be called the source (or sink) terms (5). They are defined for each dependent variable $\phi$ in table (A). $\Gamma_{eff,\phi}$'s are the exchange coefficients defined below.

TABLE A : The source terms of the conservation equation

| $\phi$ | $S_\phi$ |
|---|---|
| u | $-\frac{\partial p}{\partial x} + \frac{\partial}{\partial x}(\mu_{eff}\frac{\partial u}{\partial x}) + \frac{1}{r}\frac{\partial}{\partial r}(\mu_{eff}\,r\,\frac{\partial v}{\partial x}) - \frac{\partial}{\partial t}(\rho u)$ $-(\rho - \rho_m)g_x$ |
| v | $-\frac{\partial p}{\partial r} + \frac{\partial}{\partial x}(\mu_{eff}\frac{\partial u}{\partial r}) + \frac{1}{r}\frac{\partial}{\partial r}(\mu_{eff}\,r\,\frac{\partial v}{\partial r}) - 2\mu_{eff}\frac{v}{r^2}$ $-\frac{\partial}{\partial t}(\rho v)$ |
| k | $G_k - \rho\varepsilon - \frac{\partial}{\partial t}(\rho k)$ |
| $\varepsilon$ | $\frac{\varepsilon}{k}(C_1 G_k - C_2\rho\varepsilon) - \frac{\partial}{\partial t}(\rho\varepsilon)$ |
| h | $-\frac{\partial}{\partial t}(\rho h)$ |

Notes on Table A.

- The source term $G_k$ for k and $\varepsilon$ is given by:

$$G_k = \mu_{eff}\left\{2\{(\frac{\partial u}{\partial x})^2 + (\frac{\partial v}{\partial r})^2 + (\frac{v}{r})^2\} + (\frac{\partial u}{\partial r} + \frac{\partial v}{\partial x})^2\right\} \quad (2)$$

- The source of h implicitly assumes that the shear terms are neglected, and the flow is incompressible; this assumption is justified by the low speed of the flow.

- $\rho_m$ is an average density over the radial direction at each axial station.

- $C_1$, $C_2$ are constants in the $(k,\varepsilon)$ turbulence model given in (2, 3, 4).

In addition to solving a transport equation for each of the dependent variables, one must also solve the equation of conservation of mass.

$$\frac{\partial \rho}{\partial t} + \frac{\partial}{\partial x}(\rho u) + \frac{1}{r}\frac{\partial}{\partial r}(\rho r v) = 0 \qquad (3)$$

In the present work the above equation is manipulated in the way suggested by Spalding and co-workers (6, 7, 8) to yield an equation for pressure, which then becomes an additional dependent variable.

## DEFINITIONS AND AUXILIARY RELATIONS

The equation set is completed by the following algebraic relations:

- The effective viscosity is obtained from, $\mu_{eff} = \mu + \mu_t$ where $\mu_t = C_\mu \rho \frac{k^2}{\varepsilon}$; $C_\mu$ is a constant (2, 3, 4).

- The local effective exchange coefficients, $\Gamma_{eff,\phi}$ for the transport of scalar property $\phi$, are calculated from:

  $\Gamma_{eff,\phi} = \frac{\mu}{\sigma_{l,\phi}} + \frac{\mu_t}{\sigma_{t,\phi}}$, where $\mu$ and $\sigma_{l,\phi}$ are

  the molecular viscosity and the laminar Prandtl/Schmidt number respectively, and $\mu_t$ and $\sigma_{t,\phi}$ are their turbulent counterparts, to take account for the effect of turbulence on mixing.

  The value of the Prandtl number for $\varepsilon$ is obtained from the relation $K^2/\sqrt{C_\mu}(C_2-C_1)$ where $K$ is the Von Kármon constant (=0.42).

- In the problem under consideration, temperature or concentration differences bring about differences in density. It is then necessary, to include buoyancy forces caused by changes in volume which are associated with the temperature or concentration differences, in the equation of motion. These forces are treated here as impressed body forces, and appear in the source term of the u-momentum equation as: $(\rho - \rho_m)g_x$. This is modelled as $\rho\beta\{T-T_m\}g_x$ where: $\beta$ is the coefficient of volumetric expansion, $g_x$ is the x-component of the gravitational acceleration, and $T_m$ is an average temperature over the flow field. The buoyancy term is calculated as follows:

$$\rho\beta\{T - T_m\}g_x = \frac{\rho\beta g_x}{\bar{c}_p}\{h - h_m\}$$

$$h_{m,i} = \overline{h_{i,j}} = \left[\frac{1}{M-1}\sum_{j=2}^{M} h_{i,j}\right]_{i=2,L} \qquad (4)$$

where j is the index in the radial direction and i in the axial one.

## DETERMINATION OF THE EMPIRICAL CONSTANTS

The determination of the constants involved in the turbulence model is done according to (2, 3, 4).

## WALL FUNCTIONS

The vigorous incorporation of the effects of the vicinity of a wall to turbulence proves expensive in computer time. One economical method of accounting for these effects is by way of "wall functions"(9).

These functions are based on some ideas of Spalding and are embodied in algebraic expressions which force the numerical solution to behave in a prespecified manner.

The wall functions for velocity components and for enthalpy are based on the assumption of a log-law in the vicinity of a wall. For k, a zero diffusive flux at the wall is used; this is consistent with the assumption of a fluid layer of uniform shear stress (which results in a log distribution of velocity). For $\varepsilon$ the empirical evidence that a typical length scale of turbulence varies linearly with the distance from the wall, is used to calculate $\varepsilon$ itself at the near wall point.

## EDDY DIFFUSIVITY OF HEAT ($\varepsilon_H$) AND MOMENTUM ($\varepsilon_M$) FOR LIQUID SODIUM

It is common to assume that the turbulent Prandtl and Schmidt numbers are unity, making the eddy diffusivities of mass and heat equal. The theoretical predictions, for liquid sodium and for Reynolds numbers up to $12 \times 10^4$, for the ratio $\varepsilon_H/\varepsilon_M$ obtained by various studies differ a lot (10, 11, 12).

A very comprehensive comparison among various models of predicting the relationship between the turbulent transfer of momentum and a passive contaminant such as heat or dissolved matter, is given in (13). It is shown there, that even for liquid metals (Pr<<1) $Pr_t \rightarrow 1$ for Re$\rightarrow \infty$, and this is the value used herein.

## THE SOLUTION PROCEDURE

The governing equations (1) subject to the appropriate initial and boundary conditions were solved numerically.

The solution procedure is based on an efficient and economical procedure developed by Spalding and co-workers (6, 9, 8) a key feature of which is the so called SIMPLE (for Semi-Implicit-Method for Pressure-Linked Equations) algorithm.

The grid layout used is a "staggered" grid system also used by Spalding and co-workers (6). Hybrid differencing was used (6, 9). The difference equations are solved in turn for each variable by the application of the tri-diagonal matrix algorithm. An irregular 16 x 16 finite difference grid was found to be adequate for modelling the plenum domain. As a matter of fact, it is only for the sake of the turbulence variables (k,$\varepsilon$) near the boundaries that we choose such a fine grid. A time step of 0.1 sec. for the first 100 secs (reactor time) was used and then a time step of 1 sec. Using the above steps, the variability in the outlet temperature is much less than 1%.

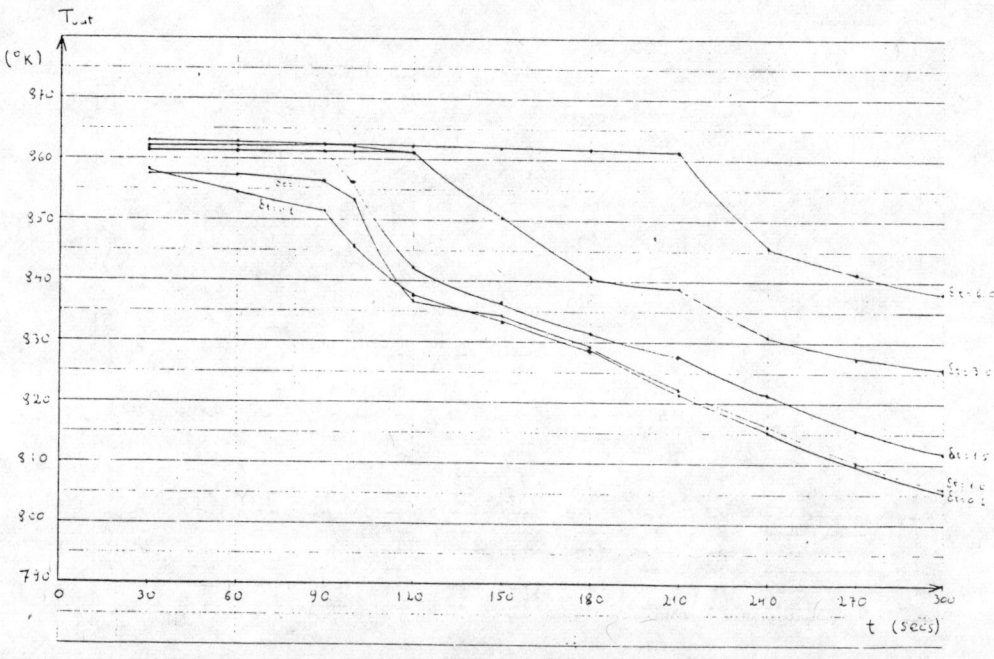

Fig. 2  Outlet Temperature versus time
Optimisation of time step δt

## RESULTS

The developed computer code was run for three different normal scram transients and for several experimental model transients. Figure 2 presents the results of some test runs to establish the proper choice of time step.

Figure 3 shows the predicted steady-state velocity field. It is indicated that under steady state isothermal conditions, the flow pattern within the plenum resembles a toroid.

In all runs, where the increase in density of the incoming fluid is accompanied by a flow coastdown to about 10% of the initial flow rate, flow stratification invariably occured. Upon entering the plenum the fluid was quickly forced downward and outward toward the exit nozzles. The predicted flow patterns are in good qualitative agreement with experiments.

Figures 4 to 8[*] present the predicted flow field for the normal scram transient every 60 secs (reactor time) after its initiation.

---
[*]We assign the maximum length not to the largest vector in the field but to a vector equal to twice the average value of all the vectors present. If a vector is larger in magnitude than this maximum, we simply represent it by a line segment of the same size, on which we print a symbol z to distinguish it.

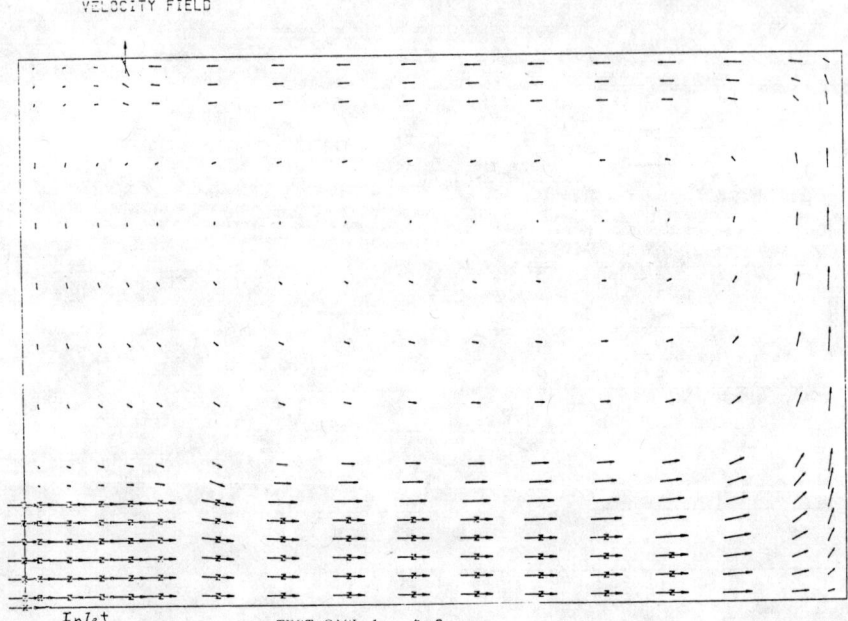

Fig. 3 Normal Scram Steady State

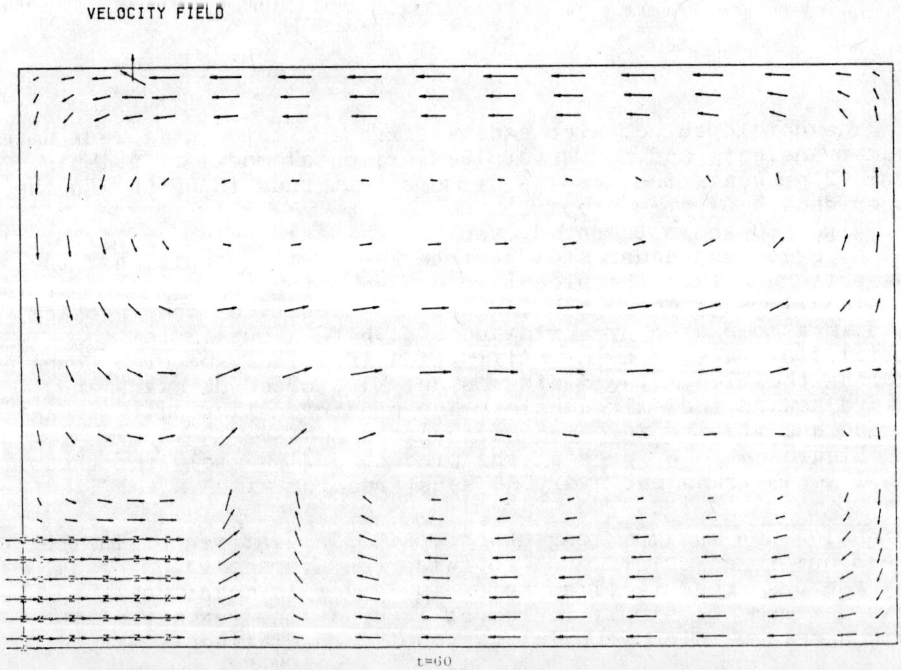

Fig. 4 Test Case 1

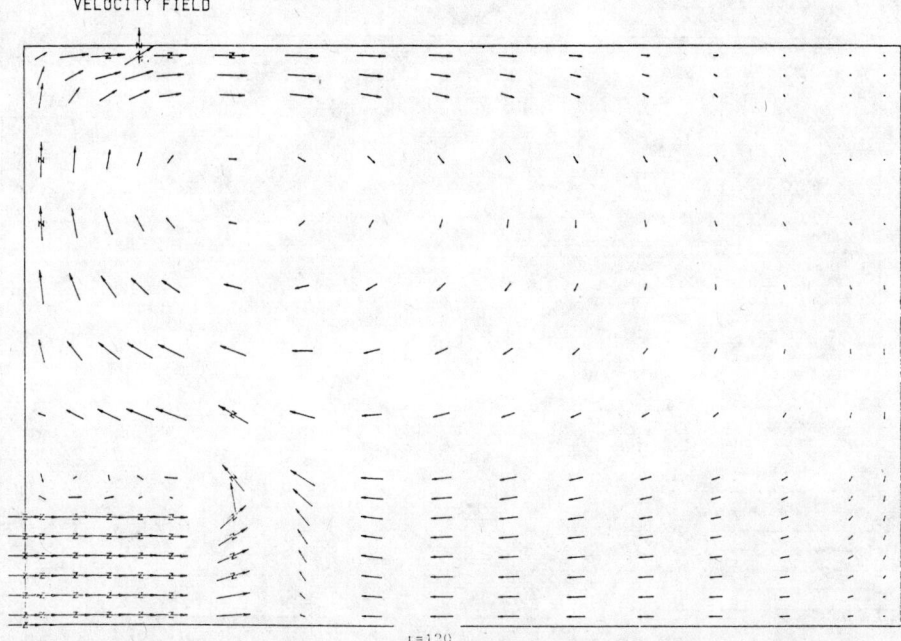

Fig. 5 Test Case 1

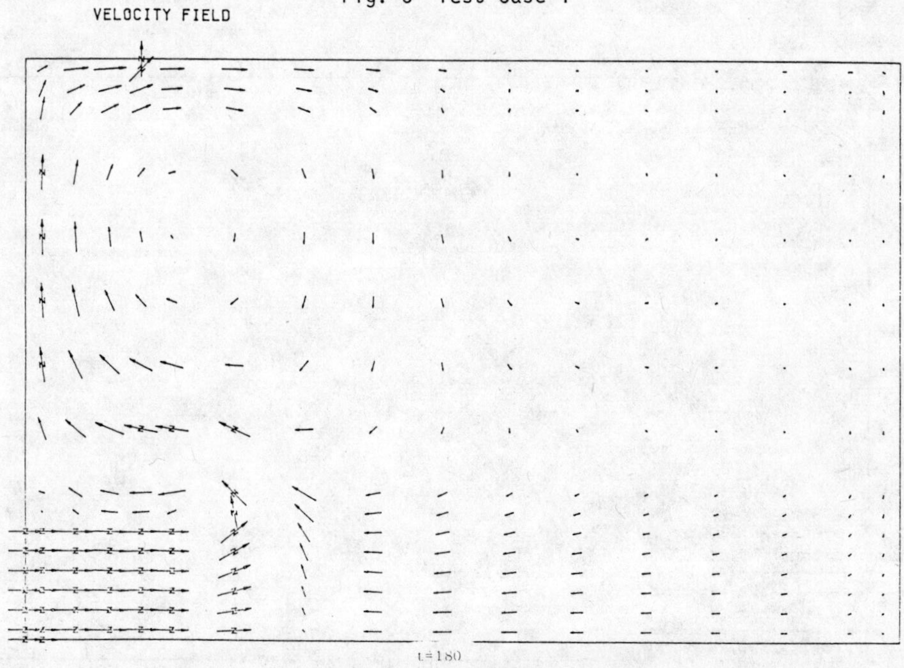

Fig. 6 Test Case 1

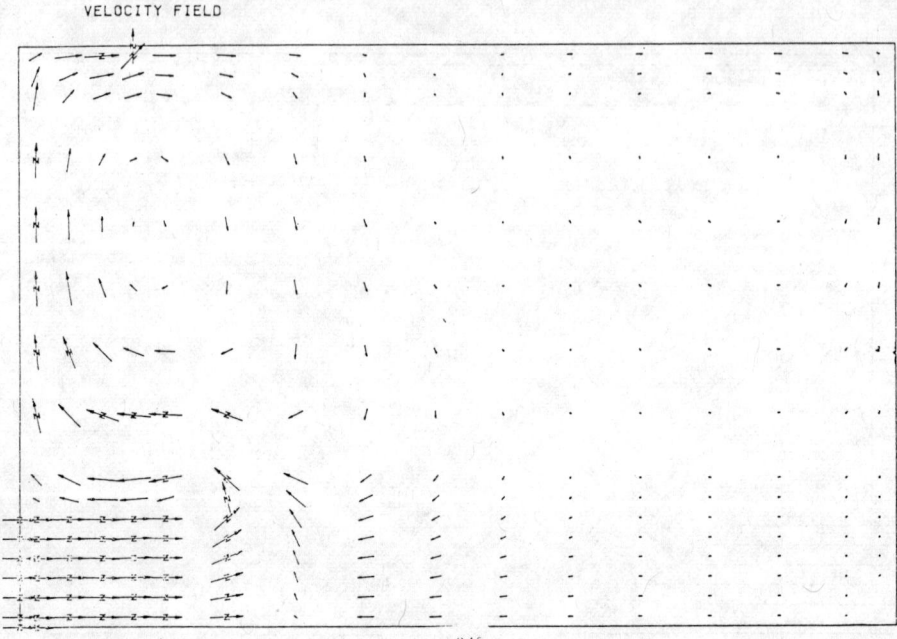

Fig. 7 Test Case 1

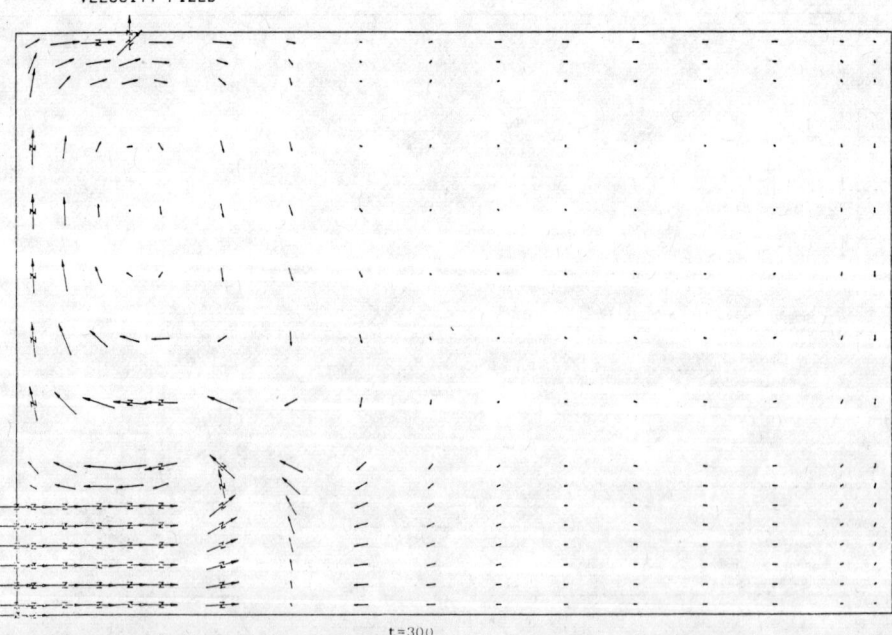

Fig. 8 Test Case 1

Figures 9 to 12** present the temperature contours for a normal scram test run, whilst figures 18 to 20 the concentration contours for the experimental model.

Considering the apparent severity of the stratified flow pattern, an abrupt outlet nozzle thermal transient might be expected. However Figure 13 indicates a relatively modest outlet transient. Therefore, a relatively large effective volume of the plenum is still active in mixing.

CONCLUSIONS

From the evidence of the computed results, it can be concluded that for all the test cases considered, the flow pattern can be described as follows:

- In the steady state, the coolant flows vertically upward, deflects horizontally at the top and then passes downward near the vessel wall toward the outlet. One large central eddy dominates the flow.
- The flow can be characterised by a Rankine toroid driven at the inner opening with a jet emerging from the reactor. The fluid in the outlet plenum is initially circulating at normal operating conditions and when the reactor is "scrammed" the core jet velocities start to decrease and a new pattern develops.

---

**The contours plotting routine is capable of producing only one curve per contour value.

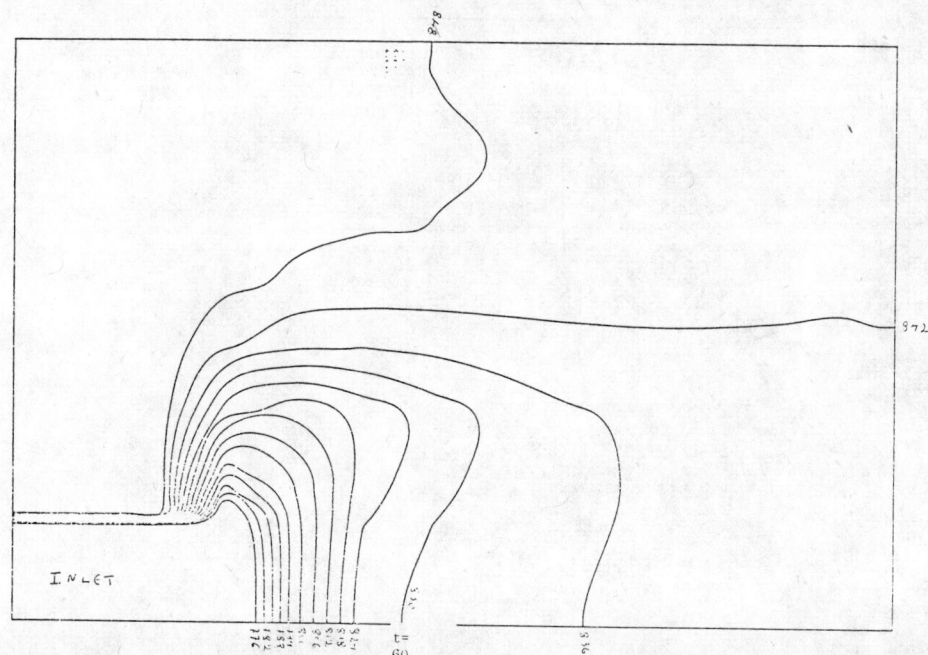

Fig. 9 Normal Scram - Test Case 3

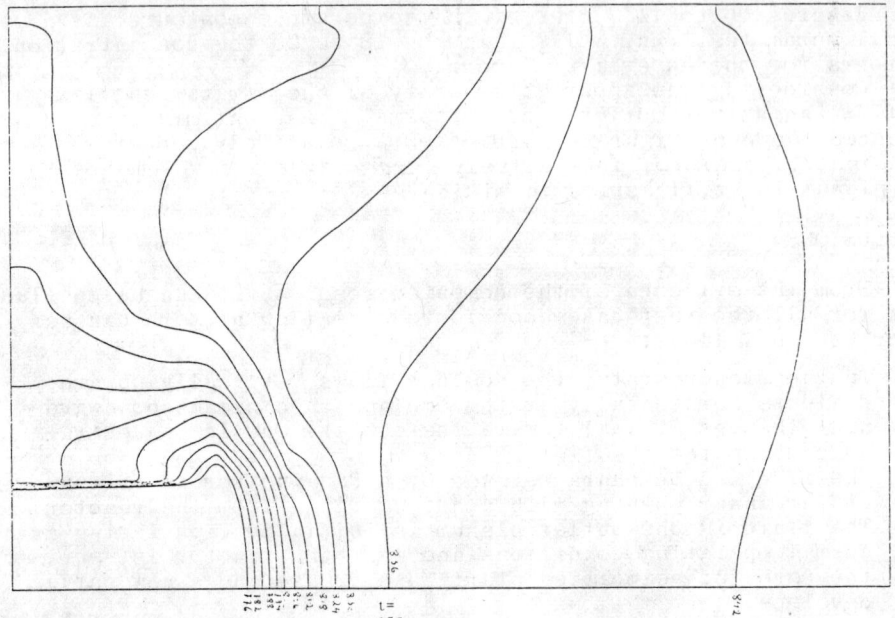

Fig. 10 Normal Scram - Test Case 3

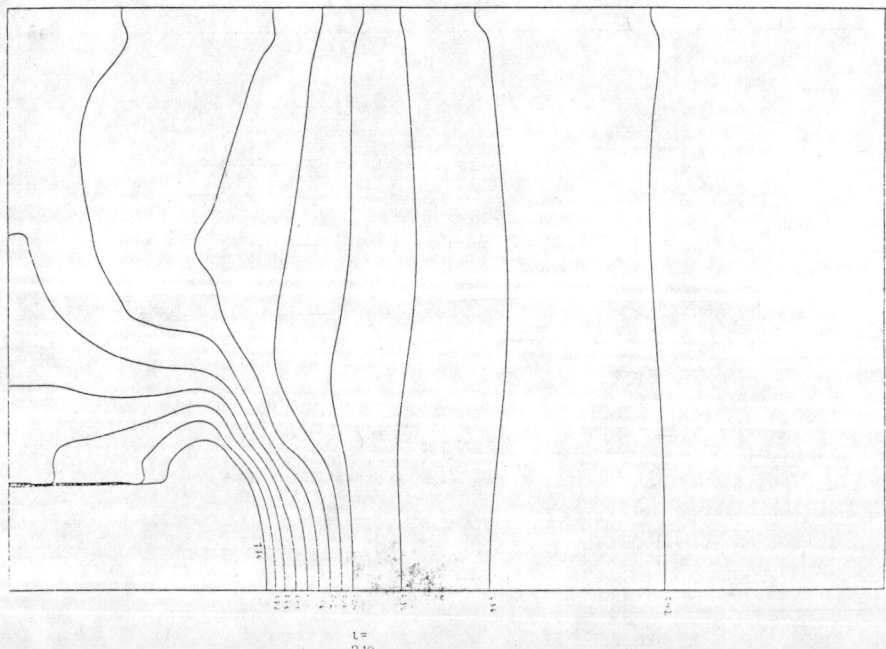

Fig. 11 Normal Scram - Test Case 3

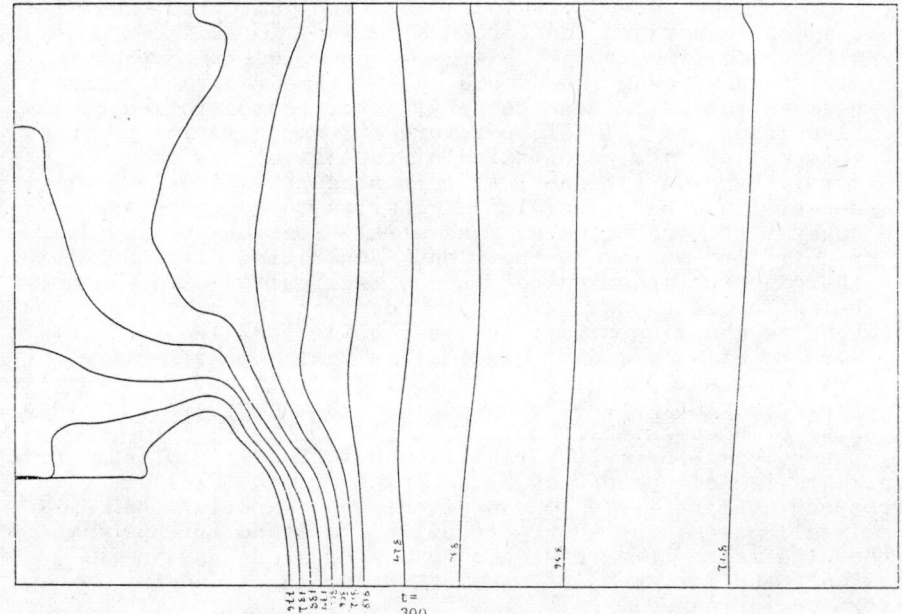

Fig. 12  Normal Scram - Test Case 3

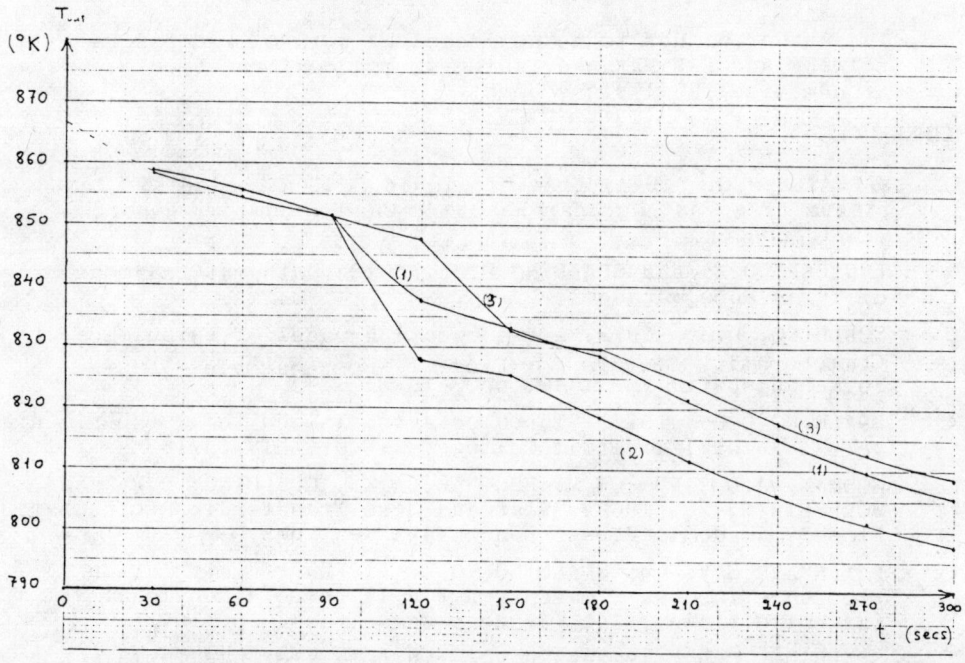

Fig. 13  Outlet Temperature Versus Time (Test Cases 1, 2 and 3.
Normal scram, different inlet transients)

- Initially the core jet velocity becomes quite low, the mixing zone collapses and the toroid slows up.
- After some time the jet energy becomes less than the static differential head introduced by the temperature difference between the mixed mean temperature or concentration of the plenum and the inlet temperature or concentration of the core outlet. The penetration of the jet decays and the toroid becomes stagnant. At this stage the lower velocity, denser fluid has insufficient inertia to overcome the negative buoyancy forces, and stratification occurs. Instead of penetrating well into the plenum and mixing with the fluid therein, the incoming jet is short-circuiting the plenum, being forced downward and outward, towards the outlet nozzles.
- Despite the flow stratification, outlet nozzle transients were relatively modest, indicating that some effective volume of the plenum remains active in mixing. For the time period investigated (300 secs) no new steady state pattern has yet been established.

The complex thermal-hydraulic behaviour of the plenum during transients is well predicted by the present model. All results correspond qualitatively to experiments and expectations.

In all cases, the short-circuiting is found irrespectively of whether the inlet temperature increases or decreases in the transient, and irrespective of whether the inlet density is uniform or increased.

It appears that the flow pattern is greatly influenced by the decrease in the incoming flow rate.

ACKNOWLEDGEMENTS

The author wishes to acknowledge the work of Professors D B Spalding and S V Patankar which forms the foundation of the present numerical analysis scheme.

REFERENCES

1. MARKATOS, N.G. (1974) "Flow of liquid Sodiun coolant in the plenum of a Fast Nuclear Reactor", CHAM Technical Report 770/1, London.
2. LAUNDER, B.E. and SPALDING, D.B. (1972) "Mathematical models of Turbulence," Academic Press.
3. SPALDING, D.B. (1972) "A two-equation model of turbulence" Commemorative Lecture for Prof. F. Boswajakovic, VDI - Forschungsheft, Vol. 546, pp 5-16.
4. SPALDING D.B. (1973) "Turbulence Models and their experimental verification", Imperial College, M.E.D., HTS/73/19.
5. GOSMAN, A.D., PUN, W.M., RUNCHAL, A.K., SPALDING, D.B. and WOLFSHTEIN, M. (1969) "Heat and Mass Transfer in Recirculating Flows", Academic Press, London and New York.
6. PATANKAR, S.V. and SPALDING, D.B. (1972) "A calculation procedure for heat, mass transfer in three-dimensional parabolic flows", Int. J. Heat Mass Trans. $\underline{15}$, 10.
7. SPALDING, D.B. (1972) "A novel-finite-difference formulation for differential expressions involving both first and second

derivatives", Int. J. for Numerical Methods in Engineering Vol. 4, pp 551-559.

8. SPALDING, D.B. (1974) "The calculation of Convective Heat Transfer in Complex Flow Systems", Proceedings of the Fifth International Heat Transfer Conference, Vol. VI, Tokyo, Japan.

9. PATANKAR, S.V. and SPALDING, D.B. (1970) "Heat and Mass Transfer in Boundary Layers", Intertext London, 2nd Edition.

10. DWYER, O.E. (1963)"Eddy transport in liquid metal heat transfer", A.I.Ch.E. Journal 9, 2, pp 261-268.

11. TYLDESLEY, J.R. and SILVER, R.S. (1968) "The prediction of the transport properties of a turbulent fluid", Int. J. Heat Mass Transfer 11, 9.

12. BORISHANSKI, V.M. and ZABLOTSKAYA, T.B. (1967)"High Conductivity fluid flow turbulent heat transfer", Atomnaya Energiya 22, 2.

13. REYNOLDS, A.J. (1975) "The prediction of turbulent Prandtl and Schmidt numbers", Int. J. Heat Mass Transfer 18, 9, p.1055.

# EXPERIMENTAL AND ANALYTICAL STUDIES OF NATURAL CONVECTION IN EBR-II

RALPH M. SINGER, JERRY L. GILLETTE, WAYNE K. LEHTO, DALE MOHR, CHARLES C. PRICE and JOHN I. SACKETT

Argonne National Laboratory, Argonne, Illinois and Idaho Falls, Idaho, U.S.A.

ABSTRACT

The EBR-II is a sodium-cooled, fast breeder reactor located in Idaho, and is designed to operate at a thermal power of 62.5 MW and an electrical generation rate of 20 MW. In a continuing program devoted to the understanding of the thermal, hydraulic, and neutronic behavior of this reactor under both normal and off-normal operating conditions, a series of natural convection tests have been conducted including both steady-state and transient operating conditions. Instrumentation utilized for the control and observation of the reactor behavior during these experiments included both the normal plant sensors as well as those located in-core within a special fueled subassembly. The results of the steady-state measurements have been compared to the predictions of an analytical model of the entire primary heat transport circuit and satisfactory agreement was obtained.

INTRODUCTION

In many types of engineering applications in which heat is removed by means of the forced circulation of a coolant, it is necessary to understand the thermal behavior of the system if forced flow is suddenly lost. This is especially important in the case of liquid-metal-cooled, fast breeder reactors, where, even when fission power generation has ceased, a low level of heat generation continues due to the decay of fission products. Thus, in the extreme situation of a loss of the operability of all of the coolant pumps, including their emergency back-up systems, the only means of cooling the reactor is by natural circulation. The ability to predict the effectiveness of this inherent mechanism of reactor cooling is obviously an important aspect of the assessment of the safety of such systems.

Although the phenomenon of natural convection has been extensively studied, little information is available concerning the behavior of systems as complex as nuclear reactors. Apparently, natural convective tests were conducted in the SEFOR reactor [1] as part of its testing program, but, to the knowledge of the authors, no formal reporting of the results has been made. The effectiveness of the emergency cooling system in the Rapsodie reactor, which involves the inducement of natural circulation in the reactor core by means of an external nitrogen loop, was investigated; some of the details of the testing was reported in [2]. In addition, a series of low fission power, natural circulation tests were conducted as an integral part of the start-up of the Phenix reactor, and are briefly reported in [3].

Although there were some early suggestions to conduct similar tests in the Experimental Breeder Reactor II (EBR-II) shortly after its start-up [4], such testing was deferred until a fully instrumented in-core subassembly became available [5]. This paper presents the results of these tests, including steady-state operation with both fission and decay power and a transient test utilizing decay power.

DESCRIPTION OF REACTOR COOLANT SYSTEM AND INSTRUMENTATION

Primary Heat Transport System

A schematic representation of the EBR-II primary vessel and the essential elements of the primary heat transport system are shown in Figure 1. From this sketch, it is apparent that EBR-II is of the "pot" design with all of the components submerged in a large pool of sodium. Primary coolant flow is provided by two centrifugal pumps operating in parallel, with one intermediate heat exchanger (IHX) transferring the energy generated in the reactor to the secondary sodium system and ultimately to the steam generators and turbogenerator. The reactor itself consists of 16 rows of subassemblies, the inner 7 constituting the active core and the outer 9 containing reflector and blanket subassemblies. Since the reactor is presently being utilized as an irradiation test facility, the core loadings are quite heterogeneous, with metal, oxide, and carbide fueled subassemblies, boron carbide elements, structural material test assemblies, and others all present in the first 7 rows.

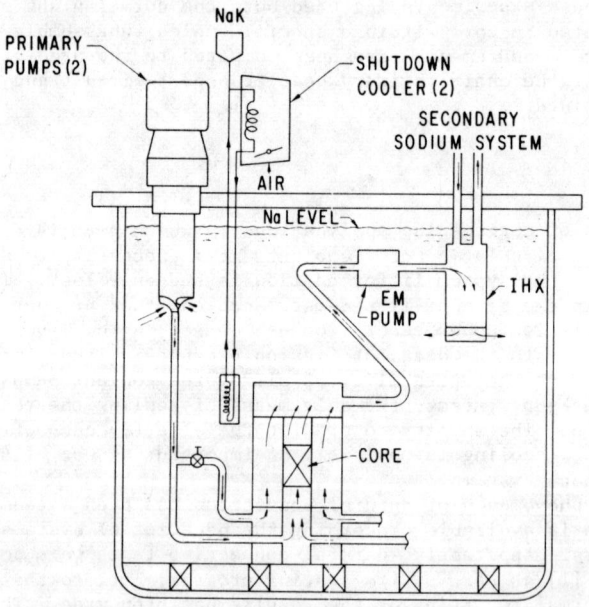

Figure 1. Schematic of the EBR-II Primary System.

The active fuel length in the core is 343 mm, with 91 elements in a metal fuel driver subassembly. These driver fuel elements are 4.42 mm in diameter, with the fuel thermally bonded to the stainless steel cladding by sodium and are contained within a hexagonal can which has a flat-to-flat dimension of 56.1 mm. The spacing of the individual fuel elements is maintained by wire wraps with a diameter of 1.24 mm and an axial pitch of 152 mm. The design operating power of the reactor is 62.5 MWt. Additional descriptive information on EBR-II is available in [6].

Instrumentation

Although EBR-II is extensively instrumented, only that instrumentation germaine to the discussion of the present study will be mentioned here. Sodium outlet thermocouples are located in the outlet plenum, 6.3 mm above the top of 20 subassemblies, 17 of which are in the core (first 7 rows). Thermocouples are also located at various points along the inlet piping and in the outlet nozzle and are used to determine the reactor inlet and outlet temperatures. Permanent magnet flowmeters are present in the inlet and outlet piping in order to measure total flow rates. The neutron power of the reactor is sensed by three fission counters and eight compensated ion chambers.

The in-core instrumentation is located in a modified driver subassembly located in a converted control-rod position in the fifth row of the core. This experimental subassembly, designated XX07, consists of 61 elements, 57 of which contain Mark-IA metal fuel. The diameter of these elements is 4.42 mm and similar to the driver fuel, are spaced with 1.24 mm spacer wire wound on a 152 mm axial pitch. The 61 elements are contained within a hexagonal can which measures 46.4 mm across its inside flats. These dimensions result in a channel hydraulic diameter of 2.75 mm. Within this subassembly, there are two inlet, permanent magnet flowmeters, 10 fuel centerline thermocouples located 21.6 mm below the top of the fuel (at beginning of life), and 13 coolant thermocouples mounted as wire-wrap spacers. Three of these coolant thermocouples are located at the core bottom, three at core midplane, five at core top (same height as the fuel thermocouples), and two at 610 mm above core top. There also are two self-powered detectors within the active fuel length which are sensitive to neutron and gamma flux respectively, and six thermal-expansion-difference monitors which passively indicate local maximum temperatures. The flowmeters were calibrated in out-of-pile testing prior to insertion in the reactor over the flow rate range of $3 \times 10^{-6}$ to $4 \times 10^{-3}$ $m^3/s$ at temperatures from 260 to 430°C. The estimated uncertainty in the measured flow rates was ±0.6%. The thermocouples were calibrated in-place by operating the reactor at essentially zero power and full flow, resulting in isothermal conditions within the instrumented fuel assembly. The estimated error in the temperature measurements is about ±1°C. Additional information on XX07 is available in [7].

DESCRIPTION OF THE NATURAL CONVECTION EXPERIMENTS

Steady-State Test with Fission Product Decay Heating

The initial conditions for this test were established by holding the reactor in a subcritical state with fully withdrawn control rods[*] for ten days. At this time, the fission product decay heat load was approximately 60 kW, of which 0.5 kW were generated in the instrumented subassembly, XX07. The main primary pumps were off and a forced flow of 0.0288 $m^3/s$ (1.69 x $10^{-4}$ $m^3/s$ through XX07) was maintained by an auxiliary electromagnetic pump, resulting in an XX07 and core temperature rise of 2°C[**]. The test was initiated by gradually

---
[*]EBR-II utilizes fueled control rods so that their withdrawal shuts the reactor down.
[**]The corresponding conditions at normal full operation are 56.9 MW, 0.517 $m^3/s$, and 102°C for the reactor and 0.43 MW, 3.33 x $10^{-3}$ $m^3/s$, and 111°C for XX07.

reducing the power supplied to the EM pump over a period of 70 minutes, at which time this pump was also shut off. All primary coolant flow was then due to natural convective effects only.

Since the power generated in the reactor was due to the decay of fission products and was quite low and essentially constant, the buoyant driving force in the coolant circuit was varied by changes in the secondary system sodium forced flow rate. An increase in this flow rate causes a reduction in temperature of the primary sodium in the intermediate heat exchanger (IHX) and thus, an increase in the density difference between the sodium in the core and the IHX; this in turn causes the convective flow rate to increase. Four different thermal-hydraulic steady-states of the system were studied in this fashion.

The dependence of the convective flow rate and core temperature rise upon the secondary system sodium flow rate (or equivalently, the rate of cooling in the IHX) under the conditions of this experiment are summarized in Figure 2, using the data from XX07. It is apparent that the secondary system conditions strongly influence the primary natural circulation flow rate. In fact, for moderate secondary sodium flow rates, the core temperature rise was reduced to about 5°C, a value only slightly larger than the initial test value of 2°C obtained under forced flow of the primary sodium.

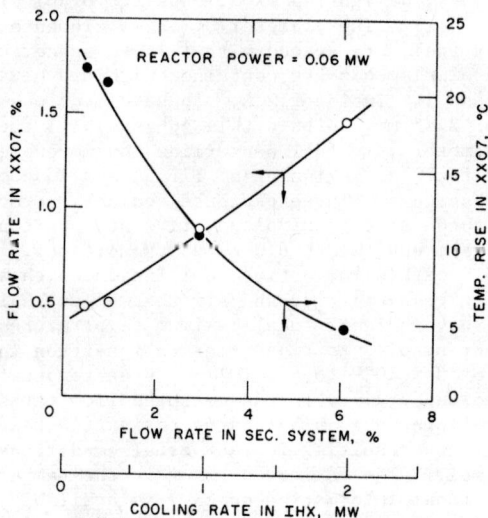

Figure 2. Response of In-core Temperature and Flow to IHX Cooling Rate (Decay Heating)

Steady-State Test with Fission Heating

This test was conducted under very similar conditions to the one previously described, the essential difference being that the reactor was now maintained in a critical condition at a low fission power level. Because of this state of the reactor, effects of reactivity feedback due to temperature changes had to be taken into account in developing the proper procedures from changing from one convective state to another. An increase in the reactor fission power will initially increase the coolant temperature in the core, thus resulting in an increase in the convective flow rate, causing a reduction in the coolant temperature rise; this in turn, because of the negative temperature coefficient of reactivity in EBR-II, causes an increase in reactivity and an increase in power. The net effect is an increase in the power-to-flow ratio resulting in a higher equilibrium core temperature rise. Although pre-test calculations had clearly shown that this phenomenon would be quickly convergent, the controlled variation of reactor fission power by control rod motion was accomplished quite slowly.

In this test, the convective flow rate could be varied by changes in both the reactor fission power level as well as the secondary system flow rate. Values of the latter parameter encompassed those used in the previous decay power test, and the combined fission plus decay power was maintained at values ranging from 0.35 MW to 0.81 MW. The results of this test are summarized in Figure 3, where the influence of reactor power upon the reactor core conditions is illustrated. From this figure, it is apparent that with a constant IHX cooling rate, an increase in the reactor power results in an increase in the convective flow rate as well as an increase in the temperature rise across the core; thus, the power-to-flow ratio increases with increases in power. The effect of the secondary system sodium flow rate upon these thermal-hydraulic parameters is represented by the points at the highest reactor power level achieved in the test (0.81 MW). This condition was reached by starting from a convective state corresponding to 0.62 MW and slowly increasing the secondary sodium flow rate without changing the control rod positions. As the primary sodium convective flow rate increased due to the reduced temperatures in the IHX, the reactor core temperatures correspondingly decreased, adding reactivity to the system. The reactor power therefore increased to 0.81 MW; the increase of 0.19 MW being due entirely to the temperature coefficient of reactivity in EBR-II. As shown in Figure 3, although the reactor power has increased substantially, the net combined effect of this power increase along with the increased IHX cooling rate has been a sufficiently large increase in the convective flow rate so as to maintain the previous in-core coolant temperatures.

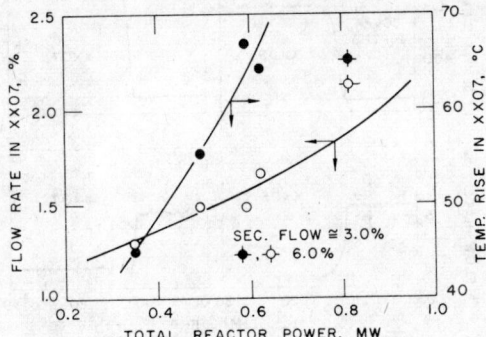

Figure 3. Response of In-core Temperature and Flow to Reactor Power (Fission Heating)

## Convective Testing with the Reactor Cover Lifted (Fuel Handling Configuration)

In order to remove subassemblies from EBR-II, the primary coolant pumps must be turned off and the reactor cover (which forms the upper portion of the reactor outlet plenum) is removed from the reactor vessel. When the cover is removed, the EM pump on the outlet coolant line no longer draws coolant through the reactor core (see Figure 1). Thus, the core is cooled only by natural convective flow. The flow path is now through the pump inlets, primary piping, and core, after which it is discharged to the primary tank directly from the top of the subassemblies. Essentially no coolant leaving the core directly enters the outlet pipe leading to the IHX. Thus, the convective flow through the core is driven only by energy generated in the core and is unaffected by cooling in the

IHX. As a special test [8] designed to investigate the in-core thermal-hydraulic conditions during this situation, the reactor cover was lifted, leaving the instrumented, fueled subassembly XX07 in place. Fortunately, during this test, another instrumented, non-fueled subassembly, XX05, was in the core loading and was also left in place. XX05 contained one flow meter and several coolant thermocouples.

The experiment was initiated about 7 hours after the start of reactor shutdown, at which time the fission product decay power level was approximately 0.33 MW. Prior to the lifting of the reactor cover, while forced flow was still being provided by the EM pump, the flow rates and temperature rises in XX05 and XX07 were $2.43 \times 10^{-5}$ m$^3$/s, 9°C and $1.75 \times 10^{-4}$ m$^3$/s, 13°C, respectively. The corresponding values at full power and flow were $4.20 \times 10^{-4}$ m$^3$/s, 37°C and $3.29 \times 10^{-3}$ m$^3$/s, 113°C, respectively. As shown in Figure 4, when the reactor cover was lifted at 11 minutes and a transition to convective flow resulted, the flow rate and temperature rise in XX05 changed to $1.26 \times 10^{-5}$ m$^3$/s and 54°C, while in XX07, they were $4.28 \times 10^{-5}$ m$^3$/s and 57°C.

From these results, several observations can be made: first, a flow redistribution has occurred under natural convective conditions resulting in the temperatures in a fueled and non-fueled subassembly becoming approximately equal. Secondly, the temperatures experienced by these subassemblies during fuel handling, i.e., when the reactor is shut down and the cover is lifted, are not insignificant; in fact, the coolant temperatures reached in XX05 exceeded those obtained under normal full power operation.

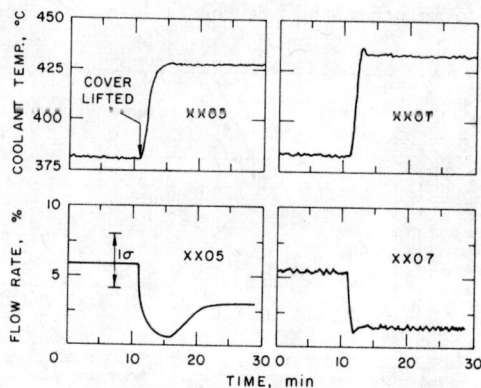

Figure 4. Response of In-core Subassemblies to the Lifting of the Reactor Cover.

## Transient Flow Coastdown to Convective Conditions

The final experiment in this experimental program was designed to investigate the dynamic aspects of a transition from forced flow cooling of the core to natural convective cooling. In order to minimize thermal shocks to the reactor system, the initial conditions were established with the reactor subcritical and the control rods fully withdrawn, but with a significant fission product decay heat load (0.91 MW), and forced flow provided only by the EM pump ($1.77 \times 10^{-4}$ m$^3$/s through XX07 and 0.0274 m$^3$/s total primary flow rate). This resulted in an initial temperature rise across the reactor of 27°C and across XX07 of 37°C. The transient was initiated by abruptly shutting off the EM pump; the primary coolant flow immediately dropped, and then gradually increased as convective flow developed. The in-core temperatures responded as would be expected from the flow transient, reaching a peak approximately 60 seconds following the initiation of the transient, and then gradually decreasing. These observations are shown graphically in Figures 5 and 6, based on XX07 measurements.

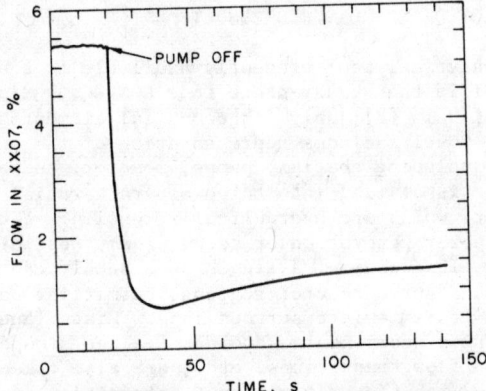

Figure 5. Flow Coast-down from a Forced Flow State to a Convective Flow State.

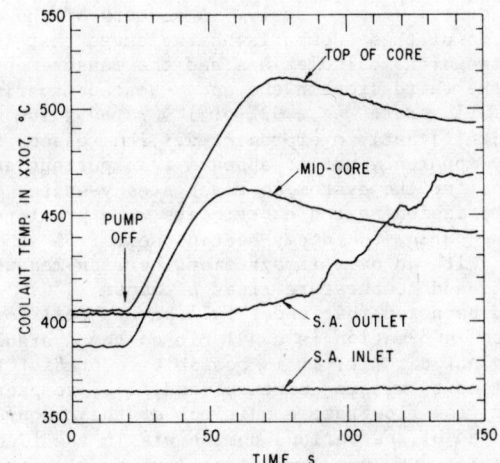

Figure 6. In-core Thermal Response to a Flow Coast-down to Natural Convection.

One of the more interesting results of this test was the appearance of a significant undershoot in flow and overshoot in temperature during the transient. The primary reason for this phenomenon is the existence of relatively cool liquid and structure in the riser portion of the heat transport circuit. In order for the final steady level of convective flow to be reached, the initially cool riser section must be heated up; during this heat-up period, the net buoyant head is less than the ultimate steady-state value and the convective flow is therefore less. Increases in the convective flow then result from the transport of heat from the core to the riser region which increases the net buoyancy driving forces. This transport process is relatively slow due to the low coolant flow rates involved and the heat capacity of the structural material. As a result of this thermal lag, the minimum flow rate dropped to about 55% of the final steady-state value, but the in-core temperatures exceeded the steady values by only about 15% due to the core heat capacity.

COMPARISON OF PREDICTED AND MEASURED RESULTS

The primary analytical tool presently available to analyze the convective conditions in EBR-II is the steady-state code CONVECT. Since this code has been described previously [8], only its essential elements will be summarized here. The physical model includes representation of the entire primary heat transport circuit, including the IHX, pumps, reactor, and all interconnecting piping. The reactor is divided into three separate radial regions (core, reflector, and blanket) which are hydraulically coupled. Each region also has its own particular power generation or decay scheme depending on its composition and structure. In addition, a single subassembly can be located within any one or more of the three reactor regions, permitting both hydraulic and thermal coupling to its immediate surroundings. Miscellaneous effects such as heat losses from various components, coolant leakage around the reactor cover, turbulent-to-laminar flow transitions, etc., are also taken into account.

Using this code with the primary input information of secondary system sodium flow rate, energy generation rate and distribution, and hydraulic resistances of the components in the flow circuit, an attempt was made to predict the steady-state natural convective conditions that existed during the first three experiments described previously. A summary of the results is listed in Table I. Inspection of these comparisons indicates that in general, reasonably good agreement between the calculations and the measurements was obtained. One exception is for the third step in the decay-heated experiment where the secondary system sodium flow rate was 2.92% of its rated value where the convective flow rates were significantly overpredicted. The reasons for this discrepancy are not completely apparent, but it appears that perhaps an insufficient length of time was allowed for the system to reach steady-state. In general, the conditions during the fission heated experiments were predicted with greater accuracy than those during the decay-heated ones, (±5% vs ±17%, excluding the third decay test), with an overall agreement between measured and calculated flows of about ±9%, and temperature rises of about ±4°C. As a basis for comparison, it should be noted that under full power, full forced flow conditions where more accurate information is available on the hydraulic characteristics of the primary coolant circuit, it is possible to predict the local flow rates and temperature rises to within about ±6% and ±3°C, respectively.

Under the very low flow rate conditions of these convective tests, the hydraulic resistances of the various components in the flow circuit have significant uncertainties due to turbulent-to-laminar flow transitions and the lack of precise calibration data. Additionally, during the decay-heated tests, the flow rates and temperature rises were quite small and due to the inherent signal noises which resulted in flow rate measurement precisions of about ±0.1% of the rated flow, it is not surprising that this particular test resulted in the largest percent deviation between the predictions and measurements. An additional factor which can influence the convective flow rate in this system is the thermal stratification that can occur in the bulk primary sodium in the reactor vessel; this effect was not included in the analytical model. During the cover-lift test, this phenomenon, as well as the existence of a thermal plume above the top of the subassembly outlets (which were open to the reactor vessel), can directly impact the in-core flow rate.

Although it was not possible to compute the observed behavior during the transient flow coastdown experiment using the steady-state CONVECT code, it was possible to calculate the asymptotic thermal-hydraulic conditions for long times after all dynamic effects have dissipated. For this test, the calculated final flow rate and temperature rise in XX07 was $4.27 \times 10^{-5}$ m³/s and 126°C, whereas the measured values were $4.27 \times 10^{-5}$ m³/s and 130°C, respectively.

Table 1. Comparison of Calculated and Measured Thermal-Hydraulic Conditions During Natural Convective Testing in EBR-II

| Test Description | Total Reactor Power (MW) | Sec. Na Flow* (%) | Primary Na Flow* | | XX07 Flow* | | XX07 Temp. Rise | |
|---|---|---|---|---|---|---|---|---|
| | | | Meas. (%) | Calc. (%) | Meas. (%) | Calc. (%) | Meas. (°C) | Calc. (°C) |
| Decay Heat | 0.06 | 0.444 | 0.61 | 0.37 | 0.48 | 0.29 | 23.1 | 27.8 |
| | 0.06 | 0.987 | 0.68 | 0.71 | 0.49 | 0.54 | 21.6 | 14.8 |
| | 0.06 | 2.92 | 1.05 | 1.87 | 0.89 | 1.38 | 11.5 | 6.7 |
| | 0.06 | 6.11 | 1.79 | 1.84 | 1.45 | 1.37 | 4.7 | 6.8 |
| Fission Heat | 0.59 | 2.51 | 1.75 | 1.79 | 1.48 | 1.36 | 67.2 | 69.8 |
| | 0.35 | 3.04 | 1.53 | 1.77 | 1.29 | 1.34 | 44.8 | 42.5 |
| | 0.49 | 3.01 | 1.86 | 1.97 | 1.48 | 1.48 | 55.8 | 53.3 |
| | 0.62 | 3.32 | 1.97 | 2.15 | 1.66 | 1.70 | 64.6 | 61.5 |
| | 0.81 | 6.0 | 2.52 | 2.41 | 2.13 | 2.04 | 65.5 | 71.5 |
| Cover Lift, Decay Heat | 0.33 | 1.0 | -** | - | 1.28 | 1.65 | 57.0 | 54.2 |

*100% flow values are: sec. Na flow, 0.366 m³/s; primary Na flow, 0.517 m³/s; XX07 flow, $3.33 \times 10^{-3}$ m³/s.
**Total flow is measured in the outlet pipe which was short-circuited during test.

SUMMARY AND CONCLUSIONS

A series of natural convective tests were conducted in the fast breeder reactor EBR-II utilizing both normal plant instrumentation as well as certain special in-core instrumentation located in a driver-fueled subassembly. These tests encompassed both steady-state and transient operation of the reactor with both fission and fission-product decay heat generation. The behavior under the fuel handling configuration of the system as well as the normal mode was studied. The effects of changing coolant density distributions along the primary circuit path were investigated by varying the rate of energy generation in the core and the rate of cooling in the IHX.

The measurements were reasonably well predicted using a natural convective code which included hydraulic and thermal interactions between parallel channels. The success of these calculations is now being used as a basis for the development of a complete dynamic code describing the behavior of the reactor system under transient operating conditions in which natural convection effects are important.

ACKNOWLEDGMENTS

This work was supported by the U.S. Energy Research and Development Administration. Also, there were many members of the EBR-II Project who contributed significantly to the success of the complex experiments described in this paper, and the authors would like to specifically acknowledge the contributions of: A. Smaardyk for the engineering of XX07, R. N. Smith and the reactor operating crews for the special operation of EBR-II, R. Hyndman and M. Tuck for the on-line data acquisition, and E. Dean and J. E. Sullivan for the data reduction.

REFERENCES

1. Natural Circulation Testing of the SEFOR Reactor Coolant System, unpublished information.

2. Chenal, J.-C., Pointer, R., and Delisk, J.-P., "Emergency Cooling of the Rapsodie-Fortissimo Core," ASME Paper No. 74-WA/HT-51, (1974).

3. Guillemard, B., "Start up of PHENIX," Nuclear Engrg. Intl., $\underline{19}$, n. 216 (May, 1974) pp. 411-414.

4. Monson, H., personal communication.

5. Gillette, J. L., "An Environmental Instrumented Subassembly for EBR-II," Trans. Am. Nucl. Soc. $\underline{17}$ (1973) p. 180.

6. Koch, L. J., et al., "Hazard Summary Report: Experimental Breeder Reactor II (EBR-II)," Argonne National Laboratory Report No. ANL-5719) (May, 1957), and ANL-5719 (Addendum), (June, 1962).

7. Gillette, J. L., et al., "Design and Fabrication of the EBR-II Environmental Instrumented Subassembly - Test XX07," to be published as an ANL Technical Report, (1976).

8. Price, C. C., Sackett, J. I., and Dean, E. M., "EBR-II Core Thermal-Hydraulic Perturbations Induced by Post-Shutdown Operations, Trans. Am. Nucl. Soc., $\underline{21}$ (1975) p. 415.

9. Mohr, D., and Gillette, J. L., "A Code for Analyzing Parallel-Channel Thermal-Hydraulics in EBR-II during Natural Convective Conditions (the CONVECT Code)," Trans. Am. Nucl. Soc., $\underline{21}$ (1975) p. 299.

# STUDY OF TEMPERATURE GRADIENTS DUE TO GAS THERMOSYPHONS INDUCED WITHIN THE PHENIX NUCLEAR REACTOR

**J. L. BOY-MARCOTTE, P. CHEVALIER, and M. JANNOT**
*Bertin & Company, 78370 Plaisir, France*

## ABSTRACT

The Phenix nuclear reactor consists of a vessel containing the core immersed in liquid sodium. The space confined between the surface of the liquid metal and the concrete top cap of the vessel is filled with argon. The thick top cap has several penetrations for heat exchangers and sodium circulators pumps. Through the annuli thus formed, a free convection-driven thermosyphon type stream develops vehiculating sodium vapors.

BERTIN & Co was awarded a study for assessing the thermal gradients generated by the thermosyphons and the stresses exerted within the metal liners.

The thermosyphon modelisation used for calculations is presented. It is a one dimensional model accounting for sodium vapor condensation. This model coupled to a finite elements two dimensional conduction programm yields the temperature chart along a plane showing upward and downward legs of the thermosyphon.

The results were confronted to measurements made after the reactor was built. The validity of the assumptions and of the model has thus been proved.

## LIST OF SYMBOLS

$C_f$ = friction coefficient
$C_p$ = specific heat of the gas at constant pressure (J/kg K)
$D_H$ = hydraulic diameter (m)
$e$ = thermosyphon thickness (m)
$g$ = gravitational acceleration (m/s$^2$)
$H$ = thermosyphon height (m)
$h$ = heat transfer coefficient (W/m K)
$i$ = sodium latent heat of vaporization (J/kg)

| | | |
|---|---|---|
| k | | = pressure drop factor |
| $M_1$ | | = molecular weight of argon (kg per mole) |
| $M_2$ | | = molecular weight of sodium (kg per mole) |
| $Nu$ | $= \dfrac{hD_H}{\lambda}$ | = Nusselt number |
| $Pr$ | $= \dfrac{\mu C_p}{\lambda}$ | = Prandtl number |
| $P_s$ | | = sodium vapor pressure (N/m$^2$) |
| P | | = gaz pressure (N/m$^2$) |
| R | | = universal constant for perfect gases (J/mole K) |
| $Re$ | $= \dfrac{\rho u D_H}{\mu}$ | = Reynolds number |
| T | | = gas temperature (K) |
| $T_{p1}, T_{p2}$ | | = thermosyphon wall temperature (K) |
| u | | = gas velocity |
| x | | = curved abscissa along the thermosyphon (m) |
| z | | = height from reference plane |
| $\delta$ | | = boundary layer thickness (m) |
| $\lambda$ | | = thermal conductivity (W/m K) |
| $\mu$ | | = gas viscosity (kg/m.s) |
| $\rho$ | | = gas density (kg/m$^3$) |
| $\psi$ | | = heat flow per unit area (W/m$^2$) |

## 1. THE PHENIX REACTOR

Phenix is a 250 MWe experimental fast neutrons reactor. Trials were performed from fall 1972 to May 1974 when nominal power rate was reached (1).

Figure 1 shows a crosscut diagram of this integrated type reactor.

The 12 meters diameter main vessel is filled with liquid sodium and houses the core. The primary sodium which cools the core flows within the main vessel. It is circulated by means of three pumps and the heat is removed through six exchangers. The secondary sodium is piped from the vessel and transfers the energy to steam generators.

The upper part of the reactor is closed by a concrete cap acting also as a neutron shielding.

The space between the top cap and the surface of the liquid sodium is filled with argon.

# THE PHENIX NUCLEAR REACTOR

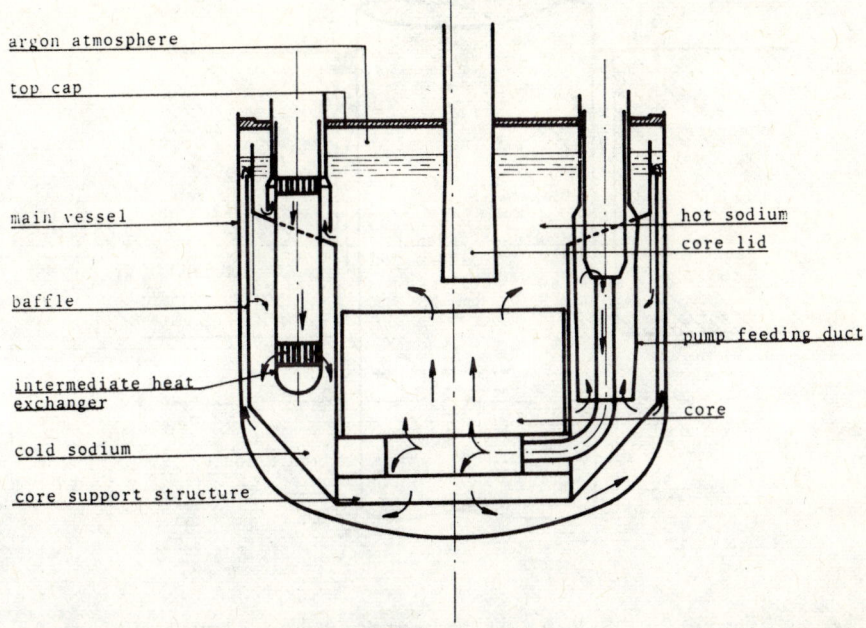

Figure 1  The Phenix breeder
Crosscut diagram along a plane including a pump and an exchanger

Several penetrations are provided within the thick top cap for :

- the rotating plug aimed to fuel rods handling
- the three pumps
- the six exchangers.

As these penetrations pass through the concrete structure, annuli are formed (figure 2) and a free convection-driven thermosyphon develops making the argon to stream in.

This stream is enhanced :

- by the fact that the concrete is cooled by water pipes
- by the pressure of sodium vapors as the sodium molecular weight (23 g/mole) is lower than that of argon (40 g/mole).

Figure 2 shows in a schematic way a penetration within the concrete structure and the argon convective stream pattern in the annulus.

This convection-driven flow generates thermal disbalance within the metal liners aimed at protecting both concrete and penetra-

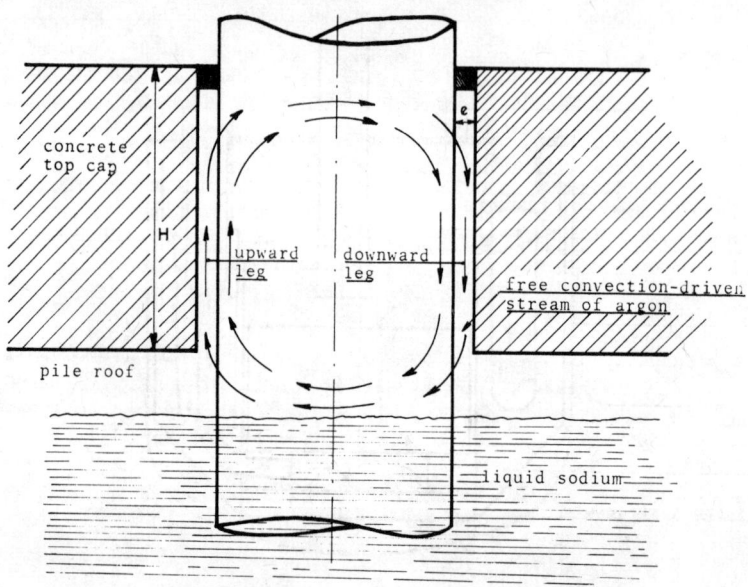

Figure 2   Schema of a thermosyphon structure

tions (pumps, exchangers, rotating plug) and which are the outer boundary of the annulus.

Thermal gradients brought on by the disbalance generate stresses within the liners.

The problem was to determine the order of magnitude of these gradients and subsequently to assess the stresses in and the mechanical behaviour of the liners.

## 2. THERMOSYPHON STRUCTURE

Calculating the heat transferred through thermosyphon is far from easy (2) and can hardly be achieved but with both straightforward configurations and boundary conditions.

Therefore we did not attempt to face up to this huge task. Instead we sought the hypotheses leading to a simple model of the phenomena so as to get a reasonable limit to actual gradients.

Such an approach is valid if we take in consideration that the actual configuration cannot be directly modelized and has to be simplified.

Boundary conditions are not known with great accurracy either.

# THE PHENIX NUCLEAR REACTOR

Especially the pile ceiling temperatures depend on the parts penetrating the concrete roof. Their values, resulting from other calculations, derive from approximate values.

The thermosyphons are 10 cm thick, the rotating plug has a 4 m diameter and pumps and exchangers have a 1.5 m diameter.

The flow within the annulus can therefore be considered as a flow between two plane walls.

Such a flow has been studied by SIEGEL and NORRIS (3) in a configuration similar to ours, i.e. between two heated walls, bottom and sides closed, the upper side open. They reported that two flow patterns can set up between such two walls.

a) If the gap is large enough so that both wall boundary layers do not mix, the flow is two dimensional, stream downward along the middle plan and upward along each heated wall.

b) If the thickness to height ratio is low, the flow turns to be unsteady and three dimensional. Both upward and downward streams fill the whole gap and the pattern exhibits upward legs and downward legs.

From SIEGEL and NORRIS work the latter pattern happens when

$$\frac{e}{H} < 0.21 \qquad (1)$$

A more general relationship can be stated relating the boundary layer thickness $\delta$ to the gap thickness e.

Experimentations with flow visualization run by our research team have shown that transition is experienced for a ratio

$$\frac{e}{\delta} \lesssim 3 \qquad (2)$$

Near this value the flow is especially unsteady and difficult to investigate.

In the present case a rough calculation of a boundary layer flowing along a plane wall equivalent to one side of the annulus, leads to the ratio :

$$\frac{e}{\delta} \ll 1 \qquad (3)$$

It can thus be seen that a two dimensional flow cannot set up, and that we are facing a configuration with a "b type" three dimensional flow, shown on the figure 2.

## 3. THERMOSYPHON MATHEMATICAL MODEL

The assumptions are the followings :

- wall temperatures are supposed to be known and depend only on the height "z" for every upward or downward leg
- we consider one upward leg and one downward leg of same cross section area and unit depth (figure 3)

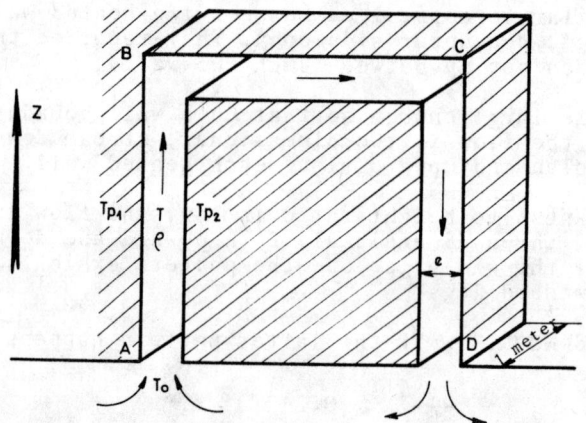

Figure 3   One dimensional thermosyphon model

- relationships for heat transfer and friction in turbulent flow between two vertical walls are used, for instance :

$$Nu = 0.023 \, Re^{0.8} \, Pr^{0.33} \quad \text{(Dittus and Boelter)} \quad (4)$$

$$C_f = 0.079 \, Re^{-0.25} \quad \text{(Blasius)} \quad (5)$$

- interaction of up and down legs is neglected and so is the conduction along the section perimeter
- in the upper region, as in the bottom one as well if the thermosyphon is closed, the gas streams horizontally. Both heat transfer and friction are neglected in these points
- the partial pressure of sodium vapor stands at the saturation mark $P_s$, given by

$$5 \log P_s = 6.45 - \frac{5544}{T} - 0.54 \log T \quad (6)$$

With such assumptions the thermosyphon model is one dimensional. The relations which govern the motion are the followings :

- Buoyancy

$$\Delta P_m = g \oint_{ABCD} \rho(T) \, dz \qquad (7)$$

- Pressure difference due to friction

$$\Delta P_f = \sum_i k_i \, \rho_i \, \frac{u_i^2}{2} + \int_A^D \frac{4}{D_H} C_f \, \frac{V^2(x)}{2} \, dx \qquad (8)$$

where the subscript i relates to the various pressure drops

- Gas heat balance

$$\rho(x) \, C_p(x) \, e \, u(x) \, dT = h \, (2 \, T(x) - T_{p1}(x) - T_{p2}(x)) \, dx \qquad (9)$$

- Conservation of mass (accounting for condensation)

$$\frac{\partial (\rho u)}{\partial x} = \frac{M_2}{RT} \, u \, \frac{\partial P_s}{\partial x} \qquad (10)$$

- Equation of state

$$\rho = \frac{1}{RT} \left[ \frac{P - P_s}{M_1} + \frac{P_s}{M_2} \right] \qquad (11)$$

- Specific heat flow $\varphi$ transferred to walls through condensation

$$\varphi = \frac{e \, i}{2} \, \frac{\partial (\rho u)}{\partial x} \qquad (12)$$

where :

$$i = 4.97 \cdot 10^6 - 1000 \, T \qquad (13)$$

## 4. SOLVING

The problem is solved in the plan of symmetry of a pump, of an exchanger or of the rotating plug. The choice of a meridian plan allows to deal separately with an upward leg and a downward leg of the thermosyphon.

In this plan, conduction heat transfer is processed by a finite elements calculation. It is completed by introducing :

- complementary terms accounting for radiation
- the heat transferred by convection in the form of heat exchange coefficients and outside temperatures.

To do this, thermosyphons are netted, the net corresponding to that of the finite elements method.

In this way relations (7) to (11) can be numerically solved.

The whole solving is carried out as follows :

A - The finite elements calculations yield approximate temperatures of the thermosyphon walls

    Aa - A certain mass flowrate is assumed to enter into the thermosyphon

    Ab - The temperature distribution $T(x)$ is computed

    Ac - The buoyancy $\Delta P_m$ is computed

    Ad - The bulk pressure drop $\Delta P_f$ is computed

    Ae - From comparing $\Delta P_m$ with $\Delta P_f$ results a closer approximation of the mass flowrate. The calculation is looped from Ab until convergence is achieved.

B - With the figures obtained for the thermosyphon the finite elements calculation is run again yielding a new set of figures for wall temperatures and the whole calculation is looped from Aa.

## 5. RESULTS

This mathematical model has been elaborated in 1969 and implemented for Phenix studies (4) up to 1973.

The measurements which were made after the reactor has entered into operation showed that only one upward leg and one downward leg actually exist in this configuration.

Values resulting from measurements compare with those from calculations on the figure 4 for an exchanger penetration and on the figure 5 for the rotating plug penetration.

In both cases the reactor was running at its nominal power rate, the liquid sodium temperature being 560°C.

Figures 4 and 5 show the part of the structure involved in the calculation. The detailed diagram shows that, in fact, there are several concentric annuli as far as exchangers are concerned, the corresponding thermosyphons were introduced in the model by clustering upward and downward legs.

Indeed, as the thermosyphons are separated but by a thin metallic cylinder , the upward leg of the most energetic thermosyphon induces an upward leg in the thermosyphon located behind the liner.

In the two thermosyphons shown on figures 4 and 5, the argon flows

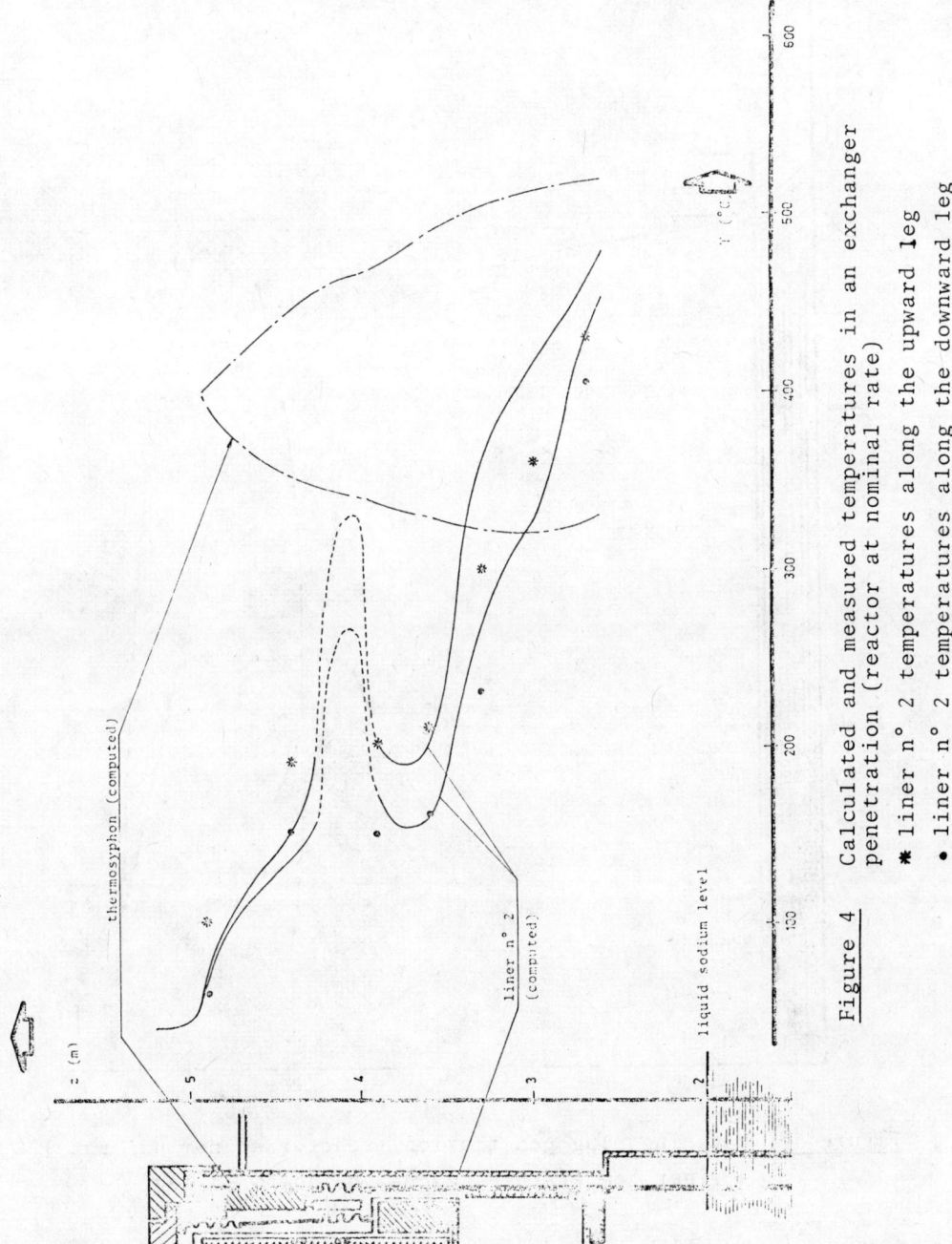

Figure 4  Calculated and measured temperatures in an exchanger penetration (reactor at nominal rate)

* liner n° 2 temperatures along the upward leg
• liner n° 2 temperatures along the downward leg

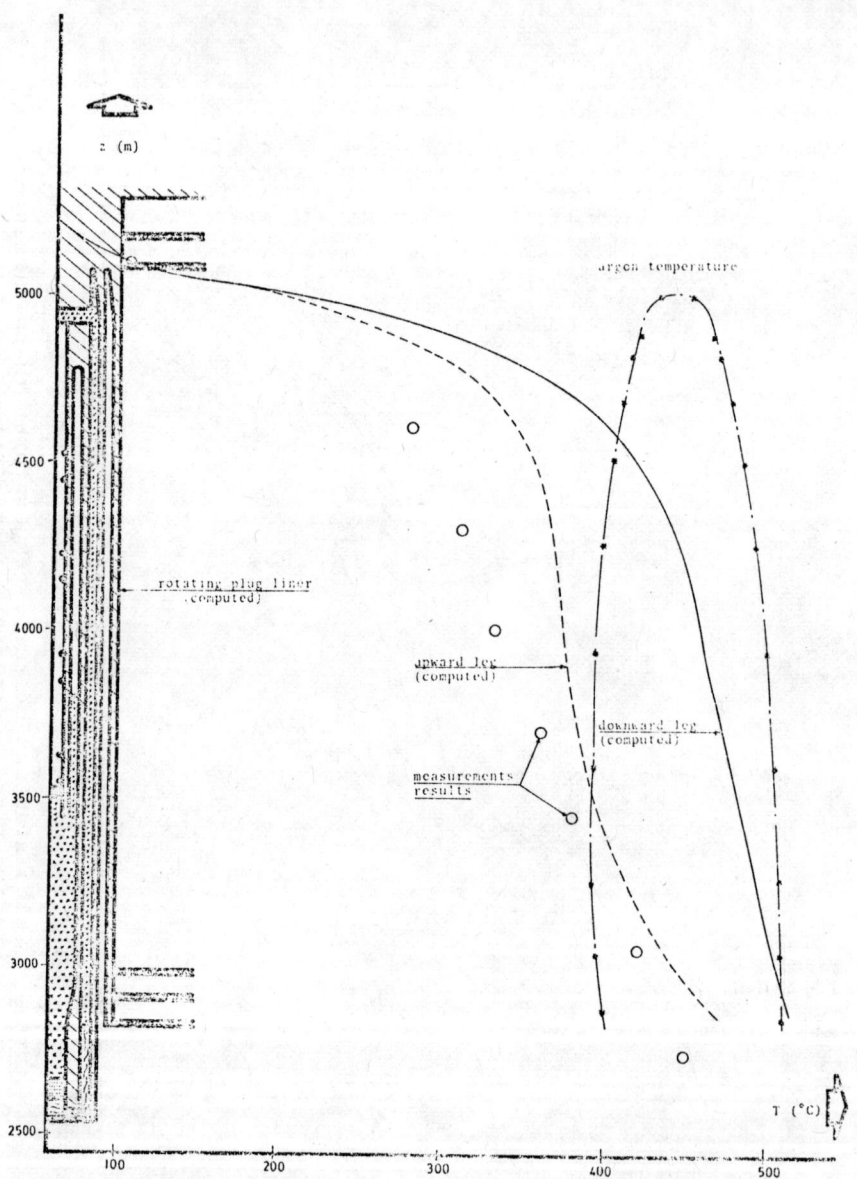

Figure 5   Rotating plug penetration (reactor at nominal rate)
• liner temperatures along the downward leg

at 2 m/s and the corresponding convective heat transfer coefficients are on the order of 5 $W/m^2°C$.

As of the exchanger the sodium condensation heat flow ranges from 1.3 to 2.2 kW when the temperatures energizing the thermosyphon range from 517°C to 547°C.

In the rotating plug case, to an entrance temperature of argon of 510°C corresponds a power transferred through sodium condensation of 2.5 kW. Compared to the power transferred through convection in the thermosyphon -about 13 kW- this value is low enough to legitimate a posteriori the simple assumption used for calculating the condensation.

On the figures 4 and 5 again the liner's calculated temperatures departs to some extent from those measured. However between two points of the same liner, one of them in contact with the upward leg and the other with the downward leg, the temperature differential is in agreement with measurement results ; this difference is function of the height z and shows a maximum of about 100°C.

## 6. CONCLUSION

The results presented above demonstrate that a calculation founded on simple hypotheses gives the engineer sufficiently accurate values, having in mind the lack of precision on boundary conditions. As far as sodium condensation is concerned the model would be easy to improve. But as for the thermosyphon a next step in development would necessarily be going three dimensional.

We believe that such a model is a useful tool easy to implement by engineers.

## REFERENCES

(1) J. BRANCHU, J. P. CROCHON, P. DEBERGH, J. DEFAUCHY, J. M. GAMA, J. MAULARD, J. C. MAZZON
B. I. S. T. 199 pages 45-60, Janv. 1975

(2) "Advances in Thermosyphon Technology"
D. JAFIKSE,
Adv. in Heat Transfer, vol. 9, Acad. Press, 1973

(3) R. SIEGEL, R. H. NORRIS
Trans. A. S. M. E., 79, I, pages 663-673, 1957

(4) M. BOURBOTTE, J. L. BOY-MARCOTTE
Dissymétrie thermique liée aux thermosyphons
Note BERTIN N. T. 73 Cg 17, 1973

# FREE CONVECTION PHENOMENA
# IN GAS-LIQUID MIXTURES

# THE CALCULATION OF FREE-CONVECTION PHENOMENA IN GAS-LIQUID MIXTURES

D. BRIAN SPALDING

*Imperial College of Science & Technology*
*Mechanical Engineering Department, Exhibition Road,*
*London sW7 2BX, UK*

ABSTRACT

The practical importance of the title subject is reviewed and exemplified.

Means of predicting the flow are described, both when the two phases slip relative to one another and when they do not.

The GALA (Gas And Liquid Analyser) technique is explained. This employs the volumetric rather than the mass continuity equation. It is shown to be advantageous both when the gas and liquid are finely interspersed and when they are separated by an interface stretching across the domain of study.

The IPSA (InterPhase Slip Analyser) technique is explained. This generalises to two-phase flows, with unequal phase velocities, the techniques of computation which have been found to work for single-phase flow.

The techniques in question are valid for multi-phase flow processes in general, not merely those implied by the title.

# 1. INTRODUCTION

## 1.1 Practical occurrence

Free-convection phenomena in gas-liquid mixtures are of great practical significance, as the following list of examples confirms:-

- The steam-air mixture in a boiler moves and circulates, as a rule, under the influence of buoyancy forces. Although there may be little "slip" between the two phases in some parts of the boiler, in other parts the "slip" is essential; without it, the bubbles could not detach themselves from the liquid.

- Many chemical reactors involve the bubbling of gases through liquids, often under circumstances in which the whole mixture is set into motion by a mechanical stirrer.

- When a liquid is subjected to electrolysis, bubbles of gas are often formed. Their rise, under the influence of buoyancy, affects the flow of electrolyte and so influences the mass-transfer resistances at the electrodes.

- Sometimes the formation, growth and movement of bubbles need to be predicted and controlled because of the mechanical damage which they can cause. Thus, if a bubble forms in superheated liquid sodium in a nuclear reactor, it may grow explosively, into voilent motion, and perhaps destroying the containment vessel.

- When tanks are emptied or filled with liquid, the movement of the interface is influenced by the relative magnitudes of the gravitational, inertial and surface-tension forces. Sometimes the surface folds over, so that gas bubbles are trapped: a gas-liquid mixture is then formed.

- Finally, the "breaking" of ocean waves may be mentioned; these represent large-scale entrapment processes, strongly influenced by gravity effects.

A procedure for prediction such phenomena quantitatively is strongly to be desired.

## 1.2 Physics

Prediction procedures must be based upon understanding of the physical processes of which the phenomena are composed.

The following list contains the most important ones associated with the behaviour of gas-liquid mixtures:-

- Bubbles embedded in a liquid, and droplets distributed through a gas, often exchange momentum, heat and mass with the media surrounding them; the processes may involve friction, vaporisation and condensation, absorption, and dissolution. The laws of such processes are well known, at least for small spherical bubbles and droplets; in every case, the process is more rapid the greater is the difference in velocity, temperature or concentration (according to type); and it is most effective when the diameter is small.

- The differences of velocity between the two phases are usually the result of the differing body forces exerted by gravity per unit volume: bubbles tend to rise, and droplets to fall. Of course, velocity differences can originate in other ways, as when liquid fuel is injected into a combustion chamber.

- The surface-tension phenomenon is simple for pure liquids; but it is strongly influenced by impurities. Some "frothy" mixtures are extremely resistant to inter-phase slip because of this effect.

- Gas-liquid mixtures often flow at such high velocities, and in such large equipment, that the flow must be supposed to be turbulent rather than laminar. Nevertheless, the turbulence is certainly different in its quantitative laws from the turbulence of a single-phase fluid.

- Often the flow of the two-phase mixture is impeded by the presence of distributed resistances such as:- baffles, tubes, packings and other equipment features. The liquid may cling to the solid surfaces; and the latter probably present different resistances to the liquid and the gas phases.

The laws governing these processes are only patchily known; and the variety of their practically-occurring manifestations is so great that, for many of them, more than ad hoc empirical investigations may never be made.

However, a general theoretical framework is needed, into which these individual formulations can be fitted; it is this which the present paper provides.

## 1.3 Mathematical aspects of the prediction problem

In recent years, numerical analysts have learned how to compute single-phase flow and heat-transfer phenomena rather well. All the main problems encountered have been solved; and improved solution procedures are continually appearing.

For two-phase phenomena, the situation is rather less satisfactory. For they exhibit, in strong measure, some of the more difficult aspects of single-phase flows; and they have special difficulties of their own. The following matters engage the attention of the numerical analyst concerning himself with free-convection phenomena in gas-liquid mixtures:-

- When slip between the phases is neglected (i.e. both gas and liquid are regarded as possessing a single, shared, velocity vector at each location and time), large density variations are encountered. Obviously, if water enters a heated pipe at one end, and leaves as steam at the other, the density change is very large indeed; moreover, it is dependent upon the pressure. Thus, two-phase mixtures exhibit the phenomena found in compressible flow of a single-phase gas, but with greater possibilities of variety.

- Density variations, in a gravitational field, entail non-uniform body forces; and the resulting motions tend to equalise these body forces. Problems of this kind, because they involve a second connexion between the momentum and the species-conservation equation (the first connexion is that via continuity), are especially prone to numerical instabilities.

- Surface-tension effects require special attention, even when the liquid phase is pure. Their accurate inclusion in the momentum equations requires accurate prediction of the location and curvature of the interface. Surface tracking is therefore an operation which the analyst must know how to perform.

- The most difficult aspect of all results from the presence, in general, of inter-phase slip. Of course, there are then more variables to compute (three velocity components at each point for each phase); but the problem is not merely quantitative: it is rather that general algorithms for solving the equations appear not to have been published. Such algorithms must take account of the fact that there are now two continuity equations to solve (one for each phase); but there is only one pressure at each point.

## 1.4 Outline of the present contribution

The present article concentrates attention on methods of formulating and solving the equations of free-convection effects in gas-liquid mixtures.

Especial attention is devoted to two connected novel analytical devices: GALA (Gas And Liquid Analyser) and IPSA (Inter-Phase Slip Analyser). These are valid for two-phase flows in general; but they are of especial utility for analysing the behaviour of gas-liquid mixtures in the practical circumstances indicated in Section 1.1.

Discussion will be brief; practical applications of the ideas will be published elsewhere.

## 2. THE QUASI-SINGLE-PHASE MODEL

### 2.1 Assumptions

If slip is neglected, the gas-liquid mixture can be regarded as a single-phase fluid, having its own peculiar thermodynamic- and transport-property relations. Thus, if water and steam are in question, the relation between specific volume and enthalpy at fixed pressure will ordinarily be expressed by a three-part formula: for pure liquid; for pure vapour; and for a mixture of the two.

The transport-property relations are more difficult to ascribe them the thermodynamic ones; indeed, with current knowledge it is possible, in most cases, to settle only the orders of magnitude. However, this difficulty will not be dwelt upon here: it is not an obstacle to the establishment of a general procedure for solving the equations, once values have been assigned to the transport properties.

Special mention may perhaps be made of distributed resistances, for example a perforated baffle plate: although the resistance of the plate to single-phase flow may be known, it is not the case that the single-phase formula, with the mixture density inserted, will yield the right pressure drop. However, "correction factors" can in principle be found; and sometimes their values are known. Once again, there is no need to dwell on the topic here.

### 2.2 Differential equations

The equations of "conservation" of mass, momentum, thermodynamic energy, turbulence quantities, etc., can all be expressed in the compact form:

$$\frac{\partial(\rho\phi)}{\partial t} + \text{div} \, (\rho \vec{u} \, \phi - \Gamma_\phi \, \text{grad} \, \phi) = S_\phi \qquad , \quad (2.2\text{-}1)$$

where the dependent variable $\phi$ can stand for any of:- 1 (for mass, irrespective of species); velocity components; stagnation enthalpy; mass fraction of a species in the mixture; turbulence energy; etc. Further significances of symbols are:- $\rho$, the mixture density; t, time; $\vec{u}$, the velocity vector; $\Gamma_\phi$, the exchange coefficient; $S_\phi$, the source of $\phi$, per unit volume, which balances the other terms.

It is through the source terms that the equations for the various $\phi$'s distinguish themselves. $S_\phi$ equals zero when $\phi$ equals unity; it contains pressure gradient, body force and solid-surface friction if $\phi$ is a velocity component; it contains radiation absorption when $\phi$ is the enthalpy of the mixture; etc.

These equations will not be developed here. They may be found, for example, in Ref. 1.

## 2.3 FINITE-DIFFERENCE EQUATIONS

The usual way of solving equation (2.1-1), in circumstances of any complexity, is to integrate it over a finite number of small control volumes, filling the space in question. By means of suitable interpolation assumptions, a set of linearised "finite-difference equations" is derived, of the form:

$$\phi_P = \frac{\Sigma a_i \phi_i + b}{\Sigma a_i - c} \qquad , (2.3-1)$$

wherein i is an index which can represent each of N, S, E, W, H, L, P- (north, south, east, west, high, low, earlier time), i.e. the seven (in a 3D transient problem) neighbour states which can influence conditions at point P.

The $a_i$ coefficients contain the effects of convection, diffusion and storage. They are all positive in a properly-formulated equation. The terms b and c are representative of the source term; c should be negative for satisfactory results.

The linearity of (2.3-1) is one of appearance rather than reality; for the $a_i$'s, and b and c, are in general functions of the $\phi$'s. The general problem is therefore non-linear; and the equations for different equation sets are linked together. Iterative solution procedures are needed.

## 2.4 Solution procedures

It is easy to <u>devise</u> iterative solution procedures for the equation set; but it is not easy to ensure that they will always (or even <u>ever</u>) converge. Since the problem of procuring convergence is most severe for the hydrodynamic equations ($\phi \equiv 1$, u, v, w, the last three being velocity components), and since their handling is the crux of GALA and IPSA, a brief mention will be made of two solution procedures which have been proved to be successful in many problems.

The first method is SIMPLE (<u>S</u>emi-<u>I</u>mplicit <u>M</u>ethod for <u>P</u>ressure-<u>L</u>inked <u>E</u>quations)(Ref.2). Its main features are:-

(i) Guess the fields of u, v, w, and of pressure p. (Usually the values in store are employed.)

(ii) Solve the equations of type (2.3-1) for new u's, v's, w's, by appropriate matrix-inversion techniques. (The $a_i$'s, b's, c's are those appropriate to the guesses of (i).)

(iii) Calculate the resulting errors in equation (2.3-1) for continuity ($\phi \equiv 1$).

(iv) Differentiate the latter equation with respect to the velocities appearing in the $a_i$'s; and relate the velocity differences to pressure differences by differentiating the corresponding finite-difference equations for u, v, w. Hence obtain a set of equations for pressure correction, containing the continuity errors of (iii) as source terms.

(v) Solve these pressure-correction equations by matrix-inversion techniques.

(vi) Apply the corresponding corrections to u, v, w so that continuity is now satisfied.

(vii) Solve the finite-difference equations for energy, etc.

(viii) Compute the corresponding new values of density, transport properties, body forces, etc.

(ix) Return to (i) and repeat until the errors at (iii) are sufficiently small.

The method called SNIP (<u>S</u>tart with <u>N</u>ew <u>I</u>ntegration for <u>P</u>ressure)(Ref.3) differs from SIMPLE in that:-

(i) The pressure field is not guessed, but deduced from the velocity field by integration over judiciously-chosen paths.

(ii) The solution for at least one of the velocity-component fields can therefore be dispensed with.

(iii) to (viii) These steps are the same as for SIMPLE; but the pressure corrections are discarded, having served only as a means of obtaining the velocity corrections.

SNIP is newer than SIMPLE; and it is characterised by less experience but greater promise. Both procedures have been applied successfully to the calculation of gas-liquid mixtures in the quasi-single-phase approximation.

## 2.5 Special practices appropriate to free convection

When significant density changes occur, and when gravitational body forces are important, steps (vii) and (viii) of the above procedure may create some difficulties; for the body-force changes may be very large, so that the changes in velocity increase at each cycle rather than diminishing: the procedure <u>diverges</u>.

It is worth mentioning a few practices that have been found to diminish divergence. They are:-

### (a) The reduced-pressure method

The pressure p is replaced by a "reduced pressure P", defined by:

$$P \equiv p + g \int_0^z \{\rho - \rho_{ref}\{z\}\} \, dz \qquad , \quad (2.5-1)$$

whereby it is presumed that the gravitational acceleration g operates in the negative-z direction. The function $\rho_{ref}\{z\}$ is an arbitrary one, chosen so that the $(\rho - \rho_{ref})$ terms are small. For example, the following definition is often employed:

$$\rho_{ref} = \frac{\int \rho \, dx \, dy}{\int \rho \, dx \, dy} \qquad , \quad (2.5-2)$$

where the integrations are carried out over the whole of the plane of fixed z.

The momentum equations for the x and y directions then contain $\partial P/\partial x$ and $\partial P/\partial y$ in place of $\partial p/\partial x$ and $\partial p/\partial y$; and the z-direction equation is changed by the substitution:

$$\frac{\partial p}{\partial z} + g\rho = \frac{\partial P}{\partial z} + g(\rho - \rho_{ref}) \qquad . \quad (2.5-3)$$

This change reduces the tendency of temporary errors in $\rho$ to promote numerical instability.

### (b) Inertial relaxation of velocity

Whenever a numerical procedure exhibits instability of the "overshoot" type, in which errors of alternating sign and increasing magnitude appear, under-relaxation is suitable. The particular type of under-relaxation which is suitable for curing instabilities of the present kind is called "inertial". Its nature is best explained by writing equation (2.3-1) in the equivalent form:

$$\phi_P = \frac{\Sigma a_i \phi_i + b + I\phi_P}{\Sigma a_i - c + I} \qquad , \quad (2.5-4)$$

where I is the inertial-relaxation parameter.

# CALCULATION OF FREE-CONVECTION PHENOMENA IN GAS-LIQUID MIXTURES

In use, this equation is employed to obtain a new value of $\phi_P$ by inserting the previous value of $\phi_P$ on the right-hand side. Obviously, if I is very large, only a small change of $\phi_P$ can occur.

The question is: what value should be ascribed to I? Here it will merely be indicated that I has, like the $a_i$'s, the dimensions of mass per unit time, and that it should be made proportional to the fluid density times the square root of the gravitational acceleration times the cell height to the power 3/2. The required proportionality factor depends upon the ratio of cell height to cell width.

(c) The use of SNIP

It has been mentioned that the SNIP procedure involves obtaining the pressure distribution by integrating the differential equations along judiciously-chosen paths. If buoyancy forces are significant and the computation prone to instability, experience shows that these paths should be vertical.

A consequence is that, whatever the distribution of density, the pressure distribution is such as to produce no change in the vertical velocities. Of course, changes do result indirectly: for the horizontal momentum equations are not in general in balance; so changes in the horizontal velocities occur, which cause continuity errors. Then the velocity-correction stage (vi) causes changes in both vertical and horizontal velocities, leading to elimination of these continuity errors.

Here it may be remarked that, in the SNIP procedure, there is no objection to making artificial modifications to the differential coefficients which represent the variations of velocity with pressure. For example, it is permissible to diminish those for the vertical components of velocity relative to those of the horizontal velocity. By such devices, which are distinct from under-relaxation, buoyancy-induced instability can be brought under control.

## 3. THE GALA METHOD

### 3.1 The main idea

The GALA (Gas And Liquid Analyser) method (Ref.4) focusses attention on the volumetric continuity equation. This expresses the fact that, in the absence of compressibility or volume sources, the volume of fluid flowing into a control volume equals that flowing out of it, even though there may be significant discrepancies between the inflows and outflows of mass.

The practical advantage of GALA is that the continuity principle is expressible in finite-difference form with little need for careful accounting for the densities of the materials crossing the control-volume boundaries, and residing within the control volume.

There is a conceptual advantage also: continuity is concerned with kinematics, not masses; therefore to think and compute in terms of velocities is actually more direct than to involve mass balances in the accounting.

### 3.2 The continuity equation

When $\phi$ is placed equal to 1 in equation (2.2-1), and $s_\phi$ is put equal to zero, the result is:

$$\frac{\partial \rho}{\partial t} + \text{div}(\rho \vec{u}) = 0 \qquad (3.2\text{-}1)$$

This can be written as:

$$\frac{1}{\rho}\frac{\partial \rho}{\partial t} + \text{div }\vec{u} + \vec{u} \text{ grad } \rho = 0 \qquad (3.2\text{-}2)$$

i.e.:

$$\boxed{\text{div }\vec{u} = -\frac{D}{Dt}\ln \rho} \qquad (3.2\text{-}3)$$

(3.2-3) is the form of the differential equation which is used in the GALA method for single-phase flow. In finite-difference form, the equation becomes:

$$\sum \vec{u}_i \cdot \vec{A}_i = \dot{V} \qquad (3.2\text{-}4)$$

where the left-hand side represents the sum of the products of outward normal velocity with the appropriate cell areas, and $\dot{V}$ represents the total rate of volume increase within the cell.

It is equation (3.2-4) which is used in SIMPLE and SNIP, first to yield the continuity imbalances (step (iii)) and secondly as the origin of the pressure-correction equation. The velocities on the left-hand side are of course influenced by the local pressure gradients; and the volume source on the right-hand side may be influenced by a pressure increase within the cell.

### 3.3 Constant-density flow

In many cases, the two fluids can be regarded, with little error, as having individually-constant densities. Thus, in an air-water flow of moderate size and low velocity, the air can be presumed, like the water, to be incompressible.

Then equation (3.2-4) becomes:

$$\sum \vec{u}_i \cdot \vec{A}_i = 0 \qquad (3.3\text{-}1)$$

which is particularly simple to use.

Of course, there is no implication that the density at any particular location is not changing with time: it *is* changing, as bubble succeeds liquid film and droplet succeeds gas. However, none of these complexities need appear.

This simplification is of the highest importance to the development of economical and accurate prediction procedures.

### 3.4 Flows with density changes

When water and steam flow through a boiler, there is a significant density change resultant upon heat transfer, even though the pressure variations are small enough for their effect on the steam density to be neglected.

This means that $\dot{V}$, in equation (3.2-4), is not zero. It is however often easy to calculate. For example, if the heat transfer into the cell is $\dot{Q}$, the volume source in the cell can be deduced from:

$$\dot{V} = \frac{v_{fg}}{h_{fg}} \dot{Q} \qquad , (3.4-1)$$

where $v_{fg}$ is the specific volume change, and $h_{fg}$ is the specific enthalpy change, for the transformation of water into steam at the prevailing pressure.

There may of course be additional contributions to $\dot{V}$, associated with pressure changes; but the term on the right of equation (3.4-1) is by far the largest; and it can be given a value, with fair certainty, as a rule, from the auxiliary relations governing the heat-transfer rate.

### 3.5 Interface tracking

When the liquid and gas are finely dispersed, the density of the local mixture, which is needed for the evaluation of terms in the momentum and energy equations, is best evaluated via an equation of state from the local enthalpy, or alternatively from a separate differential equations of which the dependent variable is the mass or volume fraction.

However, when the two phases are separated by continuous interfaces, and air takes its place, a different procedure is appropriate: one has to determine where the interface lies, and to allocate densities to control volumes according as they lie wholly on one or other side of the interface, or are intersected by it.

Two methods have been found to be satisfactory, the first being more suitable for steady flows and the second for unsteady flows.

The first method involves the solution of a conservation equation for the interface height, h, which can be written:

$$\frac{\partial h}{\partial t} + \int_{z_0}^{h} (\text{div } \vec{u}) \, dz = 0 \qquad , (3.5-1)$$

where z is the vertical height. Of course, this equation has to be put into finite-difference form, by integration over a finite time interval for a column of cells piled on top of each other. This method implies that the liquid lies always on the lower side of the interface.

The second method involves "particle tracking". This means that a series of particles are supposed to be placed, at some initial time, on the phase interphase; each is characterised by coordinates $\vec{x}_p$ where p denotes the particle. Then the variation of $\vec{x}_p$ with time is deduced from the finite-difference version of:

$$\frac{D\vec{x}_p}{Dt} = \vec{u}_p \qquad (3.5-2)$$

where $\vec{u}_p$ is the instantaneous velocity vector of the particle.

As time progresses, some of the imagined particles may become too widely separated for linear interpolation between them to define adequately the interface. When this occurs, additional particles can be interspersed.

This particle-tracking method, with the associated interpolation, provides means of adequately computing the densities in all the time-dependent, convective flux, distributed-resistance and body force terms in the finite-difference equations.

Steady flows can also be handled by particle-tracking techniques; but there are differences: the particles must be imagined as being supplied at appropriate "source points" on the interface; they trace out a part of the surface with their trajectory and then disappear.

### 3.6  Problems in which gas-phase dynamics play no part

When a water wave splashes upon a solid surface, the nearby air is moved by it; but the inertia of the air is so much less than the inertia of the water, that it might as well be non-existent.

However, the GALA method <u>does</u> treat the air dynamics as well as that of the water, but for convenience rather than necessity. It is convenient from the point of computer-code organisation to solve the finite-difference over the same grid, throughout the computation, regardless of whether the cells are occupied by air or water; and the presence of the nearly-inertialess air makes itself felt by the very small pressure differences which are computed. All that is necessary is to insert the appropriate densities; and these are a direct consequence of the interface-tracking feature of the computations.

### 4.  THE IPSA METHOD

### 4.1  Definitions

The Inter-Phase Slip Analyser (IPSA) method (Ref.5) focusses attention on the following variables:-

u,v,w,  the velocity components of the light phase;
r    ,  the volume fraction occupied by this phase;

# CALCULATION OF FREE-CONVECTION PHENOMENA IN GAS-LIQUID MIXTURES

$s$ , the volume source per unit volume of this phase;
$U, V, W, R, S$, the corresponding properties of the dense phase;
$\rho_1, \rho_2$, the densities of the light and dense phases respectively.

Average velocities, densities, etc., are defined by:

$$\bar{u} \equiv ru + RU \qquad , \quad (4.1-1)$$

$$\bar{\rho} \equiv r\rho_1 + R\rho_2 \qquad , \quad (4.1-2)$$

etc.

The pressure p is supposed to serve for both of the intermingled phases.

Inter-phase transfer coefficients will be introduced where appropriate.

## 4.2 The continuity equations

There are now two continuity equations to solve, namely:

$$\frac{\partial r}{\partial t} + \text{div}(r\vec{u}) = s \qquad , \quad (4.2-1)$$

$$\frac{\partial R}{\partial t} + \text{div}(R\vec{U}) = S \qquad . \quad (4.2-2)$$

Here $\vec{u}$ and $\vec{U}$ are the velocity vectors whose components are $u, v, w$ and $U, V, W$ respectively.

These equations can be expressed in finite-difference form in a manner analogous to (2.3-1), as:

$$r_P = \frac{a_i r_i + s}{a_i} \qquad , \quad (4.2-3)$$

$$R_P = \frac{A_i R_i + S}{A_i} \qquad . \quad (4.2-4)$$

Here the $a_i$'s and $A_i$'s are the products of appropriate velocities times cell areas, divided by the whole cell volume.

Of course, $r_P$ and $R_P$ are connected by:

$$r_P + R_P = 1 - B_P \qquad , \quad (4.2-5)$$

where $B_P$ is the proportion of the cell volume blocked by the presence of a third (solid) phase; it is a necessary element of the equation when shell-side heat-exchanger flows are being computed, for example.

It should be mentioned that the coefficients of equations (4.2-3) and (4.2-4), like those of equation (2.3-1), are calculated in an "upwind" fashion. This is part of the "proper formulation" touched upon in the comment on the latter equation; and it implies that the terms summed in the numerators of the equations are finite only for inflow and those in the denominator only for outflow. This implication can be made explicit by writing (4.2-3), for example, as:

$$r_P = \frac{\sum_{in} a_i r_i + s}{\sum_{out} a_i} \qquad (4.2-6)$$

The $a_i$'s associated with the time-dependent term, it may be mentioned, appear in both numerator and denominator.

It is wise, before proceeding, to recognise that equation (4.2-6) reduces to a familiar form for single-phase flow. Thus, if s is zero, and the r's are put equal to unity (for a domain without blockage), the result is:

$$\sum_{out} a_i = \sum_{in} a_i \qquad , (4.2-7)$$

which implies equality of inflow and outflow. Equation (4.2-3), (4.2-4) and (4.2-5) can thus be seen to extend and generalise the continuity principle, and to reduce to familiar forms in familiar circumstances.

### 4.3 The velocity-correction equation

The momentum equations for the two phases, and the other conservation equations, can all be put in differential and finite-difference forms which correspond to equations (2.2-1) and (2.3-1). The counterpart of the former, for the light phase, becomes:

$$\frac{\partial}{\partial t}(r\rho_1 \phi) + div\{r(\rho_1 \vec{u} \phi - \Gamma_\phi \, grad \, \phi)\} = s_\phi \qquad ; (4.3-1)$$

and there are corresponding finite-difference equations.

The $s_\phi$ terms contain inter-phase transport elements. For example, if $\phi$ stands for u, one component of $s_\phi$ will be $(U-u)F_x$, where $F_x$ is an interphase-friction coefficient. These expressions link the equations for the light and dense phases.

The momentum equations are important not just for themselves: they are also the source of the velocity-correction equation which, as in SIMPLE and SNIP, must be introduced to ensure that errors in the continuity equation are corrected. The nature of this equation will now be explained.

Let $r^*_P$ and $R^*_P$ be the values of the volume ratios which are computed from the finite-difference equations for a cell with the aid of an approximate velocity field; and let $r'_P$ and $R'_P$ be corrections to those ratios which are needed to satisfy equation (4.2-5). Then:

$$r'_P + R'_P = 1 - B_P - r^*_P - R^*_P \qquad . (4.3-2)$$

# CALCULATION OF FREE-CONVECTION PHENOMENA IN GAS-LIQUID MIXTURES

Let it be supposed that these primed quantities are the consequence of changes to the a's and A's in the relevant equation. Let these also be denoted by primes. Then, from equation (4.2-6):

$$r'_P = \frac{(\sum_{out} a_i)(\sum_{in} a_i' r_i) - (\sum_{in} a_i r_i + s)(\sum_{out} a_i)}{(\sum_{out} a_i)} \quad , \quad (4.3\text{-}3)$$

and there is a corresponding equation for $R'_P$. Substitution into (4.3-2) gives the obvious result.

This is not the end of the matter; for the primed coefficients have to be connected with pressure differences. This is easy; for every a' is connected with a velocity change; and every velocity change can be connected with pressure changes at adjacent points. Thus, for example:

$$u' = \frac{\partial u}{\partial p_{left}} p'_{left} + \frac{\partial u}{\partial p_{right}} p'_{right} \quad . \quad (4.3\text{-}4)$$

The differential coefficients result from differentiation of the finite-difference momentum equations.

The result of combining equations of the type of (4.3-3) and (4.3-4) with (4.3-2) is an equation of the form:

$$p'_P \sum \alpha_i - \sum (\alpha_i p'_i) = 1 - B_P - r_P^* - R_P^* \quad , \quad (4.3\text{-}5)$$

where the p''s are adjustments to the grid-point pressures. Here the $\alpha_i$'s are coefficients which express the ways in which the $a_i$'s depend upon the velocity changes, and in which these depend upon the pressure changes.

Equation (4.3-5) is the generalised pressure-correction equation, the solution of which can be substituted in (4.3-4) to produce velocity corrections. When these velocity corrections are substituted in (4.3-3) and the corresponding equation for $R'_P$, equation (4.3-2) will be satisfied.

## 4.4 Solution procedure

The complete system of equations can now be solved by way of either the SIMPLE or the SNIP procedures. The differences as compared with the single-phase problems can be seen by comparing the following steps with their counterparts in Section 2.4:-

(i) Guess the fields of u, v, w, r, U, V, W, R and of pressure p (usually the values in store are employed).

(ii) Solve finite-difference equations for all the above variables, except p, by matrix-inversion techniques.

(iii) Calculate the right-hand sides of equation (4.3-5) for all cells.

(iv)   Calculate the coefficients $\alpha_i$ in the equation (4.3-5).

(v)    Solve these pressure-correction equations by matrix-inversion techniques.

(vi)   Apply the corresponding corrections u', v', w', r', U', V', W', R', so that equation (4.3-5) is satisfied for all cells.

(vii), (viii), (ix)   As in Section 2.4

Evidently, there is more work to do, when inter-phase slip is to be considered; but the pattern of the computational procedure is very similar to that for single-phase flow.

## 4.5   Discussion

### (a)   Errors in the continuity equations

When a cycle of computations (steps (i) to (ix)) has been carried out for single-phase flow, the velocity fields satisfy continuity (but not, until convergence has been attained, momentum). This is not quite the case when inter-phase slip is being analysed; and the reason is worth considering.

The expression (4.3-3) for $r'_p$, and the corresponding one for $R'_p$, are obtained by a differentiation which postulates that the neighbouring r's and R's are unchanged; but, in reality, they <u>will</u> be changed. It is not practicable to proceed otherwise; but a consequence is that the r's and R's which result from a computation cycle, although they will satisfy (4.2-5), will not quite satisfy (4.2-3) and (4.2-4).

Of course, as the iterations succeed one another, these errors diminish. The situation may be recognised as similar to that which prevails in respect of the momentum equations, even for single-phase flow. The $\partial u/\partial p$ terms are obtained by a differentiation which neglects the influences of simultaneous changes of nearby u's; consequently, the series of adjustments do not, except in the converged state, result in velocity fields which satisfy the momentum equation.

### (b)   Interactions between light-phase and dense-phase velocities

When inter-phase friction is significant, convergence of the calculation scheme will be slow unless the velocity interactions between the phases are taken into account in the correction equations. Thus, equation (4.3-4) should be extended, to become:

$$u' = \frac{\partial u}{\partial p}\bigg|_{left} p'_{left} + \frac{\partial u}{\partial p}\bigg|_{right} p'_{right} + \frac{\partial u}{\partial U} \cdot U' \quad ; \quad (4.5\text{-}1)$$

and there is a corresponding equation deducible by differentiating the finite-difference equation for U, namely:

$$U' = \frac{\partial U}{\partial p}\bigg|_{left} p'_{left} + \frac{\partial U}{\partial p}\bigg|_{right} p'_{right} + \frac{\partial U}{\partial u} u' \quad , \quad (4.5\text{-}2)$$

Obviously, these equations can be combined so that u' and U' each appear separately. It is these resulting expressions for u' and U' which should be employed in the derivations of the coefficients of the pressure-correction equation.

It is not only momentum interchanges which should be accounted for in this way. The same is true of heat transfer and mass transfer: the interactions require to be accounted for in the adjustment equations, if convergence is to be rapid, whenever the interaction coefficients are large.

These assertions have been made without proof, for which space is lacking; but they may be readily verified by either analysis or computational experience.

## 5. CONCLUSIONS

It has been shown in this article that the methods of predicting single-phase flow phenomena can be extended also to problems of gas-liquid flow, including those in which free convection is present. The argument requires to be supplemented by practical demonstrations, to carry conviction; but these will be forthcoming.

This being so, the questions which it is natural to ask are: How long will it be before detailed and realistic calculations are being made of the radial velocity and void-fraction distributions of, for example, steam-water mixture in straight pipes? How long before flow in curved pipes will be predicted? And what about shell-side flows in turb-heated boilers?

These questions cannot be answered precisely. However, the calculations can <u>already</u> be performed; the task is to perform many of them, and to make comprehensive comparisons with experiment.

Of course, accurate predictions depend upon the existence of sufficiently accurate formulae for such phenomena as inter-phase friction and heat transfer. These are not available at present. However, just as turbulence theory suffered a re-birth, once computational procedures were available for solving the turbulence-model equations, so may it also be with two-phase flows.

What appears to be needed, therefore, is a bold demonstration of the predictive power of the new technique, based on the best available auxiliary relations. The latter can be improved, in the course of time, as comparisons with experiment proceed.

The present article has been written so as to point the way to, and to encourage, these developments.

## 6. NOMENCLATURE

All symbols are defined in the text. The following list is provided to assist the reader to find the definitions.

| Equation | Symbols introduced |
|---|---|
| 2.2-1 | $\rho$, $\phi$, t, $\vec{u}$, $\Gamma_\phi$, $s_\phi$. |
| 2.3-1 | $a_i$, $\phi_i$, b, c. |

| Equation | Symbols introduced |
|---|---|
| 2.3-1 (contd.) | subscripts N, S, E, W, H, L, $P_-$. |
| 2.5-1 | $P$, $p$, $g$, $\rho_{ref}$, $z$. |
| 2.5-2 | $x$, $y$. |
| 2.5-4 | $I$. |
| 3.2-4 | $A_i$, $V$. |
| 3.4-1 | $v_{fg}$, $h_{fg}$, $Q$. |
| 3.5-1 | $\vec{h}$, $\vec{z}_0$. |
| 3.5-2 | $x_p$, $u_p$. |
| Section 4.1 | $\underline{u}$, $v$, $w$, $r$, $s$, $U$, $V$, $W$, $R$, $S$, $\rho_1$, $\rho_2$. |
| 4.1-1 | $\underline{u}$. |
| 4.1-2 | $\underline{\rho}$. |
| 4.2-5 | $B_p$. |
| 4.3-2 | $r'_p$, $R'_p$, $r^*_p$, $R^*_p$. |
| 4.3-3 | $a'_i$. |
| 4.3-4 | $p'_{left}$, $p'_{right}$. |
| 4.3-5 | $\alpha_i$. |
| 4.5-1 | $u'$, $U'$. |

## 7. REFERENCES

1. **SPALDING D B**

   "THIRBLE: Transfer of heat in rivers, bays, lakes and estuaries".

   Imperial College, London, Mechanical Engineering Department Report HTS/75/4, 1975.

2. **PATANKAR S V & SPALDING D B**

   "A calculation procedure for heat, mass and momentum transfer in three-dimensional parabolic flows".

   Int.J. Heat & Mass Transfer, Vol. 15, pp 1787-1806, Pergamon Press, London, 1972.

3. **SPALDING D B**

   "Basic equations of fluid mechanics, and heat and mass transfer; analysis of convective flows".

   Imperial College, London, Mechanical Engineering Department Report HTS/76/6, 1976.

4. **SPALDING D B**

   "A method for computing steady and unsteady flows possessing discontinuities of density".

   CHAM Ltd., 86 Burlington Road, New Malden, Surrey, Report No. 910/2, December 1974.

# EXPERIMENTS IN TURBULENT THERMAL CONVECTION DRIVEN BY INTERNAL HEAT SOURCES

J. C. RALPH and R. McGREEVY
UKAEA, AERE Harwell, Oxon, England

R. S. PECKOVER
UKAEA Culham Laboratory, Oxon, England

ABSTRACT

Thermal convection in enclosed spaces driven by volumetrically distributed heat sources is of interest in safety studies of sodium cooled fast breeder reactors. In particular the design of retention devices for containing core material in liquid form requires a knowledge of internal heat transfer coefficients in horizontal layers of heat generating fluid cooled through the top and bottom bounding surfaces.

Analogue laboratory experiments have been performed using tanks filled with weak chloride solutions. The internal heating in the primary experiments has been volumetric Joule heating with heat extraction taking place through the temperature controlled horizontal bounding surfaces. Operating conditions in the tests extended well into the turbulent régime and temperatures within the convecting fluid were found to be strongly time dependent. Mean temperature profiles obtained using traversing thermocouples and flow visualization films indicated a relatively thick boundary layer above the bottom cooling surface with a thinner layer below the top cooling surface from which plumes intermittently break away.

Comparison with tests in which the lower boundary was insulated showed that the heat transfer characteristic is only weakly dependent on the lower surface boundary condition. Our tests did show however that the properties of the cooling surfaces can have a significant effect on the time dependent nature of the temperature oscillations of the fluid and on the resulting mean heat transfer coefficient through the top plate. This effect, not reported before in connection with turbulent convection, is important particularly for fluids which may form solid crusts of variable thickness.

1. INTRODUCTION

Free convection heat transfer in which heat enters the system through a boundary or immersed surface has been extensively studied both theoretically and experimentally. When fluid is heated from below, through a horizontal surface, and cooled from above, the natural instability of the fluid due to buoyancy effects leads to bulk fluid motion. The motion is initially fairly regular and exhibits a distinct cellular pattern first reported by Bénard [1] and studied theoretically by Rayleigh [2]. The subject is summarised in books [3,4] and articles; that of Stuart [5] for example offers a relatively comprehensive review. A recent paper which extends the studies into the turbulent regime is that of Chu and Goldstein [6].

The present study is concerned with a similar heat transfer situation to the above but differing in that the heat enters the system only by uniform generation throughout the volume of the fluid. The fluid is bounded on all sides by surfaces which are essentially adiabatic except the top and bottom horizontal surfaces which are cooled and are approximately isothermal.

There has been relatively little study of thermal convection driven by internal heat generation. Theoretical studies of systems of interest have been made by Roberts [7] and Thirlby [8]. These theories are however limited to the prediction of the conditions for the onset of steady laminar convection and to the behaviour of fluid near this condition. Tritton and Zarragar [9] and Schwiderski and Schwab [10] have done largely qualitative experiments in the laminar region with a single cooled surface, and Roberts and Ralph [11] have extended this more quantitatively into the turbulent regime. The experiments of Kulacki and Goldstein [12] and Jahn and Reineke [13] with both upper and lower boundaries cooled have also extended from the onset of convection into the turbulent region.

It is well known that for free convection systems driven by an applied temperature gradient, heat transfer is a function of the Grashof (Gr) and Prandtl (Pr) nondimensional groups [5]. For a wide range of fluids having moderate Prandtl numbers and where inertial forces do not completely dominate the viscous forces the heat transfer is in fact a function only of the product (Gr.Pr) which is the Rayleigh number (Ra), where

$$Ra = \frac{\beta \rho^2 g c}{\mu k} L^3 \Delta T \ . \tag{1}$$

For such a convecting fluid confined between horizontal surfaces the characteristic length dimension L is taken as the distance between the two surfaces and $\Delta T$ the temperature difference between them. The Nusselt number for the system is defined as

$$Nu = hL/k \tag{2}$$

where the heat transfer coefficient 'h' is defined by

$$\phi = h \Delta T \ . \tag{3}$$

Nu is then found in various ranges to be related to Ra by correlations of the form

$$Nu = c_1 Ra^n \tag{4}$$

where $c_1$ and n are positive real constants.

A similar nondimensional group to (1) can be defined for a system with internal heat generation. The dependent variable $\Delta T$ can be replaced by a temperature difference constructed from the volumetric power density. The temperature difference adopted here is the one chosen by Roberts [5] which, apart from numerical factors, has also been used by other workers in the field. The $\Delta T$ in [1] is set equal to twice the temperature difference that would occur if the fluid was purely conductive. That is:

$$\Delta T = \frac{HL^2}{k} \ . \tag{5}$$

The resulting modified Rayleigh number becomes:-

$$Ra_H = \frac{\beta \rho^2 g c}{\mu k^2} HL^5 \tag{6}$$

the suffix H denoting internal heat generation.

The weakly temperature dependent physical property group $\dfrac{\beta\rho^2 gc}{\mu k^2}$ will hereafter be replaced by $c_f$ such that

$$Ra_H = c_f HL^5 \quad . \tag{7}$$

Our experiments have concentrated on obtaining actual measurements of the temperature profiles within the fluid and of the heat transferred to the cooling surfaces for Rayleigh numbers in the range $3.7 \times 10^8 \leq Ra_H \leq 1.08 \times 10^{12}$.

## 2. EXPERIMENTAL

### 2.1 General

The arrangement of the convection chamber and associated instrumentation is shown schematically in Fig 1 (a). The basic chamber consisted of upper and lower horizontal cooling plates (Fig 1 (b)) bolted to a glass cylinder of radius 22 cm. Cylinders of various lengths were used in order to alter the depth of the convection cell. Two water inlets, four outlets and a porous water distribution plate ensured uniform water flow and temperature across the platinumized titanium cooling surfaces, which also acted as electrodes.

Thermocouples were mounted through access points in the cooling plate, eight such points being available. Some of the thermocouples were capable of being moved accurately over a range of 1 cm in the boundary layer, while others were set in fixed positions. All thermocouples were electrically insulated with a thin coating of a silicone rubber compound.

The electrolyte used in the convecting cell was a weak ($\simeq$ N/100) aqueous solution of either hydrochloric acid or sodium chloride. These were chosen because of their low (0.02) temperature coefficient of electrical conductivity. However temperature variation in the cell caused up to 40% variation in power density, for which a correction had to be made.

Power was supplied to the electrodes via a variable 50Hz A.C. transformer and measured with an accurate wattmeter. The heat removed through each cooling plate was calculated from measurements of water flow rate and temperature rise across the plate. The flow rates were measured by rotameters and the temperature rise by differential thermocouples, the cooling water being supplied to each plate from a separate temperature controlled loop accurate to within ± 0.1C over long periods.

The whole experimental cell was well lagged with fibre glass insulation.

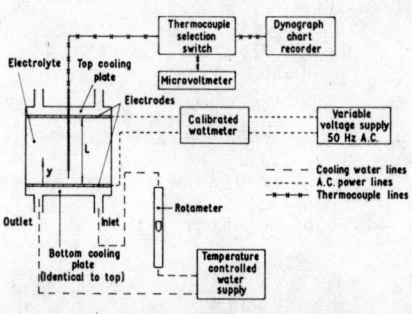

Fig. 1(a). Schematic diagram of experimental apparatus

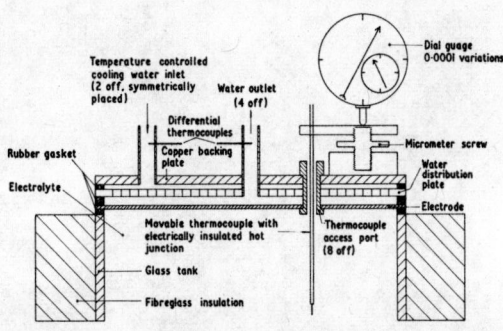

Fig. 1(b). Simplified diagram of top cooling plate

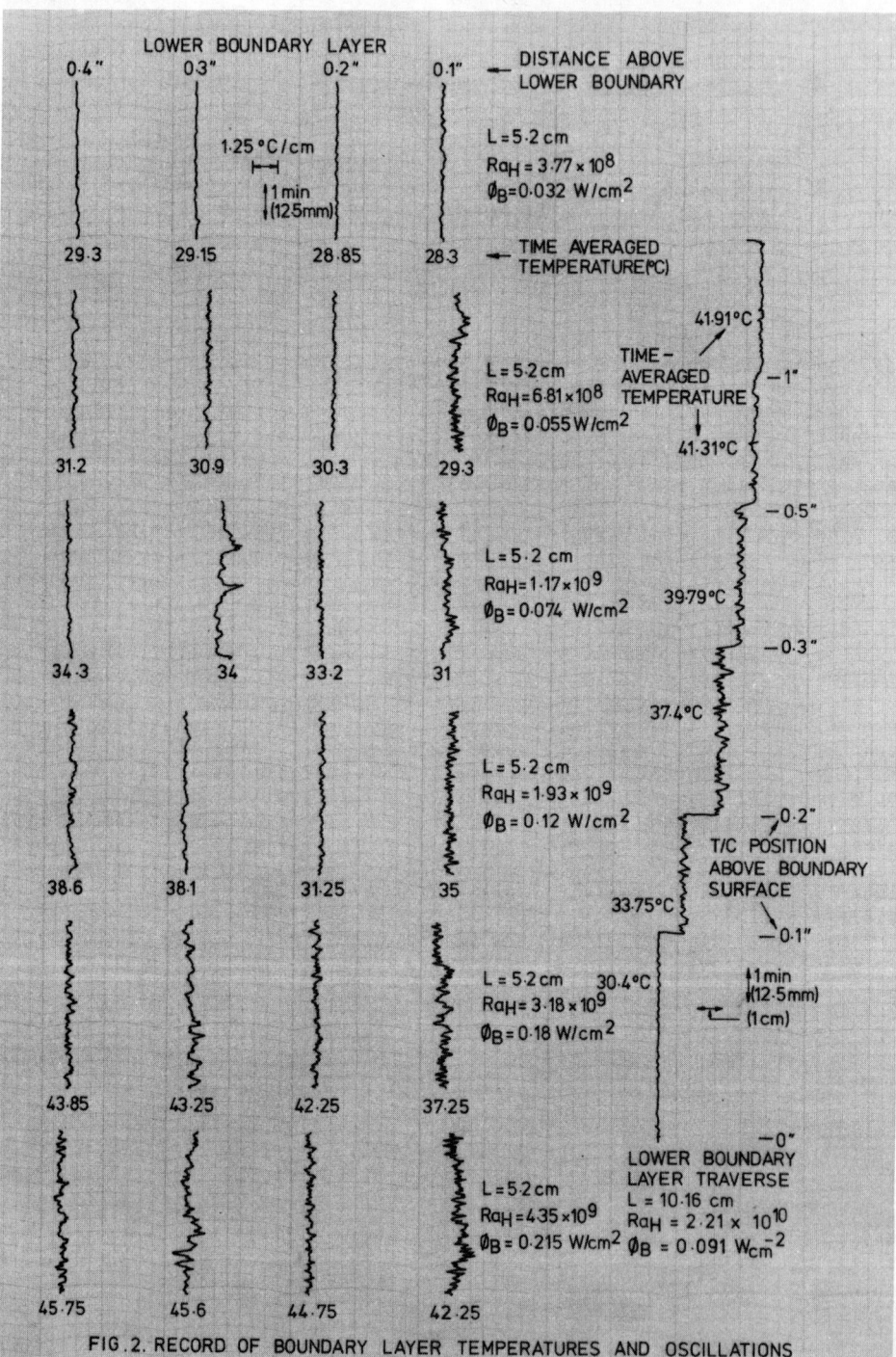

FIG. 2. RECORD OF BOUNDARY LAYER TEMPERATURES AND OSCILLATIONS

The particular experimental arrangement used, with the horizontal electrodes acting as cooling surfaces, was chosen instead of a vertical electrode cell because of problems encountered in the electrical insulation of cooling surfaces. Other workers [12,13] have not encountered any problems, presumably because of the low power densities and surface heat fluxes used. However it was found in experiments previous to those reported here that a large variety of such surfaces were rapidly attacked by current concentration at any porous spots. Thus it was felt preferable to use the present configuration rather than to carry out extensive work on testing of electrically insulating surfaces in order to find one that could be made thin enough so that it did not act as a thermal insulator.

The electrodes used, platinumized titanium, were chosen because of their wide availability, and in order that the formation of electrolysis bubbles be prevented. It is also considerably cheaper than platinum foil.

## 2.2 Test Procedure

The convection cell was set up with a known cell depth and filled with electrolyte through one of the thermocouple access points. Power was then applied for a time to drive dissolved air from the electrolyte, any gas bubbles thus trapped being voided with the aid of a hypodermic needle. Thermocouples were then fixed in known positions and power applied to the electrodes. Cooling water flow rate and temperature were also set, and periods of over six hours were then allowed for stabilisation of the system.

Equilibrium conditions having been reached measurement was made of total electrical power input, and of cooling water flow rate, inlet temperature and differential temperature rise for both plates. The temperatures of the fixed thermocouples in the electrolyte were measured and recorded on a multichannel recorder, as were the temperatures of those thermocouples that were moved through the boundary layers in accurately known steps. Periods of two to three minutes were allowed at each step in order to obtain a mean temperature.

Three different cell depths were used in the experiment, each being tested over a wide range of power inputs.

## 3. RESULTS

### 3.1 Basic

At all Rayleigh numbers encountered in the experiment it was found that temperatures in the convection cell were largely time dependent and that at any position in the cell they oscillate within a certain fixed band.

Chart records exhibiting such oscillations at fixed positions in the cell are shown in Fig 2 together with the record of a boundary layer traverse. It can be seen that the magnitude of the oscillations initially increases as the boundary is traversed, and then decreases as the bulk fluid temperature is approached. The apparent frequency of the oscillations is constant at all positions in the fluid but increases with Rayleigh number, as does the magnitude of the oscillations. A typical temperature profile illustrating the band of oscillations is shown in Fig 3. Mean temperature profiles are obtained by time averaging of the oscillation band, and a typical family of such profiles can be seen in Fig 4. It will be noted that the upper boundary layer is generally something like half as thick as the lower boundary layer, but that no significant change in boundary layer thickness is evident over the range shown.

### 3.2 Heat Transfer Results

The initial parameters measured from the basic results are $\Delta T$, (defined

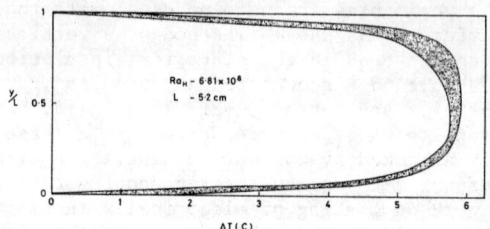

Fig.3. Typical Temperature Profile with Oscillation Band

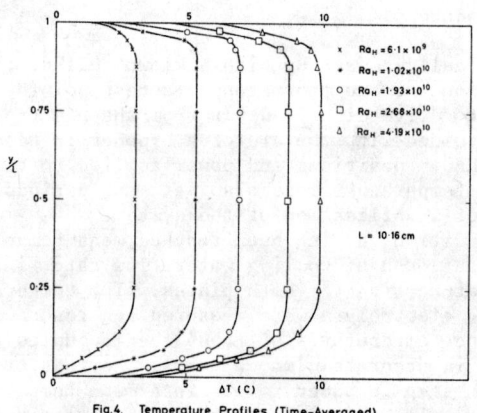

Fig.4. Temperature Profiles (Time-Averaged)

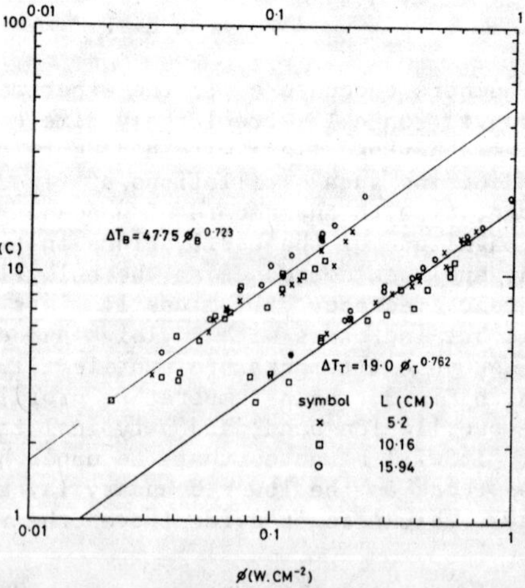

Fig.5. Fluid Temperature as a Function of Surface Heat Flux

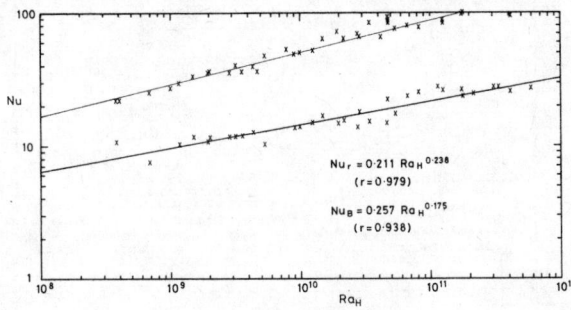

Fig.6. Correlation Between Heat Transfer Coefficients and Rayleigh Number

as the temperature difference between the cooling surfaces and the time averaged bulk fluid temperature) and $\phi$, the surface heat flux (subscripts T and B indicating heat fluxes for the top and bottom cooling surfaces respectively). A good correlation of the form

$$\Delta T = a \cdot \phi^b \qquad (8)$$

is illustrated in fig 5; that for the upper surface being almost independent of, and the lower surface being only slightly dependent on, the cell depth. As cell depth is decreased the tendency is for the surface heat flux ratio ($\phi_T/\phi_B$) to also decrease for a particular $\Delta T$.

As mentioned above a correlation of the form

$$Nu = c_1 Ra_H^n \qquad (4)$$

is assumed. Logarithmic plots of Nu vs $Ra_H$ are made in Fig 6. Single linear regressions of $\ln(Nu)$ vs $\ln(Ra_H)$ through 40 data points have been made and are also plotted.
This yields correlations of

$$Nu_T = 0.211 \, Ra_H^{0.238} \qquad (r = 0.979) \qquad (9)$$

for the upper surface and

$$Nu_B = 0.257 \, Ra_H^{0.175} \qquad (r = 0.938) \qquad (10)$$

for the lower surface.

In investigation of thermal convection with external heating from below a number of workers (Malkus [14] Willis and Deardorff [15] Krishnamurti [16] Chu and Goldstein [6] ) have investigated the occurence of discrete transition in the energy transport at the boundaries. Though the work reported here gives a good correlation for a single line there has not been sufficiently detailed examination of the effects of small changes in a narrow range of Rayleigh numbers to investigate these transitions. However work done as an extension of this report, dealing with the effect on heat transfer coefficients of differing upper and lower boundary temperatures, has indicated the existence of such transitions.

The correlations reported here (9) (10) represent a significant variation from those reported by other workers [12][13] with a similar convective arrangement. Although none of these have extended their results quite so far into the turbulent region extrapolation of their correlations reveals a factor of almost 1.5 difference in heat transfer coefficients measured at the same Rayleigh

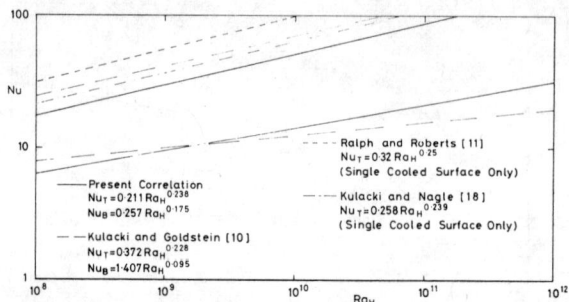

Fig.7 Comparison of Present Correlations with those of other Experimenters

number. In Fig 7 comparison is made with the extrapolated results of Kulacki and Goldstein [12] whose correlations (when expressed in terms of the Rayleigh number defined by (6)) were

$$Nu_T = 0.372\ Ra_H^{0.228} \qquad (11)$$

$$Nu_B = 1.407\ Ra_H^{0.095} \qquad (12)$$

The correlation of Roberts and Ralph [11] for a single cooled surface

$$Nu_T = 0.32\ Ra_H^{0.25} \qquad (13)$$

is also shown, as is that of Kulacki and Nagle [18]

$$Nu_T = 0.258\ Ra_H^{0.239} \ . \qquad (14)$$

In Fig 8 a comparison of temperature profiles and oscillation bands has been made with that of Ralph and Roberts [11] the cylindrical two cooled surface convection cell having been arranged to act as if with one cooled surface. Ralph and Roberts [11] report oscillations of up to 50% of the total ΔT, but no such oscillations have been encountered in the present work, these being of the order of 5 or 10%.

A secondary experiment was set up to investigate these differences.

## 4. SECONDARY EXPERIMENT

### 4.1 Secondary Experiment

In order to explain the variation in heat transfer coefficients mentioned above it was necessary to simplify the system by setting up a cell with only the upper boundary surface cooled, and an adiabatic lower surface. As the cell used by Ralph and Roberts [11] was available (hereafter referred to as the rectangular vertical cell, the electrodes being vertical) a cell of the same dimensions was constructed, but with an electrode implanted in the lower surface, and another electrode as the upper cooling boundary surface. This will be referred to as the rectangular horizontal cell (Fig 9(b). The experimental procedure was similar to that using the cylindrical cell.

Heat transfer results obtained are illustrated in Fig 9(a) together with the correlation for the upper cooled surface from the cylindrical cell. These indicate that the presence of a lower cooled surface has little or no effect on heat transfer at the upper boundary and that the change in aspect ratio is also unimportant. Work done as an extension of that reported here, and already

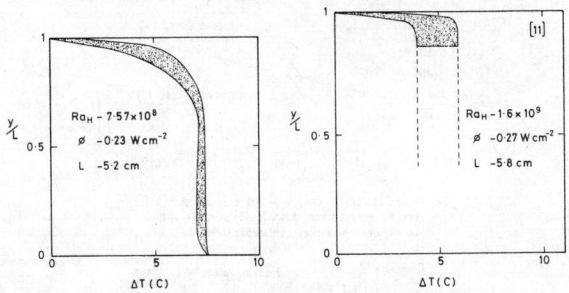

Fig.8 Comparison of Temperature Profile and Oscillation Band with Data from Ralph and Roberts [11]

mentioned in regard to discrete heat flux transitions, has confirmed this.

The two parameters in the convection cell that could affect heat transfer are (a) electrical and (b) boundary surface properties. The presence of any simple calculational error was disproved by direct repetition of some of the experiments of Ralph and Roberts [11].

The electrical effect can be divided into two categories, electrochemical and directional. The presence of an electrolytic recombination layer at the electrodes, and therefore in the case of the rectangular horizontal cell, at the cooling surface, could cause high power density and non-uniform heating in the boundary layer. An electrolyte that forms no such recombination layer (equimolar solutions of potassium ferri- and ferro-cyanide) was used but produced no alteration of heat transfer results in either of the rectangular cells.

A directional electrical effect, perhaps magnetohydrodynamic forces acting in the boundary layer, was considered because in the case of the rectangular horizontal cell current direction is vertical, and in the rectangular vertical cell, horizontal. To investigate any such effect a variety of small and large electrodes were used, in both rectangular cells, in various positions. Small variations in heat transfer coefficients measured were due to current concentration close to the upper boundary surface, but were not comparable with those variations under investigation.

The first cooling boundary surface property investigated was thermal conductivity. In the rectangular vertical cell a thin coating of stove enamel, and thus a thermal insulation, had been used to electrically insulate the aluminium cooling plate. To simulate the presence of such a thermal insulation while using the cooling surface in the rectangular horizontal cell as an electrode a porous cloth was attached to the electrode, in order to trap a layer of electrolyte. This trapped layer acted as an electrical conductor while having poor thermal conductivity. This experiment produced no change in heat transfer results, nor did similar experiments carried out on the lower electrode (though adiabatic) and on the vertical electrodes in the rectangular vertical cell.

4.2 Internal Heat Generation Simulation

Due to the limitations placed on the system by the presence of an electrical current an experiment was devised to simulate internal heat generation. Work carried out by Roberts with liquid metals had revealed that a horizontal grid of heaters provided such a simulation. Further tests carried out in water revealed that as long as a single horizontal grid heater is placed below the centre of the cell the heat transfer results obtained are in good agreement with those for internal heat generation.

Use of a grid heater in the rectangular horizontal cell (with lower electrode removed and upper electrode acting only as a cooling surface) and in the

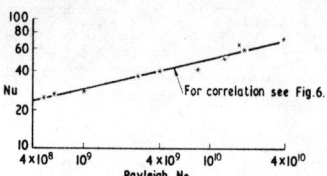

Fig. 9 (a). Comparison of cylindrical cell upper cooling surface heat transfer correlation with rectangular horizontal cell experimental data

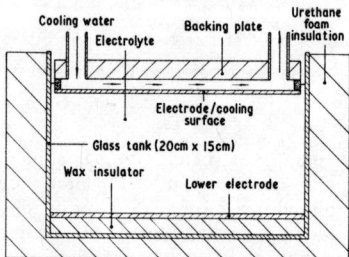

Fig. 9 (b). Rectangular horizontal convection cell

rectangular vertical cell (with both electrodes removed) confirmed the absence of electrochemical and directional effects. Experiments in both cells with thermally insulating layers also confirmed that such (thin) layers introduced no unexpected effect on the heat transfer.

A series of cooling boundary surfaces of varying properties (aluminium, copper, titanium, stainless steel and between 12.7 mm thick and 0.6 mm thick were tested using the grid heater. Heat transfer results obtained are illustrated in Fig 10. These results indicate that the properties of the cooling surfaces have a significant effect upon the heat transfer even in such turbulent conditions. Table 1 provides a key to the materials and thicknesses of the cooling plates used.

If a correlation of the form $Nu = \Gamma \cdot Ra_H^{0.24}$ is assumed for all plate thicknesses then it is clear from figure 10 that $\Gamma$ takes on a different value for each different plate. In figure 11, $\Gamma$ is plotted as a function of the plate thickness, by plotting the average $\Gamma$ for a given plate together with the observed spread for that plate. Despite the considerable scatter in figure 10, a remarkably good correlation emerges from figure 11:-

$$Nu = 0.345 \left(\frac{L_p}{L_{cell}}\right)^{1/11} Ra_H^{0.24} \quad . \tag{15}$$

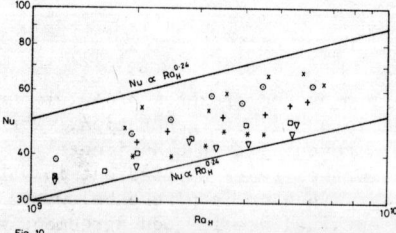

Fig. 10

Heat generation using immersion heater; heat transfer coefficients as a function of Rayleigh number for various boundary cooling surfaces.

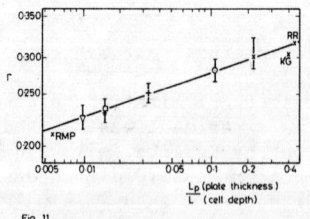

Fig. 11

The dependence of $\Gamma \equiv (Nu)/(Ra)^{0.24}$ on the ratio of thickness of top cooling plate to cell depth. For symbols see figure 10 and table 1. Each data point in this figure is the average of those in figure 10.

TABLE 1

| Material | | $L_p$ (mm) | $L_p^2/\kappa_p$ (sec) | $k_p/L_p$ (Wcm$^{-2}$K$^{-1}$) | $\Gamma$ | $\Gamma \cdot \left(\dfrac{L_p}{L_{cell}}\right)^{-1/11}$ |
|---|---|---|---|---|---|---|
| x | Al | 12.7 | 1.985 | 1.58 | 0.299 | 0.345 |
| ⊙ | Cu | 6.35 | 0.360 | 6.06 | 0.282 | 0.347 |
| + | Cu | 2.0 | 0.0357 | 19.2 | 0.254 | 0.347 |
| ▫ | Al | 0.9 | 0.0099 | 22.33 | 0.235 | 0.345 |
| * | Ti | 0.9 | 0.0836 | 2.56 | 0.230 | 0.338 |
| ∇ | SS | 0.6 | 0.0383 | 6.33 | 0.224 | 0.341 |
| RR | Al | 19.1 | 4.491 | 1.05 | 0.320 | 0.345 |
| KG | Cu | 24.0 | 5.143 | 1.60 | 0.303 | 0.332 |
| RMP | Ti | 0.9 | 0.0836 | 2.56 | 0.211 | 0.338 |

RR = Roberts & Ralph   KG = Kulacki & Goldstein   RMP = Ralph, McGreevy & Peckover (this paper)

This holds good also for the experiments using Joule heating by Roberts and Ralph [11], Kulacki and Goldstein [12], and that given earlier in this paper (eqn. 9). The difference in $\Gamma$ between its value for the thick aluminium plate (x) and the thin one (▫) indicates that $L_p$ must be an explicit correlation parameter.

However it is not the Biot number which is important, for, as table 1 shows, $(k_p L/kL_p)$ does not correlate at all with the change in Nusselt number for different cooling plates. At first sight this is very surprising. However the heat transfer coefficient of the cooling water configuration above the cooling plate is only 0.2 W cm$^{-1}$ K$^{-1}$ so that the sum of the thermal resistances of cooling plate and cooling water is dominated by the cooling water contribution.

The heat transfer in these highly turbulent convection cells may be expected to be a function of the Fourier parameter $L_p^2/\kappa_p$. For aluminium and copper, the results in table 1 are consistent with this grouping. Titanium and stainless steel are not consistent however. The thermal capacity $(\rho c)_p$ of the plates only varies by 60% so that for example $[(\rho c)_p]^{1/11}$ could be included in the correlation without significantly perturbing it. $(L \rho c)_p$ is the thermal capacity of the plate and it may be the key quantity in this fluctuating heat transfer configuration.

## 5. CONCLUSION

This paper reports work in progress. Heat transfer correlations are presented for layers heated internally by Joule heating and by an immersion heater. It is found that the upward heat flux is a function of thickness of the overlying cooling plate. The dependence on cooling plate thickness is purely empirical and further work is planned to establish this on a firmer analytical basis.

ACKNOWLEDGMENT   —   This work was carried out under the auspices of the Safety & Reliability Directorate, Culcheth, UK.

## REFERENCES

[1] H Benard. Les Tourbillons cellulaires. Ann. Chim. Phys. 23 62-144 (1901)

[2] Lord Rayleigh. On convective currents in a horizontal layer of a fluid when the higher temperature is on the underside. Philos. Mag. 32, 529-546 (1916)

[3] S Chandrasekhar. Hydrodynamic and Hydromagnetic Stability, Clarendon Press, Oxford (1961)

[4] J S Turner. Buoyancy Effects in Fluids. C.U.P. (1973)

[5] J T Stuart in Laminar Boundary Layers, L Rosenhead (ed), section IX O.U.P. (1963).

[6] T Y Chu and R J Goldstein. Turbulent convection in a horizontal layer of water. J.Fluid Mech. 60, 141-159 (1973).

[7] P H Roberts. Convection in horizontal layers with internal heat generation. Theory J.Fluid Mech. 30, 33-44 (1967).

[8] R Thirlby. Convection in an internally heated layer. J.Fluid Mech. 44, 673-693 (1970).

[9] D J Tritton and M N Zarraga. Convection in horizontal layers with internal heat generation. J.Fluid Mech. 30, 21-31 (1967).

[10] E W Schwiderski and H J A Schwab. Convection experiments with electrolytically heated fluid layers. J.Fluid Mech. 48, 703-719 (1971).

[11] J C Ralph and D N Roberts. Free Convection heat transfer measurements in horizontal liquid layers with internal heat generation. UKAEA Report. AERE-R7841 (1974).

[12] F A Kulacki and R J Goldstein. Thermal Convection in a horizontal fluid layer with uniform volumetric heat sources. J.Fluid Mech. 55, 271-287 (1972).

[13] M Jahn and H H Reineke. Free convection heat transfer with internal heat source, calculations and measurements. 5th Int.Heat Transf.Conf. paper NC2.8 Tokyo (1974).

[14] W V R Malkus. Discrete transitions in turbulent convection. Proc. Roy.Soc. A 225, 185-195 (1954).

[15] G E Willis and J W Deardorff. Confirmation and renumbering of the discrete flux transitions of Malkus. Physics Fluids 10, 1861-1866 (1967).

[16] R Krishnamurti. J.Fluid Mech. 42, 295-320 (1970). 42, 295-320 (1970).

[17] E M Sparrow, R J Goldstein and V K Jonsonn. Thermal Instability in a horizontal fluid layer: effect of boundary conditions and non-linear temperature profile. J.Fluid Mech. 18, 513-528 (1963).

[18] F D Kulacki and M E Nagle. Natural convection in a horizontal fluid layer with volumetric energy sources. J.Heat Transfer 204-211 (1975).

NOMENCLATURE

for convecting fluid:

$L$ = depth of layer = $L_{cell}$
$\rho$ = density
$c$ = specific heat
$k$ = thermal conductivity
$\Delta T$ = maximum temperature drop
$\mu$ = dynamic viscosity
$H$ = volumetric heating rate

for upper cooling plate:

$L_p$ = thickness of plate
$\rho_p$ = density
$c_p$ = specific heat
$k_p$ = thermal conductivity
$\kappa_p$ = thermal diffusivity

β = coefficient of thermal expansion
$c_f$ = physical properties group [eqn.7]
g = gravity
h = heat transfer coefficient [eqn.3]
φ = flux through upper cooling plate

$c_1$ ⎫
a    ⎬ constants
b    ⎪
n    ⎭

Pr = Prandtl number = $\mu/\rho k$
Ra = Rayleigh number [eqn.1]
$Ra_H$ = modified Rayleigh number [eqn.6]
Bi = Biot number = $k_p L/k L_p$
Nu = Nusselt number [eqn.2]
$Nu_T$ = Nu for upper plate
$Nu_B$ = Nu for lower plate

# MAXIMUM ATTAINABLE SUPERHEAT AND EXPLOSIVE BOILING-UP OF LIQUIDS

**V. P. SKRIPOV**
*Department of Physico-Technical Problems of Energetics, Sverdlovsk*

## ABSTRACT

Quick heating or abrupt expansion of liquid should make it highly superheated (metastable). The resulting explosive (shock) boiling-up is discussed in the paper. Maximum attainable superheat boundary of the number of liquids under different pressures is determined from independent experiments. A good agreement of the experimental results and the predictions of the homogeneous nucleation theory is observed. The explosive boiling-up occurs near the attainable superheat boundary. That enables us to treat the explosive boiling-up regime as a physically defined extreme case of intensive boiling.

## NOMENCLATURE

| | |
|---|---|
| $J$ | nucleation rate ($cm^{-3} sec^{-1}$) |
| $k$ | Bolzmann constant (J/K) |
| $W$ | critical nucleus formation work (J) |
| $v$ | specific volume ($cm^3/g$) |
| $\sigma$ | surface tension ($J/cm^2$) |
| $p$ | pressure ($J/cm^3$) |
| $N_1$ | number of molecules per unit volume ($cm^{-3}$) |
| $\Omega, \Omega_A$ | density of boiling centres ($cm^{-3}, cm^{-2}$) |
| $a$ | thermal diffusivity ($cm^2/sec$) |
| $\tau, t$ | time (sec) |
| $X$ | thermal relaxation length (cm) |
| $\rho$ | density ($g/cm^3$) |
| $c$ | heat capacity at constant pressure (J/gK) |
| $V$ | volume ($cm^3$) |

r        bubble radius (cm)
κ        exponent of bubble growth law
q        specific heat-evolution (W/cm$^3$)
L        heat of vaporization (J/g)

Subscripts

s        at the saturation line
*        homogeneous nucleation
c        liquid-vapor critical point
A        surface
′        liquid
″        vapor

We shall discuss the extreme boiling-up regimes, for example the nonstationary process of a very quick heating of liquid or the nonequilibrium flow of hot liquid through a short nozzle. The common feature of these processes is the possibility of high superheat of liquid $\Delta T = T - T_s$ with respect to saturation temperature $T_s$ at the given local pressure. It may seem that the quantitative description of nonstationary nonequilibrium processes is the insoluble problem, since even the stationary heat transfer of boiling liquid is not described analytically. However, it is possible in our case to analyze a definite boiling-up regime which we shall call shock or explosive boiling-up. The possibility of adequate description is determined by the fact that the vaporization occurs mainly on spontaneous (fluctuation) centres. The frequency of the inception of such centres (nuclei) in the unit volume of liquid $J = J(T,p)$ is a sufficiently steep function of temperature and pressure and is known [1] from the Volmer-Döring-Zeldovich-Frenkel theory (VDZF) with satisfactory accuracy:

$$J = N_1 B e^{-G}, \qquad (1)$$

where
$$G = \frac{W_*}{kT}, \qquad W_* = \frac{16\pi\sigma^3}{3(p-p_s)^2(1-v'/v'')^2}; \qquad (2)$$

$N_1$ - is the number of molecules per unit volume of liquid, $B \simeq 10^{10} - 10^{11}$ sec$^{-1}$ is the kinetic factor, weekly dependent on superheat magnitude.

Further we shall try to demonstrate the suitability of

equation (1) by its comparison with the experimental data on nucleation kinetics.

Now let us discuss in details the situations, for which we can expect the shock boiling-up regime.

1. <u>The decrease of "ready centres" number in the system</u>.
The indications of such decrease are observed for liquid distillation and for boiling of alkali metals [2] and cryogenic liquids. Boiling becomes nonstationary and the periods of quiet convection interchange the periods of intensive boiling. The superheat of Na reaches 300°C at the temperature of metal 900°C. In the experiments on organic liquids [3] the superheats up to 90° were observed near the wire heater at the absence of boiling. However, the estimates show that the superheats, obtained during quasistatic increase of heat load are usually not sufficient for the inception of fluctuation nuclei in liquid. If we denote the density of ready centres number by $\Omega$ and the characteristic time of vaporization cycle repetition on one centre by $\tau$ , then the product $J \cdot \tau_1$ will be the analog of $\Omega$ for fluctuation nucleation. Figure 1 is a schematic representation of the dependence of these values on the superheat $\Delta T$ under fixed pressure.

If we take $\Omega \simeq 10^8 cm^{-3}$, $\tau_1 \simeq 10^{-2}$ sec, for the intercept of the curves we obtain $J \simeq 10^{10} cm^{-3} sec^{-1}$. For benzene under atmospheric pressure such nucleation rate corresponds to the superheat $T - T_s = 150°$ and for water to $\Delta T = 208°$.

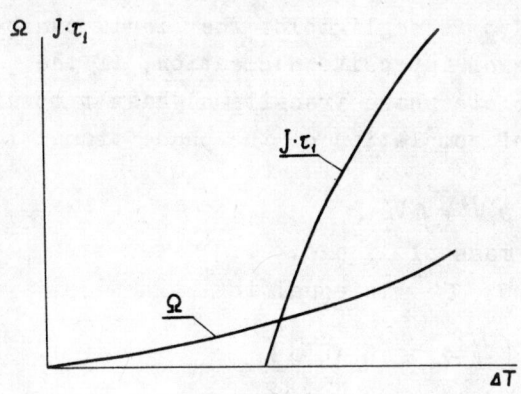

Fig.1. Volume density of ready centres $\Omega$ vs $\Delta T = T - T_s$ (schematic).

Such high superheats are not obtained during gradual increase of volume or surface heat-evolution in large installations, and also for the repeated delays of boiling.

2. <u>Quick admission of heat into the liquid</u>. It can be accomplished by means of laser or infrared (non-monochromatic) beam absorption and also by means of powerful pulse of current passing through conductive liquid. If the heat-evolution is considerably higher than the heat discharge for vaporization in ready centres, the significance of the latter as temperature regulators becomes negligible. The temperature regime of liquid considerably changes when there appears a great number of fluctuation nuclei.

We shall formulate the criterion of shock boiling-up regime, that is obtain the conditions of heating of liquid up to the temperature $T_*$ of intensive spontaneous nucleation [4,1]. We assume that the pressure of liquid is constant during the heating (it means that volume of liquid-vapor system increases). If we denote the heat-evolution in a unit of mass of liquid per unit time by $q$, then during time $\tau$ the liquid far from the bubbles (farther than the length of thermal relaxation $X \sim \sqrt{a'\tau}$ ) is heated up to the temperature $T$ :

$$(T-T_s)\langle c' \rangle = \int_0^\tau q\,dt ; \qquad (3)$$

it is assumed that the heating starts from the temperature of boiling $T_s$. Under low pressures the bubble size $r \gg X'$ and the bubble absorbs heat from the liquid layer, which is thin as compared with the bubble size. Therefore the heat influence of bubbles on liquid between them is negligible. The liquid can be heated up to the temperature of intensive nucleation, if the coalescence of bubbles (complete phase transition) hasn't occurred earlier. The mean time of completion of the phase transition is obtained from the condition

$$\rho''V'' = \rho'V_0' , \qquad (4)$$

where $\rho'V_0'$ is the initial mass of liquid.

Mass of vapor at the time $\tau$ is equal to

$$\rho''V'' = \rho''(V_0'\widetilde{\Omega})\frac{4}{3}\pi\left[\int_0^{\widetilde{\tau}} \dot{r}\,dt\right]^3. \qquad (5)$$

where $\widetilde{\Omega}$ is the specific density of ready centres of vaporization, corresponding to saturation line. Bubble growth rate is

expressed in the following way:
$$\dot{r} = K\tau^{K-1}\psi(T-T_S). \qquad (6)$$

From the equations (3)-(6) it is easy to derive the expression for maximum attainable temperature of liquid between the bubbles

$$T - T_S = (3\rho'/4\pi\rho''\widetilde{\Omega})^{1/3K} \langle q \rangle / \langle c' \rangle \langle \psi \rangle^{1/K}. \qquad (7)$$

The following designations are used:
$$\langle q \rangle = \frac{1}{\tau}\int_0^\tau q(t)dt \; ; \qquad \langle \psi \rangle = \frac{1}{\tau^K}\int_0^\tau \psi(t) K t^{K-1} dt \; .$$

It is convenient to introduce the dimensionless temperature
$$\widetilde{T} = \frac{1}{L}\int_{T_S}^T c'dT = \frac{\langle c' \rangle}{L}(T-T_S)$$

The numerical estimates show that for low pressures $\widetilde{T} \sim 1$ corresponds to the temperature of intensive nucleation $T_*$. Taking into account the equation (7) we can write the condition of reaching the temperature of intensive fluctuation nucleation in the form:

$$\frac{\langle q \rangle}{L \langle \psi \rangle^{1/K}} \left( \frac{\rho'}{\rho''} \cdot \frac{3}{4\pi\widetilde{\Omega}} \right)^{1/K} \gtrsim 1 . \qquad (8)$$

If this inequality is valid we must take into account the contribution of centres of spontaneous origin for adequate description of such system.

For pressure $p$, which is higher than half of the critical one, at the maximum attainable superheats we have $r \ll \chi'$. According to Labuntsov and others' approximation (Schraiver's data) the bubble growth rate under these conditions is

$$\dot{r} = \sqrt{\frac{a'\rho'}{2\rho''}} \left( \frac{\widetilde{T}}{1-\widetilde{T}} \right)^{1/2} \tau^{-1/2} .$$

We can write the criterion of the shock regime for the high pressure case in the form (8), if we assume $\langle \psi \rangle = (a'\rho'/2\rho'')^{1/2}$, $K = 1/2$.

Up to this point we discussed the volume heat-evolution. Discussing the nonstationary boiling at the heat evolving surface it is convenient to introduce the rate of surface heating $\dot{T}$ as a parameter. We assume that almost all ready centres are on the surface. We denote their density, corresponding to the temperature $T_S$ by $\widetilde{\Omega}$, cm$^{-2}$. The thin liquid layer, adjoining the wall has the temperature of the heat evolving surface. The maximum time for heating of liquid $\tau_A$ is determined by the condition of bubble coalescence $A''/A \simeq 1$ . (It is assumed that during this time the bubbles do not leave the surface). Then the condition of reaching the shock regime ( $T = T_*$ ) is expressed in the form:

$$\frac{\langle \dot{T} \rangle}{\langle \psi \rangle^{1/K} [\pi \widetilde{\Omega}_A]^{1/2K}} \gtrsim T_* - T_s \qquad (9)$$

Table I presents minimum values of characteristic parameters $q$, $\dot{T}$, ensuring the shock boiling-up regime, $J(T_*) \gg 1$, for methanol and water.

The densities of ready centres $\widetilde{\Omega} = 10^2 \text{cm}^{-3}$, $\widetilde{\Omega}_A = 10^2 \text{cm}^{-2}$ were taken for the calculations. It follows from the table that a very high power is required to ensure the shock regime for low pressures. The actual number of ready centres in the system rapidly increases with the temperature increase. By the time when the temperature $T_*$ is reached it can be several orders higher than the value of $\widetilde{\Omega}$ .

The data presented in the table implies the essential "softening" of the conditions of reaching the shock regime for pressures $p/p_c \gtrsim 0.5$. It is interpreted in terms of lower bubble growth rates in the high pressure case for the given value of superheat. The interesting feature of the experiments [5] on carbon dioxide for $p/p_c > 0.6$ is the proximity of the temperature departure from the steady nucleate boiling (the onset of the boiling crisis) and the temperature of intensive fluctuation nucleation $T_*$ .

Boiling crisis under quasistationary conditions has the pronounced features of hydrodynamic crisis [6] . During gradual heat load increase the stability of vapor and liquid flows, coming in the opposite directions, decreases. It results in the radical reconstruction of the two-phase layer, adjoining the wall. Vapor forms a film, separating liquid from the heating wall. Usually the density of ready centres is sufficiently high to ensure such crisis. However, the pressure increase improves mass transfer in the layer, adjoining the wall, due to the vapor density increase and the decrease of bubble departure diameter.

The mentioned proximity of the temperatures $T_i$ and $T_*$ indicates that for high pressures the density of ready centres is not sufficient for the onset of boiling crisis. Nucleate boiling is maintained until the inception of additional nuclei of fluctuation origin near the wall.

Table 1. Conditions for reaching the shock boiling-up regime.

| $p$, bar | $T_* - T_s$, °C | $q$, Wg$^{-1}$ volume boiling | $\dot{T}$, °C·sec$^{-1}$ surface boiling |
|---|---|---|---|
| Methanol, $p_c$ = 79,6 bar, $T_c$ = 240,0°C | | | |
| 1,0 | 130 | 6,0·10$^3$ | 5,0·10$^5$ |
| 3,5 | 100 | 4,7·10$^3$ | 2,0·10$^5$ |
| 6,5 | 80 | 8,8·10$^2$ | 2,5·10$^4$ |
| 11 | 63 | 1,4·10$^2$ | 2,8·10$^3$ |
| 17 | 45 | 1,3·10$^1$ | 1,8·10$^2$ |
| 56 | 5 | 2,3 | 2,0·10$^{-2}$ |
| Water, $p_c$ = 221 bar, $T_c$ = 374,1°C | | | |
| 1,0 | 205 | 1,1·10$^6$ | 1,3·10$^8$ |
| 10 | 135 | 1,3·10$^4$ | 3,4·10$^5$ |
| 25 | 91 | 1,7·10$^3$ | 2,5·10$^4$ |
| 47 | 64 | 2,2·10$^2$ | 1,9·10$^3$ |
| 130 | 18 | 1,5·10$^2$ | 2,3 |
| 190 | 3 | 7,3·10 | 2,5·10$^{-2}$ |

The proximity of $T_i$ and $T_*$ was also observed for $n$-pentane, $n$-hexane [7]. Probably for different liquids under pressures, higher than 0.6 $p_c$, the onset of boiling crisis occurs at the temperature close the temperature of intensive fluctuation nucleation in the liquid layer adjoining the wall.

The thermodynamic approach to boiling crisis, based on consideration of stability of phase state [8] seems to be useful for the analysis of extreme regimes (determination of the upper limit of temperature drop, determination of heat flows at quasistationary nucleate boiling, intensified by different external effects [9] ). The application of the fluctuation nucleation theory to nonstationary processes, leading to boiling crisis has proved to be more useful. The forming of vapor film on a rapidly heated wall is caused by the bubble coalescence in the layer, adjoining the wall. That enables us to avoid the consideration of the complicated problem of unsteady flows near the wall. The crisis condition is expressed [10] in terms of parameters, describing nucleation ( $J, G$ ) and bubble growth rate in superheated liquid ( $\langle \psi \rangle$, $K$ ). This condition can be also used for the experimental verification of the fluctuation nucleation theory. However the expressions [11,1] , relating the nucleation rate with the value and the temperature of thermal perturbation of a quick-response heater for the shock boiling-up regime, are more convenient for verification of the theory. In the experiments of P.A.Pavlov [11,12] the platinum wire (diameter 0.02 mm and length 1 cm) served as a heater and as a temperature-sensitive element. Powerful pulses of current (pulse width varied from 25 to 1000 $\mu$sec) passed through the wire, immersed into the studied liquid. The electronic temperature indication was used. The experiments on different liquids for pressures, varying from atmospheric to the vicinity of critical were carried out. A good agreement between the VDZF theory and the experimental data is observed for nucleation rates $J$ in the range $10^{12} - 10^{22} cm^{-3} sec^{-1}$. The largest discrepancy is observed for water under pressures less than 70 bar. The results for water are shown in fig.2.

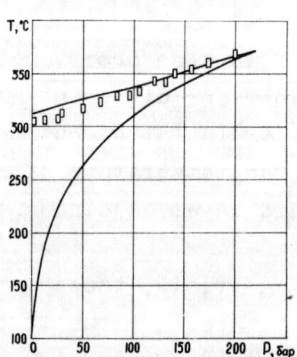

Fig.2. Comparison of the experimental (points) and the theoretical (upper line) values of temperature of explosive boiling-up of water under different pressures. Pulse heating technique. The lower line is the saturation curve.

3. <u>Flow of boiling-up liquid through short nozzles</u>. The explosive boiling-up of liquid can be attained by its quick expansion. For temperatures of stable liquid $T \gtrsim 0.9\,T_c$, a quick drop of external pressure down to atmospheric will cause the fluctuation inception of vapor nuclei in the entire volume of liquid. Vapor phase grows due to excess enthalpy (accumulated heat) of superheated liquid. This situation takes place in the case of seal failure in a high-pressure hot water line. The operation of a bubble chamber is based on the same phenomena.

The process of expansion can be made quasistationary, but highly nonequilibrium, if liquid flows into the atmosphere through a short cylindrical nozzle. The diameter of the nozzles used in the experiments [13] on $n$-pentane and $n$-hexane was 0.5 mm and the nozzle length $\ell$ varied from 0.4 to 10 mm. For $\ell = 0.7$ mm and the chamber pressure 25 bar the time of passing through the nozzle is $1.3 \cdot 10^{-5}$ sec. This value is considerably smaller than the time required for a vapor bubble to grow up to the size of the nozzle diameter. Temperature dependence of liquid flow rate was studied at the given pressure in the chamber. The results for $n$-pentane, obtained with the 0.7 mm nozzle are shown in fig. 3. The sharp flow rate decrease, observed in the definite temperature range, is caused by vaporization in the nozzle. The pressure in the nozzle was estimated by two independent techniques, which led to consistent results. Lines 1 - 4 in fig.4 show the temperature dependence of pressure in the nozzle for four values of chamber pressure. The degree of nonequilibriuty of liquid in the nozzle is determined by ordinate difference of the saturation line 5 and curves 1-4. Homogeneous nucleation region, corresponding to (in accordance with the VDZF theory) nucleation rates $J = 10^2 - 10^{15}\,cm^{-3}sec^{-1}$ is between two dotted-dashed lines. Comparing fig.3 with fig.4 we conclude that the "locking" of the nozzle occurs in the region of pressures and temperatures, where the intensive vaporization can be maintained by the flow of fluctuation centres. That enables us to treat boiling-up in short nozzle as a realization of the shock regime. The estimate of ready centres density, required for observation of a noticeable decrease of flow rate in the nozzle, gives the values $J \sim 10^8\,cm^{-3}$. This value seems to be somewhat overestimated.

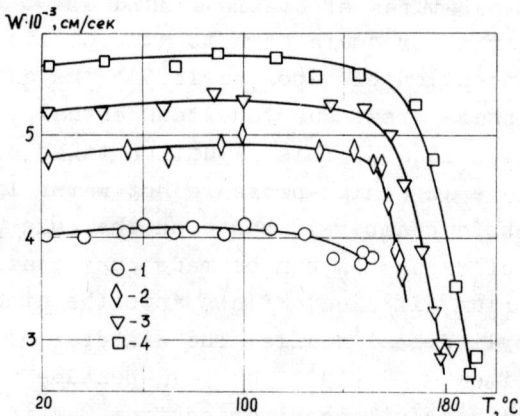

Fig.3. Temperature dependence of specific volume flow rate $w$ of $n$-pentane (parameters $T$, $p_0$ pertain to liquid in the chamber) for different pressures $p_0$: 1 - 16 bar; 2 - 21; 3 - 26; 4 - 31. Liquid flows into the atmosphere through the nozzle.

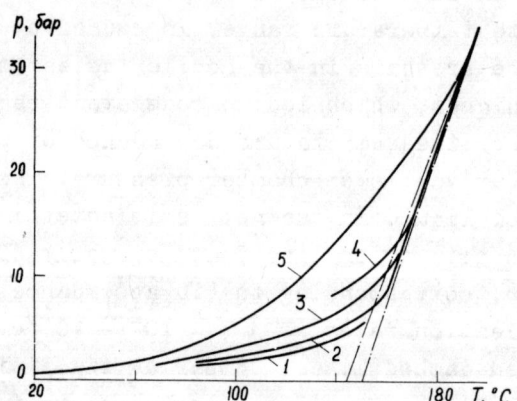

Fig.4. The temperature dependence of pressure in the cylindrical nozzle for different chamber pressures 1 - 4 (see fig.3); 5 - saturation line of $n$-pentane; K - critical point; the boundaries of homogeneous nucleation region ($J = 10^2$, $J = 10^{15} cm^{-3} sec^{-1}$) are shown by dotted-dashed lines.

### Discussion of the agreement of the homogeneous vapor nucleation theory with experiments.

We have already noted a satisfactory agreement of the predictions of the explosive boiling-up theory with the experiments if the expressions (1), (2) are used for calculation of the number of spontaneous centres. But we cannot consider such comparison as a correct verification of the VDZF theory for some reasons. Firstly, we deal with the nonequilibrium process of boiling-up, which is determined by several factors. Secondly, the highly superheated thin liquid layer adjoins the heater wall, and we cannot assert apriori that the wall influence on the conditions of nucleation is negligible. That is why other independent techniques are required for the verification of the theory.

In 1922 Wismer [14] obtained the outstanding results for maximum superheat of diethyl ether in a glass capillary tube with pressure control. In 1958 Wakeshima and Takata described [15] the simple technique for determination of the maximum attainable superheat under atmospheric pressure. Droplets of studied liquids (saturated hydrocarbons) came to the surface in a high column of sulphuric acid and were superheated due to the stationary temperature distribution along the column. The explosive evaporation of droplets was observed at the end of the way. Later on both experimental techniques were improved [1]. They allow the study of the functional dependence $J = J(T, p)$ (in a definite range of $J$ values) under quasistatic conditions and the comparison of experimental results with the VDZF theory. Consistency of the results, obtained by means of three independent techniques for many liquids in a wide pressure range and a good agreement between the experimental data and the calculations based on (1), (2) are cogent arguments in favour of the accuracy of conclusions of the VDZF theory. Table 2 presents the superheat data $Ar$[16] corresponding to nucleation rate $J = 1.3 \cdot 10^2$ $cm^{-3} sec^{-1}$.

Table 2. Temperatures of liquid argon superheat ( $T_*$ ) in a glass tube, corresponding to $J = 1.3 \cdot 10^2 \text{cm}^{-3} \text{sec}^{-1}$; $n_*$ is the number of molecules in the nucleus (bubble) of critical size.

| p, bar | $T_s$, K | $T_*$, K | | $n_*$ |
| --- | --- | --- | --- | --- |
| | | experiment | theory | |
| 1,9 | 93,6 | 131,2 | 131,0 | 260 |
| 3,6 | 101,3 | 131,9 | 131,7 | 270 |
| 8,1 | 113,2 | 133,3 | 133,3 | 400 |
| 11,0 | 118,3 | 134,4 | 134,5 | 510 |
| 14,0 | 122,7 | 135,4 | 135,6 | 670 |

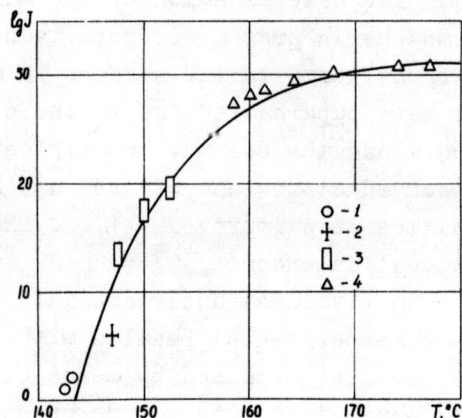

Fig.5. Verification of the fluctuation nucleation (solid line) for diethyl ether under atmospheric pressure in a very wide range of nucleation rate $J$. The experimental values: 1 - bubble chamber technique [1]; 2 - superheat of droplets [15] ; 3 - pulse heating technique [11,1] ; 4 - initiated boiling [17] ; $T_s = 34.5°C$.

Fig.5 shows the theoretical curve $J = J(T)$ for diethyl

ether at the atmospheric pressure and the experimental results 1 - 4, obtained by different techniques and overlapping a wide range of $J$ values. Points 4 correspond to the boiling-up of superheated liquid, triggered by $\gamma$-quanta. The data treatment in this case (calculation of $J$, $T$) is described in [17].

A large number of other examples of the agreement between the VDZF theory and the experimental results can be found in [1].

In conclusion we wish to remark that high superheats of liquids and explosive boiling-up can be encountered in engineering practice. There is a hope that these phenomena will be understood quite well.

REFERENCES

1. V.P.Skripov. Metastabil'naya Zhidkost'. "Nauka", 1972; "Metastable Liquids", New York-Chichester, Wily; Jerusalem-London, Israel Progr. Sci.Transl., 1974.
2. V.I.Subbotin, D.N.Sorokin, D.M.Ovetchkin, A.P.Kudrjavtsev, Teploobmen pri Kipenii Metallov v Uslovijach Jestestvennoi Konvektsii. "Nauka", 1969.
3. V.P.Skripov, N.V.Bulanov. Zh.Ingen. Phyz., 1972, 22, N 4, 614.
4. P.A. Pavlov, V.P.Skripov. Trudi Uralskogo Politechnitcheskogo Instituta col. N 189, Sverdlovsk, 1971, p 55.
5. V.P.Skripov, E.N.Dubrovina. Zh.Ingen. Phys.1971, 20, N 4, 725.
6. S.S.Kutateladze. Osnovi Teorii Teploobmena. "Mashgiz",1962.
7. G.P. Nikolajev, V.P. Skripov, E.N. Budin. Trudi CKTI, 1965, v. 62, p 137.
8. V.P. Skripov. Col. Teplo- i massoperenos, v 2, ed. BSSR Acad.of Sci. Minsk, 1962, p.60.
9. D.A.Labuntsov. Teplophyzika Visokich Temperatur, 1972,10, N 6, 1337.
10. P.A. Pavlov. Dissertation. Sverdlovsk, 1968.
11. V.P. Skripov, P.A. Pavlov. Teplophyzika Visokich Temperatur, 1970, 8, N 3, 579.
12. V.P. Skripov, P.A.Pavlov. Teplophyzika Visokich Temperatur, 1970, 8, N 4, 833.

13. N.A. Shuravenko, O.A. Isaev, V.P. Skripov. Teplophyzika Visokich Temperatur, 1975, 13, N 4, 896.
14. K.L. Wismer, J.Phys.Chem, 1922, 26, 301.
15. H.Wakeshima, K.Takata, J.Phys.Soc. Japan, 1958, 13, 1398.
16. V.G. Baidakov, V.P. Skripov, A.M. Kaverin, Zh.Exper.Theor. Phys; 1973, 65, N 3(9), 1126.
17. P.A. Pavlov, E.N. Sinitsin, V.P. Skripov, Col."Equations of State of Gases and Liquids", "Nauka", p.251.

# HYDRODYNAMIC AND THERMAL STUDY OF THE STABILITY OF BOUNDARY LAYER IN THE CASE OF FILM BOILING

**F. MOREAUX, J. C. CHEVRIER, and G. BECK**

*Laboratoire de Métallurgie L. A. 159, Parc de Saurupt, 54042 Nancy Cedex, France*

## ABSTRACT

This is a study on the stability of the vapor-film which appears during the cooling-down of a metallic sample quenched in water at a temperature of 100°C.

The impossibility to destabilize the vapor-film by mechanical means is proved either experimentally and theoretically.

Only a thermal resistance which is placed on the surface of the quenched sample will allow film-boiling to be destabilized and replaced by larvate boiling, which is caracterized by an alternating effect of wetting and non-wetting of the solid surface by the liquid. There are two conditions necessary to create larvate boiling :
- an adequate thermal resistance
- little superficial effusivity.

## NOMENCLATURE

| | |
|---|---|
| c | specific heat |
| D | diameter of a cylindrical sample |
| $E = \sqrt{\lambda c \rho}$ | , thermal effusivity |
| ∅ | heat flux |
| g | acceleration of gravity |
| h | heat transfer coefficient |
| H | height of a sample |
| λ | thermal conductivity |
| μ | absolute viscosity |
| m | mass of a sample |
| ν | kinematic viscosity |
| $Pr = C\mu/\lambda$ | , Prandtl number |
| q | heat quantity |
| R | radius of a sample |
| ρ | density |
| S | surface of a sample |
| t | time |
| $\theta_c$ | temperature in the center of a sample |
| $\theta_L$ | liquid temperature |
| $\theta_S$ | surface temperature |
| $\theta^*$ | dimensionless temperature |
| Δθ | temperature difference |

Copyright © 1976 by Hemisphere Publishing Corporation

## INTRODUCTION

When a metallic sample at a high temperature is quenched into a volatile liquid, two successive modes of vaporization are observed : film boiling and nucleate boiling.

The first mode is characterized by a film of vapor which covers the whole solid surface and insulates it from the liquid. Consequently there is very little heat transfer. The second is characterized by direct contact between the solid and the liquid, and heat transfer is important.

The transition temperature between these two modes of vaporization was determined when quenching was performed in water and in aqueous solutions of thermally stable salts (I).

Film boiling is an inconvenience generally, particularly in the case of the quenching of a metallic object, and a thorough research of the hydronamic characteristics of this vaporization mode was undertaken to find out ways to destabilize it. Studies were made on the other hand to try to destabilize film boiling by thermal means, and the solide surface is briefly wetted by the liquid.

## I. DESCRIPTION AND CHARACTERISTICS OF FILM BOILING

The liquid used for this experiment was water at a temperature of 100°C, to avoid any heating effects. Oxidised Nickel cylinders whith diameter D =16 mm and height H = 48 mm were used. They were quenched vertically into the water while the cooling-rate was being recorded. The liquid and the sample were kept absolutely still.

### I.1 - Description of film boiling

When quenching is performed from 800°C, film boiling lasts quite a long time and two different stages are clearly differenciated :

- when liquid-vapor interface is highly disturbed at the beginning of quenching when the metal surface and the liquid have the highest temperature difference. There are big vapor bubbles coming up to the surface of the liquid which cause great disturbances. The vapor film can be as thick as 10 mm when boiling water is used.

- when liquid vapor interface is little disturbed , which appears as soon as film boiling is over on the stem of the sample. There is less and less disturbance of the interface as film boiling is getting near to nucleate boiling. At this stage the interface is the site of regular waves which come up from the lower end of the sample. Then, the vapor film thickness is less than 1mm.

### I.2 - Heat transfer coefficient

During film boiling, it is possible to determine the heat transfer coefficient by using the cooling rate curve.

Newton law is $\emptyset = hS (\theta_S - \theta_L) = dq/dt$.
and $\emptyset = - mcd\theta/dt$.

By integrating these equations the heat transfer coefficient can be determined with $\text{Log} (\theta_S - \theta_L)$ versus time. The curve slope is proportional to $h = mc/S$ (slope). If the curve is a straight line, cooling follows Newton-law and the cooling rate curve is an exponential. This is what happens during film boiling, the heat transfer coefficient is $h = 264$ $W.m^{-2}.°C^{-1}$ in the case of boiling water.

The coefficient doesn't change as long as the vapor round the sample and the liquid are in physical balance. When mineral oils are used the cooling rate is different from Newton's law because of the chemical changes of the liquid.

I.3 - Study of the vapor-liquid interface

I.3.1 - Stability of the vapor film

Although the waves of the vapor liquid interface may have a very high amplitude, they don't bring the solid and the liquid into contact.

This was proved by a method for finding out whether there is an electrical contact between the sample and water (2). The water was made conductive by adding to it 1 % of NaCl, a platinium anode was dipped into it and the sample acted as a cathode in this cell for electrolysis. The electrical current was shown to go through thanks to a cathodic oscillograph fixed to the terminals of a resistance which is into the circuit. During film boiling, there was no current passing through, which shows that the interface instabilities didn't allow any even instant wetting of the metal by the liquid.

These observations are similar to BRADFIELD's (3) who noticed that it is necessary to quench a polished chromium copper sphere into water at a temperature of 80°C to obtain discontinuous contacts between the water and the sample during film boiling just before the transition stage of film boiling-nucleate boiling. At the beginning of quenching from 800°C there was no contact noticeable even in sub-cooled water.

Two types of experiments were carried out to increase the amplitude of the interface instabilities :

- in the first experiment, another sample was fixed to the first cylindrical one and the whole setup was heated to a temperature of 800°C and quenched. The second sample was meant to infuse more vapor into the vapor film round the sample and thus cause greater instabilities at the vapor interface. Then the vapor film round the cylindrical sample appeared to be thicker with more diturbances but no contact between liquid and solid was noticeable.

- in the second experiment a gas was infused into the vapor film. Heated nitrogen was used and injected into the vapor film either with a hypodermic needle of small diameter, or through the sample itself, through a little hole drilled in the solid surface. In both cases the vapor film remained stable.

These two experiments demonstrate how stable film boiling is and that is impossible to increase the instabilities of the liquid vapor interface and get the sample surface wetted.

I.3.2 - Determination of the characteristic parameters of the interface movements

Ultra-fast cinematography makes it possible to analyse the movements of the interface when the sample surface is at a temperature of 100°C above the film boiling-nucleate boiling transition which occurs at 250°C.

Under these conditions it is possible to measure the frequency of the waves which run at the interface, as well as the wavelength and the velocity of propagation. The results are listed below :

| D mm | H mm | $\Delta\theta$ °C | wavelength mm | Wave frequency Hz | wave velocity m/s |
|------|------|-------|---------------|-------------------|-------------------|
| 16   | 48   | 200   | 14 à 17       | 62 à 71           | 0,87 à 1,20       |

I.3.3 - Nature of the waves of the interface

The waves always start from the lower end of the cylinder at the stagnation point of the flow.

The measurements which characterize these waves make it possible to know for certain that they are capillarity waves. The results correspond to BRADFIELD's (3) who studied a similar case with water at 75°C and LEVICH's (4) who studied the capillarity waves which occur on the surface of a liquid.

## II. THEORETICAL STUDY OF FILM BOILING

The experimental study allowed to define a number of characteristics : newtonian heat transfer, existence of capillarity waves at the liquid-vapor interface, although it was not possible to dermine by experiment the nature of the flow and the velocity and temperature profiles in the vapor layer.

A number of authors have proposed different models for describing film boiling on a vertical wall in steady state of heat transfer : BROMLEY (5), HSU and WESTWATER (6), GREITZER and ABERNATHY (7). These latter mainly tried to measure heat transfer during film boiling and particularly gave expressions of the heat transfer coefficient h, but the hydrodynamic aspect of the different models was little developped.

On the other hand, KOH (8) proposed an analytical solution to the question of film boiling, both on the heat transfer and hydrodynamic level. He completely solved the problem of film boiling on a vertical surface in natural convection by employing the theory of laminar boundary-layer. The equations of boundary-layer concerning vapor and liquid were both taken into account. KOH also studied the case of zero flow velocity at the vapor-liquid interface, which made calculations slightly simpler. The velocity profiles calculated by KOH don't have any inflexion, which proves that the boundary-layer is quite stable.

In the case of quenching, it is possible to consider the sample surface at constant temperature when liquid vapor interface is little disturbed, that is around 300°C, and to apply KOH's model. KOH however didn't take into account the variations of the physical properties of the vapor in the film, which might alter the velocity profiles.

The model is completed (figure 1) with vapor physical properties variations and zero velocity flow at the interface (no liquid boundary layer). These considerations don't change the velocity profile but they provide too low the calculated heat transfer (8).

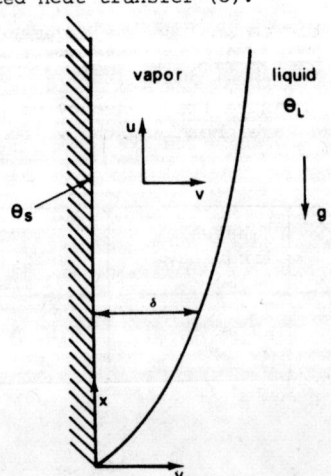

FIGURE I

Physical model and co-ordinates

The conservation equations for vapor are:

Continuity: $\dfrac{\partial u}{\partial x} + \dfrac{\partial v}{\partial y} = 0$

Momentum: $u\dfrac{\partial u}{\partial x} + v\dfrac{\partial u}{\partial y} = \nu_v \dfrac{\partial^2 u}{\partial y^2} + g\dfrac{(\Theta_S - \Theta_L)}{\Theta_L}\cdot \Theta^* + g\cdot \dfrac{\rho_L - \rho_V}{\rho_V}$

Energy: $u\dfrac{\partial \Theta}{\partial x} + v\dfrac{\partial \Theta}{\partial y} = \dfrac{\lambda_V}{\rho_V Cp_V}\dfrac{\partial^2 \Theta}{\partial y^2}$

with: $\Theta^* = \dfrac{\Theta - \Theta_L}{\Theta_S - \Theta_L}$

The boundary conditions are:

at $y = 0$ (vapor phase)

$u = 0$

$v = 0$

$\Theta = \Theta_S$

at $y = \delta$ (vapor-liquid interface)

$- u_V = u_L$

$(\mu \dfrac{\partial u}{\partial y})_V = (\mu \dfrac{\partial u}{\partial y})_L$

$(u\dfrac{d\delta}{dx} - \rho v)_V = (\rho u \dfrac{d\delta}{dx} - \rho v)_L$

$- \Theta = \Theta_L$

When $y \to \infty$ (liquid phase), $u \to 0$.

By use of stream functions, the momentum and energy equations of the vapor layer are transformed into the following ordinary differential equations:

$$\begin{cases} \xi''' + 3\xi\xi'' - 2\xi'^2 + \dfrac{\Theta_S - \Theta_L}{\Theta_L}\cdot \dfrac{\rho_V}{\rho_L - \rho_V}\cdot \Theta^* + 1 = 0 \\ \\ \Theta^{*''} + 3\,Pr.\,\xi.\,\Theta^{*'} = 0 \end{cases}$$

The equations are numerically resolved by Runge Kutta method, using a computer.

The velocity and temperature profiles are calculated for a surface at 300°C on which there is film boiling in the case of water at 100°C (figures 3 and 4).

There is no inflexion on the velocity profile which proves that the vapor film is very stable. Since, according to RAYLEIGH's theorem (9), inflexion on a velocity profile shows instability which will tend to increase. This is the reason why the capillarity waves which develop at the liquid-vapor interface cannot possibly modify the layer stability. Just as any outside disturbance cannot change the vapor flow in natural convection.

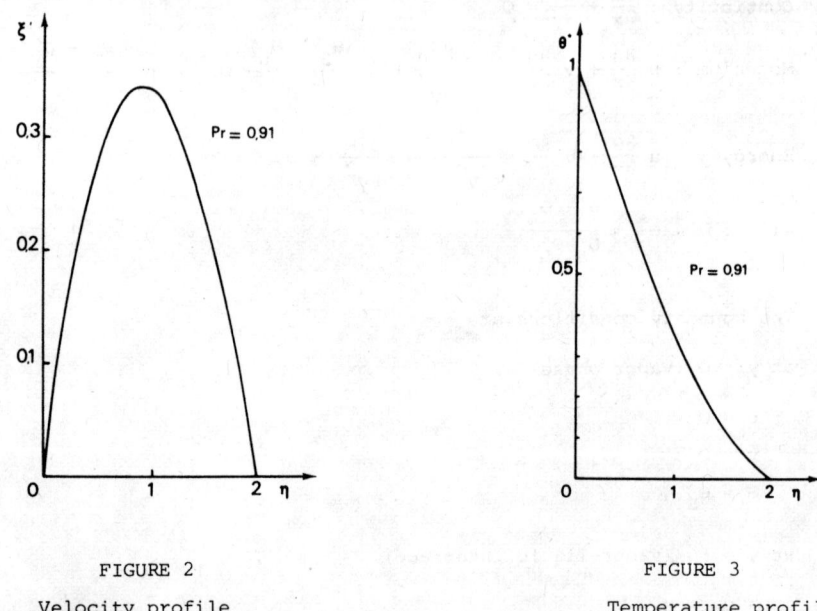

FIGURE 2  
Velocity profile

FIGURE 3  
Temperature profile

III. DESTABILIZATION OF FILM BOILING BY MEANS OF A THERMAL RESISTANCE. LARVATE BOILING.

It is possible to destabilize film boiling by setting a thermal resistance which consists of a layer of a low thermal conductivity material between the metal and the liquid. This coating may be produced during the cooling down stage, when quenching is performed into an aqueous solution of thermally stable salt (for instance NaCl) (I) (2) or by coating the sample before quenching (with zirconium oxide for instance) (1) (2).

### III.1 - Description of larvate boiling

When there is an insulating coating on the surface of a metallic sample, the cooling-down can be obtained by a special mode of vaporization, larvate boiling, which is characterized by an alternating effect of wetting and non-wetting of the coating by the liquid.

Ultra-fast cinematography together with electrical contact technic (2) (where only a small part of the coating is made conductive) allow to describe larvate boiling : it is similar film boiling when the vapor-liquid interface is the site of amplitude instabilities sufficient to cause a liquid and solid surface contact. There is a pulsed movement of the vapor film. Solid surface and liquid contacts last from 5 to 20 ms and their frequencies range from 20 to 100 Hz.

### III.2 - Mechanism of larvate boiling

When the sample and the liquid suddenly come into contact, the outside surface of the coating which is of low effusivity $E = \sqrt{\lambda c \rho}$ is instantly cooled-down. If the temperature it reaches is lower than the transition temperature, then nucleate boiling occurs. If it isn't, then film boiling sets out and the coating gets insulated from the liquid by a vapor film (figure 4). This new thermal screen decreases the heat flux transmitted to the liquid, which accor-

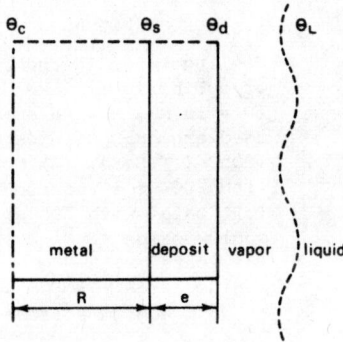

FIGURE 4

Scheme of a sample with coating of thickness e. Notation of different temperatures.

ding to Fourier's law, allows the superficial coating temperature to rise whereas the superficial sample temperature remains more or less the same.

At the initial instant of quenching, most of the temperature difference is localized in the coating which consequently stops acting as a thermal screen. But as soon as a vapor film is formed the temperature difference localizes in it. The inner and outer surface of the insulating coating are no longer at temperatures different enough and the heat flux density which passes through it is not sufficient to maintain film boiling. The liquid can come back and wet the coating surface again.

The new direct contact between liquid and coating brings the surface temperature down again and, according to Fourier's law again, the heat flux increases, which causes intense vaporization. Film boiling reappears,and is destabilized again by the same process. The whole cycle takes about 20 ms.

Therefore larvate boiling allows the wetting of a surface at the temperature of film boiling. This results in an important heat transfer.

### III.3 - Influence of superficial effusivity of the coating

The proposed mechanism of larvate boiling is only due to conduction in the insulator. The superficial effusivity of the coating however is very important and larvate boiling will be only possible when it reaches certain values.

When boiling water is used it appears from the comparison of the different cooling rates of the sample whose surface characteristics change and which is quenched from 850°C that the superficial effusivity of the coating plays an important part (figure 5). The cooling rate of the sample coated with 200 μm zirconium oxide ($E = 2230 \ W.m^{-2}.s^{1/2}.°C^{-1}$) and that of the same sample with 50 μm silver coating are very different. The thin silver coating makes cooling down considerably slower since the curve is not much different from that of a bare sample. Silver effusivity is important ($E = 31574 \ W.m^{-2}.s^{1/2}.°C^{-1}$) and when liquid and solid come into contact during larvate boiling the temperature of the coating surface does not decrease sufficiently to localize in the coating a temperature difference which allows an important heat flux to go through.

If a low effisivity material like sodium silicate ($E = 2091$) is laid over the silver film it produces larvate boiling again (curve 4, figure 5), which proves the proposed explaination.

It is possible to make such experiments in all kinds of vaporizable liquids, particularly liquid nitrogen, as shown in the curves of figure 6.

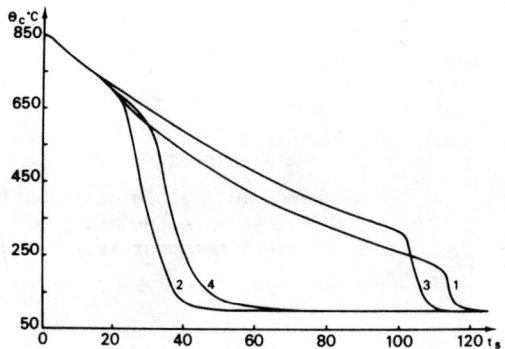

FIGURE 5

Cooling rates in the center of cylindrical nickel sample (D = 16 mm, H = 48 mm) quenched from 850°C in water at 100°C (1) Bare. (2) Coated with 200 μm zirconium oxide. (3) Coated with 200 μm zirconium oxide + 50 μm silver. (4) Coated with 200 μm zirconium oxide + 50 μm silver + 10 μm sodium silicate.

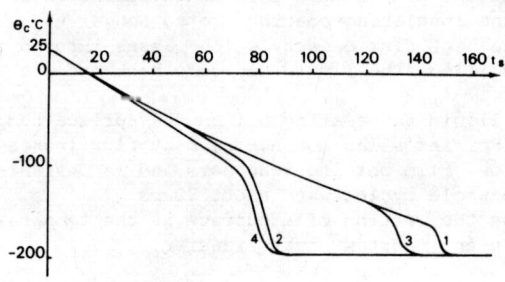

FIGURE 6

Cooling rates in the center of a cylindrical nickel sample (D = 16 mm, H = 48 mm) quenched from 25°C in liquid nitrogen (I) Bare. (2) Coated with 200 μm zirconium oxide. (3) Coated with 200 μm zirconium oxide + 50 μm silver. (4) Coated with 200 μm zirconium oxide + 50 μm silver + 10 μm polystyrene.

Two conditions are therefore necessary to obtain larvate boiling :
- on one hand a sufficient thermal resistance consisting of a coating of low conductivity material over the surface of the sample.
- on the other hand, low superficial effusivity to produce an important cooling down of the solid surface when it comes into contact with the liquid.

CONCLUSIONS

Film boiling is a mode of vaporization which is very stable and cannot be destabilized by increasing the amplitude of the capillarity waves at the liquid-vapor interface.

The laminar boundary layer theory makes it possible to calculate a velocity profile in the vapor film without any inflexion point. This allows to explain the important stability of film boiling.

A thermal resistance placed between the metal and the liquid allows to replace film boiling by larvate boiling which is characterized by an alternating effect of wetting and non-wetting of the solid surface by the liquid. Two conditions are necessary : a thermal resistance of sufficient value and small superfical effusivity.

REFERENCES

(1) J.C. Chevrier, F. Moreaux et G. Beck 1972, Int. J. Heat Mass Transfer, 15, 1631-1645.
(2) F. Moreaux, J.C. Chevrier et G. Beck, 1975, Int. J. Multiphase Flow, 2, 183-190.
(3) W.S. Bradfield 1965, Proceedings of Symposium on Two Phase Flow University of Exeter 2, A301-330.
(4) V.G. Levich 1962, Physicochemical Hydrodynamics Prentice-Hall, Inc.
(5) L.A. Bromley, 1950, Chem. Eng. Progr. 46, 221.
(6) Y.Y. Hsu et J.W. Westwater, 1960, Chem. Eng. Prog. Symp. Ser. 30, 56, 15.
(7) E.M. Greitzer et F.H. Abernathy 1972, Int. J. Heat Mass Transfer, 15, 475.
(8) J.C. Y. Koh 1962, J. Of Heat Transfer 84, 55-62.
(9) H. Schlichting 1960, Boundary Layer Theory Mc Graw Hill.

# THE COLLAPSE OF TURBULENT SHEAR LAYERS IN DENSITY STRATIFIED FLOW

VINCENT H. CHU

*McGill University, Montreal, Quebec, Canada*

## ABSTRACT

The internal hydraulics of turbulent shear layers in density stratified flow is considered theoretically through a set of integral equations of motion. The flow is found to depend on upstream and downstream conditions in a manner similar to open channel hydraulics. The general properties for supercritical flow and subcritical flows were found to depend on an "equation of state".

Guided by the theory the result of several experiments of jets and wakes in stable density stratified flow were reanalysed. Collapse of large scale turbulent motion was found to occur at a minimum gradient Richardson number of about 0.21.

## NOMENCLATURE

| | |
|---|---|
| B | buoyancy force per unit volume |
| c | wave speed of small disturbance |
| E | entrainment coefficient |
| F | buoyancy flux |
| G | buoyancy force per unit area |
| g | acceleration of gravity |
| h | mean thickness of the turbulent shear layer |
| M | flow force excess |
| Q | volume flux excess |
| q | transfer of buoyancy at the boundary, $z = 0$ |
| Ri | gradient Richardson number |
| t | time |
| U | mean velocity in x-direction |
| $W_e$ | entrainment velocity in z-direction |

x, z     Cartesian coordinates

$\alpha_U$, $\alpha_{UU}$, $\alpha_B$, $\alpha_{BU}$, $\alpha_{Bz}$     shape parameters

$\delta$     half-width of the shear layer defined at $U = \tfrac{1}{2} U_m$

$\theta$     angle of inclination

$\rho$     fluid density

$\tau$     shear stress at the boundary, $z = 0$

Subscripts

a     ambient flow

m     maximum or minimum

o     condition at $x = 0$

INTRODUCTION

In a stable density stratified flow, turbulent motion is often observed to be confined in isolated and nearly horizontal layers. The formation of these layers could be due to, for example, injection of buoyant fluid into another fluid of different density, towing of an obstacle in a density stratified fluid. or intensification of local shear by gravitational convection. The initial development of these turbulent shear layers is usually rather unaffected by density stratification. Buoyancy effects become increasingly more important as the turbulent layers spread out laterally. Eventually, as certain critical condition is reached, the turbulent motion will collapse and relaminarization will begin.

A typical example of such flow is the horizontal buoyant jets considered by the classical experimental studies of Townsend [1] and Ellison and Turner [2]. Turbulent entrainment in buoyant jets was observed to cease at a certain critical Richardson number. Collapse of turbulent motion of similar nature was also observed in turbulent wakes by Schooly and Stewart [3], Prych, Harty and Kennedy [4] and Pao [5] and in a mixing layer by Thorpe [6]. More detailed measurements of velocity and density distributions, including some correlations of turbulent fluctuations were made in atmospheric boundary layer by Businger et al. [7], and in buoyant jets with and without ambient stream by Hopfinger [8] and Vanvari and Chu [9, 10].

In this paper an attempt will be made to identify some of the common hydraulic properties that exist among stable straitified turbulent shear flows. A set of integral equations will be derived. The condition for critical flow is established by considering the propagation of small amplitude long waves. The general hydraulic properties for subcritical and supercritical flows are determined from an "equation of state". The results of several experiments of jets and wakes in density stratified flow (see Fig. 2) will be reanalysed in order to establish a critical condition for the collapse of turbulent shear layers.

INTERNAL HYDRAULICS

The internal hydraulics of turbulent jets was considered by Ellison and Turner [2], Wilkindon and Wood [11], Koh [12] and Baddour and Chu [13]. The development of a turbulent jet was found to depend on upstream condition as well as well as downstream condition in a manner similar to open channel hydraulics.

A set of integral equations, generally applicable to shear layers such as jets and wakes, will be derived here. When turbulent entrainment into the shear layer is ngeligible these equations reduce to the hydraulic relationships in open channel flow.

The general problem is defined in Figure 1. A stably stratified turbulent shear layer is shown flowing down an inclined plane. The angle with respect to the horizontal is $\theta$. The ambient density and velocity are $\rho_a$ and $U_a$ respectively. Cartesian coordinates are used with x-axis and z-axis parallel to and normal to the incline plane respectively. The mean velocity in the x directions are U. The buoyancy force per unit volume of fluid is $B = g(\rho - \rho_a)$.

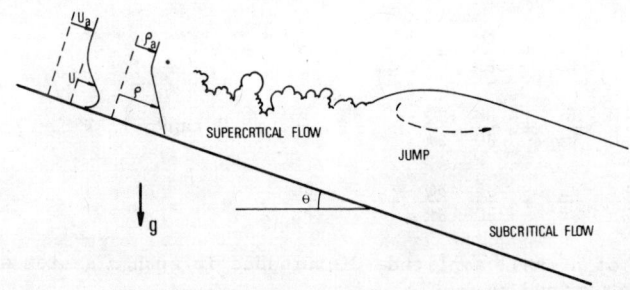

Figure 1. Definition sketch.

It is assumed that the fluid is incompressible, the layer is thin compared to its longitudinal length scale and the density variation is small. Making use of the boundary layer and Boussinesq approximations, the equations of motion are integrated across the layer; this gives

$$\frac{\partial h}{\partial t} + U_a \frac{\partial h}{\partial x} + \frac{\partial Q}{\partial x} = W_e \qquad (1)$$

$$\frac{\partial Q}{\partial t} + \frac{\partial M}{\partial x} = G \tan\theta + \tau \qquad (2)$$

$$\frac{\partial G}{\partial t} + \frac{\partial F}{\partial x} = q \qquad (3)$$

where q and $\tau$ are the transfer of buoyancy and momentum from the boundary at $z = 0$, $W_e$ is an entrainment velocity defined as the advancement of the turbulent front relative to the ambient irrotational flow and h is the mean depth of the turbulent shear layer. The remaining variables, Q, M, G and F, are dependent on the mean distribution velocity:

$$Q = \int_0^\infty U dz = \text{excess volume flux} \qquad (4)$$

$$M = \int_0^\infty [\rho_a U(U + U_a) + B \cos\theta \, z] \, dz = \text{flow force} \qquad (5)$$

$$G = \int_0^\infty B \cos\theta \, dz = \text{buoyancy force per unit area} \qquad (6)$$

$$F = \int_0^\infty B \cos\theta \, (U + U_a) \, dz = \text{buoyancy flux} \qquad (7)$$

"flow force" is defined here as excess momentum flux plus excess hydrostatic pressure.

If the development of the layer is sufficiently gradual it is possible to assume that the local structure of the turbulent shear layer to maintain, or at least approximately, at an "equilibrium" condition such that lateral distributions of mean velocity and mean buoyancy are similar. Once the mean velocity and mean buoyancy distributions are specified (see Eqs. 13 and 14) the local mean state would be uniquely determined by a mean velocity scale, a mean buoyancy scale and a characteristic thickness of the shear layer. These three basic scales may be considered as "state variables" which define the "mean state". Alternatively, any other group of three variables that depend on the three basic scales may be used as the state variables. For example, (h, Q, G) may be chosen as the state variables. In that case Equations 1, 2 and 3 can be rewritten as follows:

$$\frac{\partial h}{\partial t} + U_a \frac{\partial h}{\partial x} + \frac{\partial Q}{\partial x} = W_e \tag{8}$$

$$\frac{\partial Q}{\partial t} + \frac{\partial M}{\partial h}\frac{\partial h}{\partial x} + \frac{\partial M}{\partial Q}\frac{\partial Q}{\partial x} + \frac{\partial M}{\partial G}\frac{\partial G}{\partial x} = G \tan\theta + \tau \tag{9}$$

$$\frac{\partial G}{\partial t} + \frac{\partial F}{\partial h}\frac{\partial h}{\partial x} + \frac{\partial F}{\partial Q}\frac{\partial Q}{\partial x} + \frac{\partial F}{\partial G}\frac{\partial G}{\partial x} = q \tag{10}$$

The propagation of a small amplitude disturbance in such a system will have a wave speed c, determined by

$$\begin{vmatrix} (-c + U_a) & 1 & 0 \\ \frac{\partial M}{\partial h} & (-c + \frac{\partial M}{\partial Q}) & \frac{\partial M}{\partial G} \\ \frac{\partial F}{\partial h} & \frac{\partial F}{\partial Q} & (-c + \frac{\partial F}{\partial G}) \end{vmatrix} = 0 \tag{11}$$

It is now possible to define a critical condition in which a small disturbance will not propagate upstream. This critical condition is obtained by setting c = 0 in Equation 11; this gives

$$U_a J\left(\frac{M,F}{Q,G}\right) + J\left(\frac{M,F}{G,h}\right) = 0$$

which is equivalent to

$$\left(\frac{\partial Q}{\partial h}\right)_{F,M} = -U_a \tag{12}$$

LOCAL EQUILIBRIUM AND EQUATION OF STATE

A turbulent shear layer is said to be in "equilibrium" if lateral distributions of velocity and buoyancy would remain similar and depend on local scales only. Such an equilibrium condition is observed to achieve approximately in experiments of horizontal buoyant jets by Vanvari and Chu [9, 10] (see Fig. 2a). Their measurement indicates that mean velocity can be represented by a Gaussian profile

$$\frac{U}{U_m} = \exp\left\{-\left(\frac{\gamma z}{\delta}\right)^2\right\}, \quad \gamma = 0.833 \tag{13}$$

and mean buoyancy approximately by a linear profile

$$\frac{B}{B_m} = \begin{cases} 1 - 0.5 \frac{z}{\delta} &, z < 2 \\ 0 &, z > 2 \end{cases} \quad (14)$$

where $U_m$ and $B_m$ are the maximum mean velocity and maximum mean buoyancy and $\delta$ is the half-width of the shear layer defined such that $U = \frac{1}{2} U_m$ at $z = \delta$.

Similarity in velocity and buoyancy distributions are also observed in the outer layer of a turbulent wall jet with ambient flow by Hopfinger [8]. The buoyancy profile in his experiment is also approximately linear but is slightly wider than the velocity profile. Fitting his experimental data gives

$$\frac{B}{B_m} = \begin{cases} 1 - 0.433 \frac{z}{\delta} &, z < 2.31 \\ 0 &, z > 2.31 \end{cases} \quad (15)$$

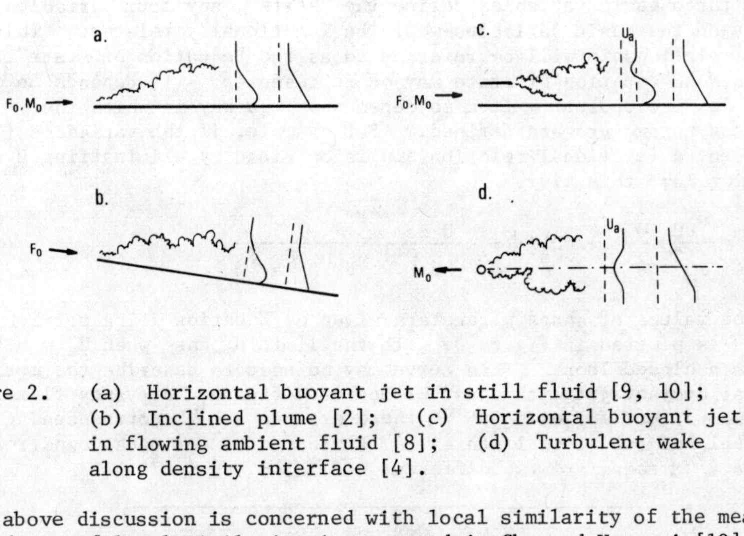

Figure 2. (a) Horizontal buoyant jet in still fluid [9, 10];
(b) Inclined plume [2]; (c) Horizontal buoyant jet in flowing ambient fluid [8]; (d) Turbulent wake along density interface [4].

The above discussion is concerned with local similarity of the mean flow. Other evidence of local similarity is reported in Chu and Vanvari [10].

If similarity in velocity and buoyancy distributions is assumed, other variables such as volume flux, buoyancy flux, flow force, and buoyancy force can be expressed in terms of the three basic "state variables" ($U_m$, $B_m$, $\delta$) as follows:

$$Q = \alpha_U U_m \delta$$

$$F = \alpha_{BU} U_m B_m \delta + \alpha_B U_a B_m \delta$$

$$M = \alpha_{UU} U_m^2 \delta + \alpha_U U_m U_a \delta + \alpha_{Bz} B_m \delta^2 \quad (16)$$

$$G = \alpha_B B_m \delta$$

where the shape parameters which depend on the lateral distributions of mean velocity and buoyancy are defined by

$$\alpha_U = \int_o^\alpha \frac{U}{U_m} d\frac{z}{\delta} , \quad \alpha_{UU} = \int_o^\infty (\frac{U}{U_m})^2 d\frac{z}{\delta}$$

$$\alpha_B = \int_o^\infty \frac{B}{B_m} d\frac{z}{\delta} , \quad \alpha_{BU} = \int_o^\infty \frac{B}{B_m}\frac{U}{U_m} d\frac{z}{\delta} \quad (17)$$

and

$$\alpha_{Bz} = \int_o^\infty \frac{B}{B_m}\frac{z}{\delta} d\frac{z}{\delta}$$

For velocity and buoyancy distributions given by Equations 13 and 14 the numerical values of these shape parameters are:

$$\alpha_U = 1.06, \quad \alpha_{UU} = 0.753, \quad \alpha_B = 1.00, \quad \alpha_{BU} = 0.707, \quad \alpha_{Bz} = 0.667 \quad (18)$$

Since three basic variables define the "state", any four variables which depend only on the basic variables would be functionally related. This functional relationship will be referred to as the "equation of state". There are many ways an equation of state may be represented. It depends on which four variables are chosen; it also depends on the way in which the dimensionless parameters are defined. For example, if the variables (Q, M, F, δ) are chosen, a functional relationship is obtained by eliminatiing $U_m$ and $B_m$ from Equation 16; this gives

$$M = \frac{\alpha_{UU}}{\alpha_U^2} \frac{Q^2}{\delta} + U_a Q + \frac{\alpha_U \alpha_{Bz} \delta^2 F}{\alpha_{BU} Q + \alpha_B \alpha_U U_a \delta} \quad (19)$$

Based on the values of shape parameters given by Equation 18, a portion of Equation 19 is plotted in Figure 3. In the limiting case when $U_a = 0$ the curve forms a closed loop. This curve may be used to describe the motion of a horizontal buoyant jet with constant momentum flux and buoyancy flux (Fig. 2a). It shows that for each $Q F^{1/3}/M$ there are two $\delta F^{2/3}/M$ corresponding to supercritical and subcritical states. Furthermore, for each downstream condition $\delta F^{2/3}/M$ there are two different $Q F^{1/3}/M$.

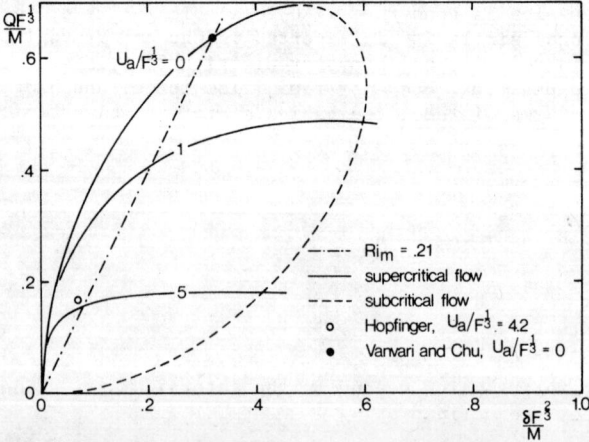

Figure 3.   Equation of state: $F(Q, \delta, M, F) = 0$.

If the upstream and downstream conditions are specified the total entrainment into the jet can be uniquely determined by taking the difference of $QF_o^{1/3}/M$ corresponding to the supercritical and subcritical flows [13].

The condition for critical flow is determined from Equation 12. When the ambient flow velocity is large the jet may remain supercritical without the influence of downstream conditions. The two experiments of Hopfinger [8], with $U_a/F_o^{1/3}$ = 4.2 and 3.5 respectively, are not affected by downstream condition.

A slightly simplified situation occurs when a jet or a wake is formed at the interface of two fluid layers of different density. In that case the suitable variables are $(Q, M, B_m, \delta)$ and the equation of state is

$$\tfrac{1}{2}M = \alpha_{UU}\frac{Q^2}{\delta} + U_aQ + \alpha_{Bz} B_m \delta^2 \qquad (20)$$

This relationship is plotted in dimensionless form in Figure 4 using the values of shape parameters given in Equation 18. The curve has two branches. The supercritical branch describes the turbulent wakes in a density interface considered by Prych, Harty and Kennedy [4] (see Fig. 2d). The subcritical branch represents jets in an opposed ambient stream.

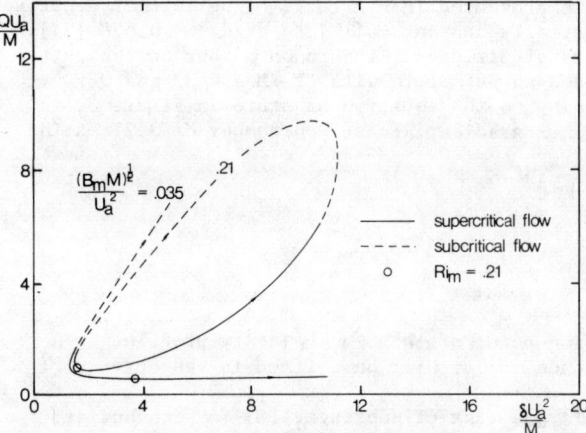

Figure 4. Equation of state: $f(Q, \delta, M, B_m)=0$.

In general the equation of state may have different functional relations at different regions of the shear layer due to changes in velocity and buoyancy distributions. This change usually occurs in an internal hydraulic jump or occurs when the turbulence begins to collapse.

When a jet is submerged by the internal jump (see Fig. 1) the velocity and buoyancy distributions downstream of the jump becomes more uniform. The subcritical branch of the curve would have to be determined by a set of shape parameters different from that given by Equation 18. The hydraulic property of jets influenced by an internal hydraulic jump is given by Wilkinson and Wood [11] and Baddour and Chu [13].

The problem concerning collapse of turbulence will be discussed in the next section.

COLLAPSE OF TURBULENCE

Collapse of turbulence is defined here as the collapse of large scale turbulent structure and the failure of mechanism by which mean flow energy is converted into turbulence. The local structure of a turbulent shear layer before collapse is observed in many ways similar to the neutrally buoyant situation [9, 10]; turbulent entrainment of ambient irrotational fluid is due to the large scale evolution of laminar and turbulent interface. Once the large scale structure has collapsed, the smaller scale turbulent motion will gradually diminish through dissipation of energy by viscous and buoyancy forces; entrainment in this region is quite negligible and is contributed mainly by sporadic breaking of internal waves.

A number of different experiments concerning the collapse of turbulence will be reviewed in this section. The flow situations under consideration are sketched in Figure 2. The purpose here is to establish a critical Richardson number in which the large scale turbulent motion collapses. The conclusion will largely be based on the experimental results of horizontal buoyant jets in a stationary ambient fluid [9, 10]. The result is shown to be consistent with other experiments of turbulent shear flows such as inclined plumes [2], horizontal wall jets in an ambient stream [8] and turbulent wakes along a density surface [4].

The experiment of Vanvari and Chu [9] was performed in a two dimensional flume by letting fresh water flow over the surface of an otherwise stationary body of saline water. The flow situation is in essence the same as the horizontal wall jet sketched in Figure 2a. Since the transfer of buoyancy and shear stress at the free surface is zero, the flow force and buoyancy flux is conserved and is equal to $M_o$ and $F_o$ at the exit. The experimental result is normalized by $M_o$ and $F_o$ and is presented in Figure 5. The initial growth of the jet is the same as in neutrally buoyant wall jet, $d\delta/dx = 0.678$ [14]. In the far field, the experiments of different Richardson number at the exit $(0.02 \sim 0.007)$ is seen to approach an asymptote with $\delta F_o^{2/3}/M \simeq 0.32$ and $Q F_o^{1/3}/M \simeq 0.64$. This asymptote, according to the equation of state in Figure 3, is corresponding to a critical minimum gradient Richardson number of 0.21. The minimum gradient Richardson is

$$Ri_m = \frac{(\frac{\partial B}{\partial z}) \cos\theta}{(\frac{\partial U}{\partial z})^2_{max}}$$

which is defined at the inflection point of the mean velocity profile. In case of a wall jet, this Richardson number will be defined in the outer part of the shear layer.

Also included in Figure 3 is one test of Hopfinger [8] for buoyant jet with ambient stream (see Fig. 2c). In this test $U_a/F_o^{1/3} = 4.2$. The jet is observed to attend a maximum thickness of $\delta F_o^{2/3}/M_o = 0.068$. This is seen in Figure 3 to agree with the critical value of minimum gradient Richardson number of 0.21. Note that the wall jet has shape parameters slightly different from the values used to calculate the curves presented in Figure 3. The comparison in Figure 3 is to demonstrate the use of the equation of state.

Entrainment velocity is also obtained by Vanvari and Chu [9] based on direct integration of mean velocity profiles. Their result is presented in Figure 6 where the entrainment coefficient $E \ (= \frac{W_e}{U_m})$ is plotted against the minimum gradient Richardson number $Ri_m$. Also included in the figure is the experimental result for inclined plume (Fig. 2b) by Ellison and Turner [2]. Despite the flow situations are somewhat different the agreement between the two sets of experiments is rather remarkable. In the limiting case when $Ri_m$ approaches zero the entrainment coefficient is found to approach a value of $E \simeq 0.36$ which is in agreement with the entrainment coefficient for a neutrally buoyant wall jet (see [10]). The critical Richardson number for the collapse of turbulent entrainment is approximately 0.21. Note that the present definitions of the entrainment coefficient and gradient Richardson number are different from the one originally used by Ellison and Turner. Their result is re-calculated here based on the data presented in Figures 4 and 6 in their paper. Ellison and Turner have also obtained some preliminary results for horizontal surface jets. These results are thought to be not reliable and are not included here (see discussion by Chu and Vanvari [10]).

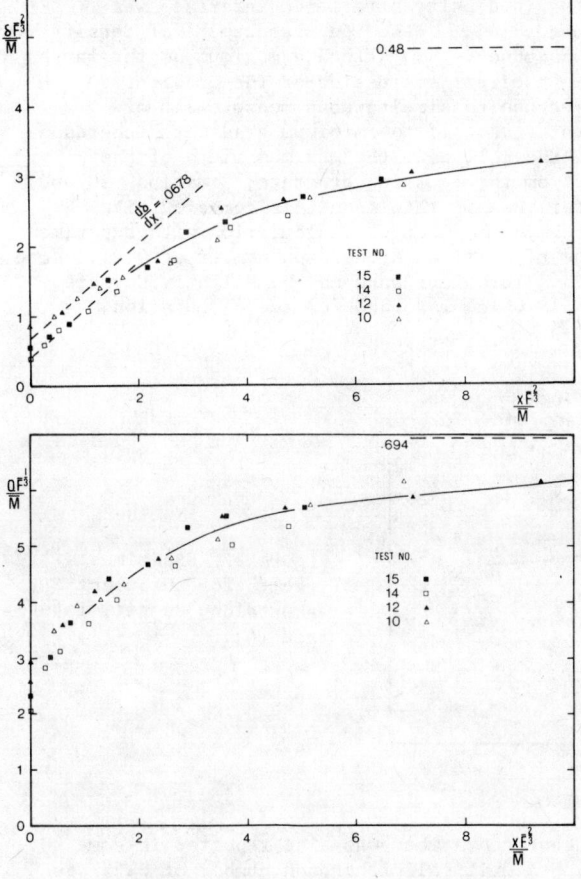

Figure 5. Experimental results for horizontal buoyant jets in still fluid [9].

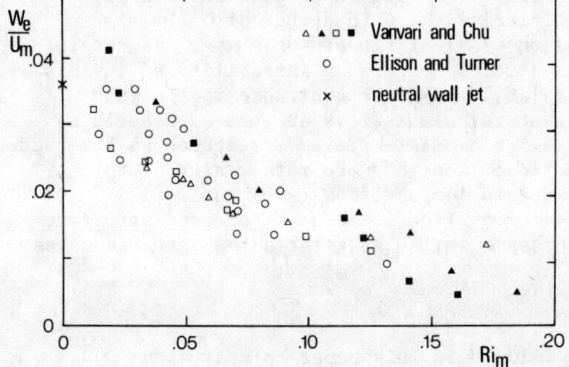

Figure 6. Entrainment coefficient

Turbulent entrainment coefficient is also obtained in a horizontal wall jet with ambient stream by Hopfinger [8]. The initial development of the jet in Hopfinger's experiment is complicated by the wake of a splitter plate (see Fig. 2c), the entrainment coefficient is generally rather higher; but despite the complexity of the flow situation, collapse of turbulent entrainment was reported to occur also at a minimum gradient Richardson number of 0.21.

Experiments of turbulent wakes in density stratified interface was performed by Prych, Harty and Kennedy [4]. Based on measurement of density profiles, the width of the wake was observed to attend a maximum as the large scale turbulent eddies collapse. A direct evaluation of the gradient Richardson number is impossible because no simultaneous measurement of velocity was made in the experiment. But if the minimum gradient Richardson number is taken to be 0.21 before the collapse, the maximum width of the turbulent wake can be determined from the equation of state, Equation 20 and Figure 4. The maximum width determined in this manner is compared with experiment in Figure 7. The agreement between the calculation with experiment gives further support to the value of critical Richardson number of 0.21. Note again that the shape parameters in a turbulent wake may be slightly deviated from the values given by Equation 18 which was used for the calculation.

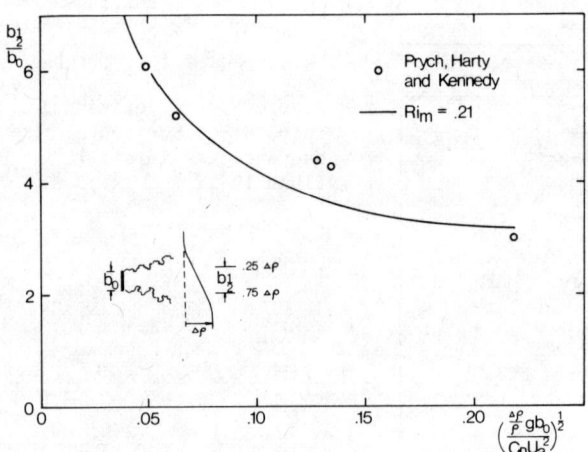

Figure 7. Experimental results for turbulent wakes along density interface [4].

Similar value of critical Richardson number was also reported in some rather different flow situations. A critical Richardson number of 0.25 was obtained by Businger et al. [7] based on measurement in atmospheric boundary layer. So [15] considered the analogy of curved shear flow and density stratified shear flow and gave a critical Richardson number of 0.21.

Almost identical value of minimum gradient Richardson number was obtained in the experimental and theoretical studies concerning instability of invisid laminar stratified flow (see Turner [16]). More recent observation in turbulent shear layers discovered that turbulent flows are not as chaotic as has been previously assumed. Large scale quasi-ordered structure was observed in fully developed turbulent shear flow; and they are rather similar to initial rollup of vortex sheet in invisid laminar flow (see Laufer [17]). It is quite possible that the buoyancy force affects the quasi-ordered structure in turbulent flow in essentially the same way as it affects the unstable waves in laminar shear flow.

CONCLUSION

A simple integral method is developed in this paper to deal with turbulent shear layers in density stratified flow. The technique is limited by the requirement of local similarity in velocity and density profiles. Since the shape parameters are not strongly depends on velocity and density distributions, the method is useful for engineering calculation and planning of an experimental study. In fact it has demonstrated in this paper to be useful in interpreting some of the previous experimental data on turbulent density stratified shear flow.

The concept of local equilibrium introduced in this paper is somewhat reminiscent to study of classical thermodynamics. The integral technique forms a theoretical basis for experimental and more sophisticated theoretical studies which are aiming to obtain information on the local processes.

REFERENCES

1. Townsend, A.A. 1957. Turbulent Flow in a Stably Stratified Atmosphere. J. of Fluid Mech. 3: 361-372.

2. Ellison, T.H. and Turner, J.S. 1959. Turbulent Entrainment in Stratified Flow. J. of Fluid Mech. 6: 423-448.

3. Schooley, A.H. and Stewart, R.W. 1963. Experiments With a Self-propelled Body Submerged in a Fluid With a Vertical Density Gradient. J. Fluid Mech. 15: 83-96.

4. Prych, E.A., Harty.F.R. and Kennedy, J.F. 1964. Turbulent Wakes in Density Stratified Fluids of Finite Extent. MIT Hydrodynamics Laboratory Report No. 65.

5. Pao, Y-H. 1970. Turbulent Measurements in Stably Stratified Fluids. Boeing Sci. Res. Lab. Document D1-82-0959.

6. Thorpe, S.A. 1973. Experiments on Instability and Turbulence in a Stratified Shear Flow. J. Fluid Mech. 61: 731-751.

7. Businger, J.A., Wyngaad, J.C., Izumi, Y. and Bradley, E. 1971. Flux-profile Relationships in Atmospheric Surface Layer. J. Atmos. Sci. 28: 181-188.

8. Hopfinger, E.J. 1972. Development of a Stratified Turbulent Shear Flow. International Symposium on Stratified Flows, Novosibirsk, 533-565.

9. Vanvari, M.R. and Chu, V.H. 1974. Two-dimensional Turbulent Surface Jets of Low Richardson Number. Fluid Mechanics Lab. Tech. Rep. No. 74-2(FML), Dep. of Civil Eng. and Applied Mech., McGill University, Montreal, Canada.

10. Chu, V.H. and Vanvari, M.R. 1976. Experimental Study of Turbulent Stratified Shearing Flow. J. of Hydraulic Div., ASCE. 102: 691-706.

11. Wilkinson, D.L. and Wood, I.R. 1971. A Rapidly Varied Flow Phenomenon in a Two-layer Flow. J. Fluid Mech. 47: 241-256.

12. Koh, R.C.Y. 1971. Two-dimensional Surface Warm Jets. J. of Hydraulic Div., ASCE. 97: 819-836.

13. Baddour, R.E. and Chu, V.H. 1975. Buoyant Surface Discharge on a Step and on a Sloping Bottom. Fluid Mech. Lab. Rept. No. 75-2(FML), Dept. of Civil Eng. and Applied Mech., McGill University, Montreal, Canada.

14. Schwartz, W.H. and Cosart, W.P. 1961. The Two-dimensional Turbulent Wall Jet. J. Fluid Mech. 10: 481-495.

15. So, R.M.C. 1975. A Turbulent Velocity Scale for Curved Shear Flow. J. Fluid Mech. 60: 43-62.

16. Turner, J.S. 1973. Buoyancy Effects in Fluids. Cambridge University Press, London, England.

17. Laufer, J. 1975. New Trends in Experimental Turbulence Research. Ann. Rev. Fluid Mech. 7: 307-326.

# FREE CONVECTION WITH HEAT ADDITION AND COMBUSTION PHENOMENA

# INVESTIGATION OF HYDRODYNAMICS AND HEAT TRANSFER AT TURBULENT NATURAL CONVECTION IN FLAT SLOTS

M. D. DIEV, S. D. KORNEEV, H. K. KURBANOV, A. I. LEONTIEV, and B. M. MIRONOV
Moscow Higher Technical School, Moscow, U.S.S.R.

ABSTRACT

Theoretical and experimental study of hydrodynamics and heat trnsfer at turbulent natural convection in flat slots is presented.

The process in thermosyphons of two types was studied. It is found that the heat flux and inclination of the thermosyphon have great influence on the transition behaviour. Flow pattern in the thermosyphon was studied at different heat fluxes and inclinations. The swing flow of the warm liquid in a single-phase thermosyphon was found to be dependent upon the heat flux and inclination. Distributions of the wall temperature in two-phase thermosyphon were registered.

The process of boiling in a slot was modelled analytically. Experiments carried out to study the influence of the gravity force showed the considerable drop of the boiling heat transfer rate while the effective gravity force is increased

Decrease of the slot thickness is followed by the increase of the boiling heat transfer rate.

NOMENCLATURE

b - slot thickness, m;
D - diameter, m;
F - area, $m^2$;
g - gravitational acceleration, $m/sec^2$;
K - constant in eq.(1);
L - length, m;

$\alpha$ - coefficient of heat transfer, $W/m^2 K$;
$\beta$ - angle of inlination, °;
$\delta$ - see eq.(1);
$\nu$ - viscosity, $m^2/sec$;
$\rho$ - density, $kg/m^3$;
$\sigma$ - surface tension, N/m;

$q$ - heat flux, $W/m^2$;
$Q$ - heat flow, $W$;
$r$ - latent heat of evaporation;
$U$ - velocity, $m/sec$;
$V$ - volume, $m^3$;
$Nu$ - Nusselt number $= \frac{\alpha b}{\lambda}$;
$We$ - Weber number $= \frac{6}{q(\rho_l - \rho_v) b^2}$.

$\tau$ - Time, sec;
$\tilde{\tau}$ - dimensionless time $= \frac{(\tau - \tau_0) q}{r \rho_v b}$;
$\varphi$ - see fig.3;
$\lambda$ - thermal conductivity, $W/mK$;
$Ga$ - Galilei number $= \frac{g b^3}{\nu^2}$;
$Re$ - Reynolds number $= D_0 \frac{q}{r \rho_v b}$;

Subscripts

$v$ - vapor;
$l$ - liquid;

$o$ - initial conditions;
$e$ - effective.

The first stage of our work was a closed thermosyphon study. Investigation of the transient behaviour was the scope of this part.

Inclination and heat flux were varied during the experimental series. The inclination was found to have a cosiderable influence on the transition. The flow pattern was more stable in inclined thermosyphon than in a vertical one. The fact, particularly, can be explained by the presence of the "swing" flow (from one side wall to another) in a vertical thermosyphon. If the heat flux is growing the frequency of such a swinging increases and some type of an internal boundary appears.

On the distance from the heater, the swing flow seems to be more quiet. Internal boundaries resolve little by little. While approaching the cooler, however, such a boundary can be seen again. Bailey and Lock /1/ noted this zone (coupling region) has to be characterized by this internal boundary.

The cooler influences on the upper part of the transport zone, which influence appears in mixing of the cold going down liquid and the warm going up liquid. The vortices can appear in this part but not as intensive as in the heater vicinity. The transition from laminar to turbulent flow can be seen in the cold liquid at higher heat fluxes.

So, we can note two parts in a slot thermosyphon where transitio occures. The laminar flow takes place between these two zones - in the adiabatic part. Tomake the flow in this part turbulent extremely high heat flux is needed.

Such flow pattern is characteristic to the vertical thermosyphon.

The flow pattern seems to be more regular if the thermosyphon is inclined.

Inclination causes the boundary layer growth on the heating surface (see fig.1a). Going up along the side wall the "warm" boundary layer meets the "cold" one. The effect of this interaction is the flow destruction. In such a case the vortices can appear: the "warm" vortex is followed by the "cold" one.

The flow near the lower side wall seems to be more quiet.

The above mentioned internal boundary can be seen in the inclined slot thermosyphon at higher heat fluxes. It appears, however, stronger than in a vertical thermosyphon. Going from the core warm liquid flows along this boundary as shown on fig.1b.

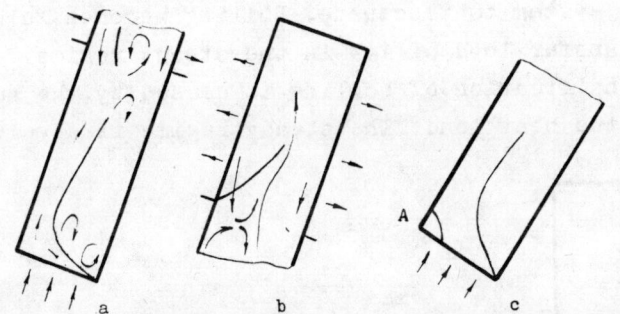

Fig. 1. Flow pattern in the slot thermosyphon.

The lower value of the heat flux is needed for the transient flow appearence if the imclination is greater than $5^o - 10^o$. Stagnation regions can be seen now at even moderate heat fluxes; the flow pattern changes essentially. The stagnation region A (see fig.1c) is more dangerous because of the considerable contact electrical resistance causing the additional heating is concentrated here. So vaporization begins if the bulk liquid temperature is $60^oC$. In this case, however, departed vapor bubbles are condenced in a surrounding liquid immediately. The fluctuations in manometer readings are noted now.

Two stable vortices are evident near heater and cooler if the thermosyphon inclination is more than $45^o$. The adiabatic zone flow is now of unstable type. Liquid is periodically thrown into the space between the two vortices. The boundary at the joint of the adiabatic and cooler zones can not be seen now.

The slot thermosyphon with 50% volume filling was used in the second series of the experiments.

At a 10W heat load the boiling of water begins in 30 minutes. It needs to be noted that the vaporization was poorly expressed, being departed the vapor bubbles rised up not more than 5mm. While the flow near the heater was fully turbulent itwas laminar near the free surface.

The process of boiling developed rather fast and in subsequent 5-10 minutes vapor bubbles rised over the heating surface on 20-25mm. The stream of the vapor bubbles starts to swing from one side wall to another causing the fluctuations of the side wall thermocouples readings. One can see now the liquid level and prssure in a system to fluctuate. Boiling becomes fully developed at a heat transfer load of 14W in the steady regime.

The intensification of boiling is caused by the subsequent increase of the heat load. The steady regime is developed now in 25 minutes.

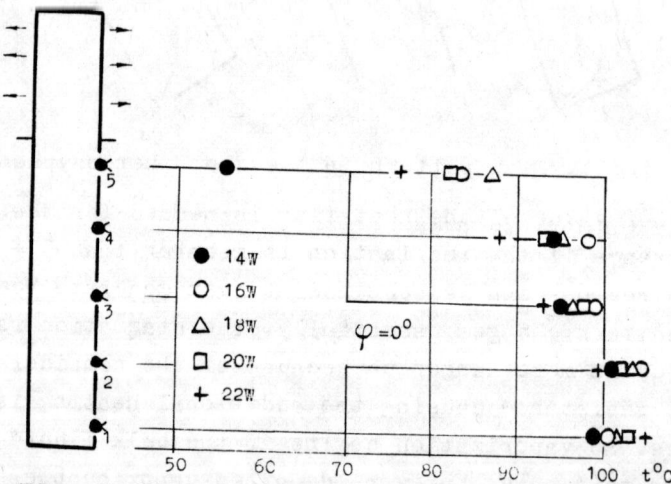

Fig. 2. Distribution of the side wall temperature.

Experiments were run in a heat trnsfer load interval from 14W to 22W and inclination from +30° to -30°. Fig.2 presents the experimental data obtained for the side wall temperature distribution in a vertical slot thermosyphon. The readings of the three lower thercouples are independent of the heat input. The condeser cold wall has great influence on the temperature of the upper part of the adiabatic zone wall; it is expressed stronger at 22W (thermocouples 4 and5). The liquid level reaches the cooler wall at this heat input and the temperature in the upper liquid layers decreases rapidly.

Heat transfer in a slot and in a pool are practically equal if the side walls are adiabatic as measurements of the heat transfer coefficient showed. It may be expected, however, that the side wall heating would increase the heat transfer rate because of the heat transfer area growth. To study this effect the special experimental unit was used.

Distilled water was used as a process liquid. Experiments were run at atmospheric pressure. To study the vapor bubbles dynamics the filming was used. While investigating heat transfer one of the glass plates was repleced by the measuring element.

The slot boiling heat transfer data were compared with correlations obtained in /2/ by A.I.Leontiev et al.

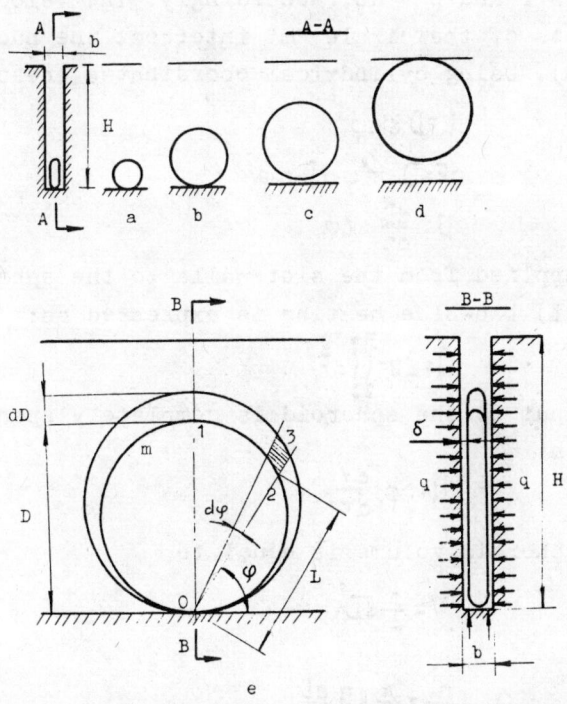

Fig. 3. Physical model of hydrodynamics and heat transfer of the process of boiling in a slot canal.

The space between bottom and slot walls was filled by the superheated liquid. There are flattened out and separated from the boundary surfaces by thin liquid layer vapor bubbles in the liquid. The heat, spent on the vapor generation, is supplied to the flattened out vapor bubbles surface from boundary walls by molecular conductivity through this liquid layer. There are two stages of the bubbles growth in a slot.

During the first stage the growing vapor bubble takes a form of an increasing spheroid and is kept in a slot by surface tension (see fig.3a, b). The thickness of a liquid layer separating spheroid from the slot wall at a little slot thickness is dependent upon the surface tension at the interface, viscosity and inertia forces. We shall assume (as in /3/):

$$\delta = K \cdot \sqrt{\sqrt[3]{\frac{\rho \varepsilon B}{\sigma U_*}}}. \tag{1}$$

The value of K has to be taken from the experimental data.

Let us compare the vapor spheroids at moments $\tau$ and $\tau + d\tau$ when its size is D and D + dD, accordingly. The velocity vectors of all the points of the circle "m" intersect the nucleation cite "O" (see fig.3e). Using cylindrical coordinates one can obtain:

$$L = D \cdot \sin\varphi; \tag{2}$$

$$dF = D \cdot \sin^2\varphi \cdot dD \cdot d\varphi; \tag{3}$$

$$U = \frac{dD}{d\tau} \cdot \sin\varphi. \tag{4}$$

The heat supplied from the slot walls to the spheroid at uniform (q=const) two-side heating is expressed as:

$$Q = 2q \int_0^{D}\int_0^{\pi} dF. \tag{5}$$

The heat input to the spheroid is completely spent on the vapor generation:

$$Q = r\rho_v \frac{dV}{d\tau}. \tag{6}$$

The vapor spheroid volume is equal to

$$V = \frac{\pi}{4} \delta D^2, \tag{7}$$

so

$$\frac{dV}{d\tau} = \frac{\pi}{2} \delta D \frac{dD}{d\tau}. \tag{8}$$

From the equivalence of (5) and (6) and taking into account (8) we can write:

$$\frac{dD}{d\tau} = \frac{qD}{r\rho_v \delta}. \tag{9}$$

By integrating (9) with $\tau = \tau_0$ and $D = D_0$:

$$D = D_0 \cdot \exp\left[\frac{q}{r\rho_v \delta}(\tau - \tau_0)\right]. \tag{10}$$

As

$$U_v = \frac{dD}{d\tau},$$

so

$$\delta = K \cdot \nu \cdot \left\{ \frac{\rho_\ell \rho_v \beta^2 r}{6 q D_0 \exp\left[\frac{q}{r\rho_v \delta}(\tau - \tau_0)\right]} \right\}^{1/3}. \tag{11}$$

The heat transfer coefficient is expressed by:

$$\alpha = \frac{\lambda \left\{ 6 q D_0 \exp\left[\frac{q}{r\rho_v \delta}(\tau - \tau_0)\right]\right\}^{1/3}}{K \cdot \nu (\rho_\ell \rho_v \beta^2 r)^{1/3}}. \tag{12}$$

This expression may be written in nondimensional form:

$$Nu = \frac{1}{K}\left(Ga \cdot Re \cdot We \cdot \exp \bar{\tau}\right)^{1/3} \tag{13}$$

During the second stage the spheroid departs the bottom and grows while moving up in a slot (see fig.3c, d).

The velocity of the vapor spheroid growth is added to the velocity of it's movement in a slot $U_v$; the according expressions can be obtained on the second stage of the vapor spheroid growth.

Fig.4 shows the comparison of the data obtained and the analytical correlation (12). It is evident that decrease of the slot thickness causes significant intensification of the heat transfer rate.

The vertical slot boiling heat transfer data generalization is shown on fig.5.

The canal plane inclination experiments were carried out with the heat transfer varied. The simulation of the effective gravity force decrease characterize this part of the experiment. The effective gravity force was defined as the gravity force projection on the canal plane: $g_e = g_0 \cdot \sin\beta$.

The relative gravity force was obtained as the ratio of the effective gravity force to the gravitational acceleration ($\bar{g}$).

The heat transfer coefficient was found to be strongly dependent upon the gravity force. Fig.6 shows the experimental data obtained for the heat transfer coefficient mesured at various heat flux and effective gravity force in a slot with 1,5mm thickness.

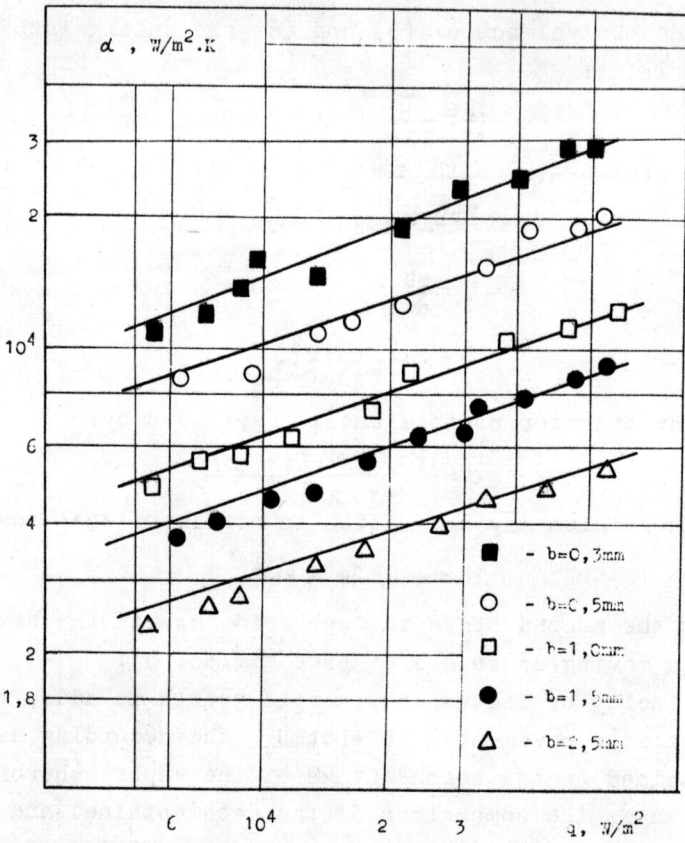

Fig. 4. Comparison of analytical and experimental data.

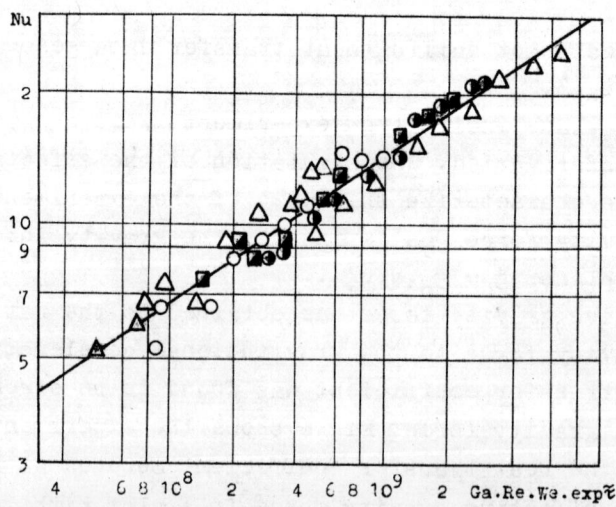

Fig. 5. Heat transfer in a vertical slot.

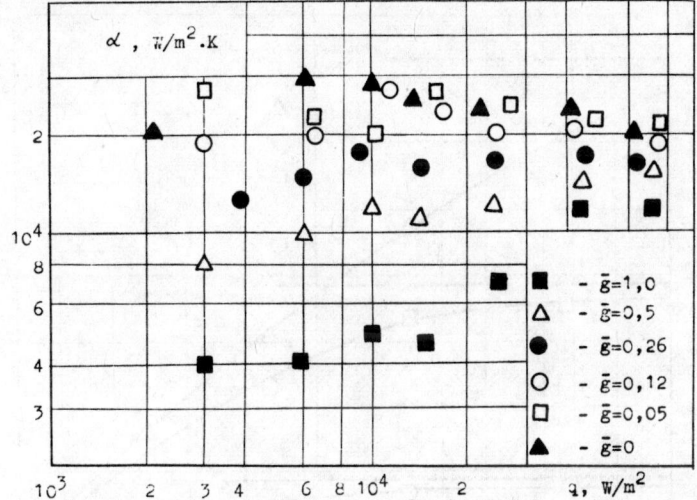

Fig. 6. Heat transfer in an inclined slot.

The gravity force influence on the heat transfer coefficient is more strong at the relatively low heat fluxes. At the heat flux $0,5 \cdot 10^4$ W/m$^2$ the decrease of the relative gravity force from 1 to 0 causes the increase of the heat transfer coefficient 7 times as much. With the heat growing the gravity force influence decreases. At the heat flux $1 \cdot 10^4$ W/m$^2$ the same gravity force decrease causes the heat transfer coefficient increase 5 times as much; at $5 \cdot 10^4$ W/m$^2$ - 2 times as much (fig.7).

Fig.8 presents the similar diagram for the slot thickness of 0,7mm. The comparison of these figures shows the greater gravity force influence on the heat transfer for the 1,5mm slot thickness.

As for the critical effects, we can note that the influence of the gravity force is similar to that of the canal thickness; as gravity force decreases the burnout heat flux decreases too.

The filming study showed the gravity force decrease followed by the slip velocity decrease. The vapor bubble diameter at the departure moment increases with gravity force decrease. This diameter, however, depends upon interaction with other bubbles. So, if the preceding bubble has not gone from the bottom as far as its diameter this causes earlier departure of the following bubble. The latter has the greater slip velocity under such circumstances. It catches up the preceding bubble and they mix. Five and more bubbles can mix by this way sometimes. Thus, the process of

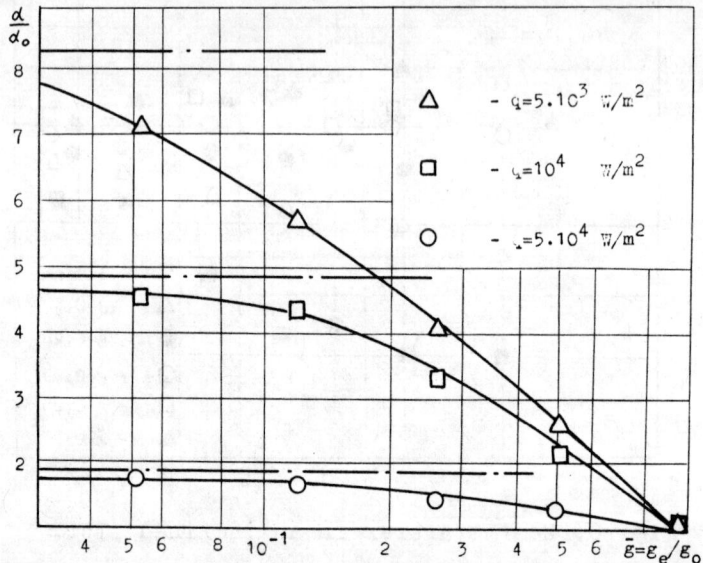

Fig. 7. Influence of the relative gravity force on the relative heat transfer coefficient.

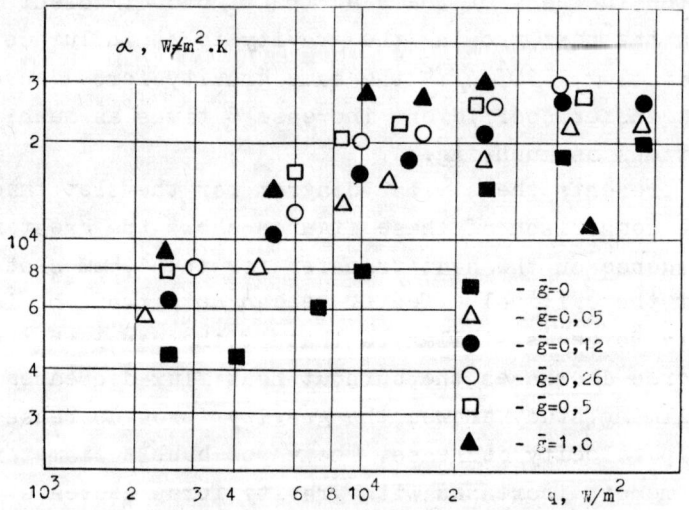

Fig. 8. Experimental heat transfer results for inclined slot canal.

bubble generation and departure at the diminished gravity force is of a cyclic type.

## CONCLUSIONS

1. The transition from laminar to turbulent natural covection in a closed slot thermosyphon is followed by the swing movement of the warm going up liquid stream; this effect is observed at relatively low Grashof numbers. Amplitude and frequency of the fluctuation grow with Grashof number increase. The transition behaviour is strongly dependent upon the inclination. The flow is more stable at little inclination.

It was found that while convection at the adiabatic zone was still laminar it was turbulent in the vucinity of heater and cooler.

At higher inclination ($45°$ and more) there are two steady vortices in the thermosyphon; they are situated near heater and cooler and flow is unstable.

2. The slot canal boiling heat transfer is strongly dependent upon the canal geometry.

3. The gravity force influence on the heat transfer coeffhcient is presented.

4. The slot canal boiling heat transfer correlations suggested are in a satisfactory agreement with the analysis.

## REFERENCES

1. F.J.Bailey and D.S.H.Lock, J. Heat Transfer, **87**, 1965.
2. А.И.Леонтьев и др., Труды МВТУ, №195, М., 1975.
3. В.Г.Григорьев и др., ТВТ, Т9, №6, 1975.

# EFFECT OF DISCRETE WALL ROUGHNESS ON FREE CONVECTIVE HEAT TRANSFER FROM A VERTICAL TUBE

**C. V. N. SASTRY, V. NARAYANA MURTHY, and P. K. SARMA**
Department of Mechanical Engineering, Andhra University, Waltair 530013, India

ABSTRACT

Experimental results are presented for the problem of free convection from a vertical tube with wires of various gauges wrapped around the tube maintaining a constant pitch (p) to diameter (d) ratio of unity. Artificial roughness thus created gives higher heat transfer rates in comparison with the turbulent free convective heat transfer results obtained on hydraulically smooth surface of a vertical tube. Further, it is shown analytically that the heat transfer coefficient is proportional to $\sqrt{Gr}$ and the constant of proportionality is a function of the roughness parameter ($K_s/L$). The experimental data are further correlated by a single dimensionless equation of the type

$$Nu = C (Gr.Pr)^{1/3}$$

$$C = 0.204$$

The index 1/3 suggests that the shear stress relationship valid for the turbulent boundary layer is equally applicable to the present study except for a change in the value of the coefficient of proportionality.

NOMENCLATURE

a     Constant in Eq.(1)

$C_p$     Specific heat

d     Diameter of the tube

f     Friction coefficient

h     Heat transfer coefficient

K     Thermal conductivity

$K_s$     Absolute roughness

L     Height of the tube

m     Exponent in eq.(1)

p   Pitch
t   Temperature
u   Velocity in x direction
v   Velocity in y direction
x   Spatial coordinate
y   Spatial coordinate

Subscripts

x   Differential with respect to x
y   Differential with respect to y
yy  Second differential with respect to y
w   Wall

Non-dimensional numbers

Gr  Grashof number  $(gL^3 \beta \Delta t / \nu^2)$
Nu  Nusselt number  $(hL/K)$
Pr  Prandtl number  $(\mu C_p / K)$
Ra  Rayleights number  $(Gr.Pr)$

INTRODUCTION

Several problems on free convective heat transfer from different body shapes and under varied thermal conditions at the solid boundary have been extensively investigated and a summary of the same can be found in the reviews of Ede[1] and Raithby and Hollands [2]. However, it can be observed that the influence of the nature of surface on free convective heat transfer is not tackled. In order to arrive at efficient and compact heat exchangers, augmentation techniques have been utilized and the methods adopted are either to create artificial roughness at the solid boundary or to provide turbulators to increase the intensity of turbulence in the fluid stream, thus, facilitating more energy transport from the hot surface. These studies are primarily devoted to forced convection inside ducts of different geometries. Salient aspects on internal flows with artificial roughness at the solid boundary are summarised by Rohsenow and Hartnett[3]. Thus, the purpose of the present paper is to present the experimental data on free convection from a vertical tube with discrete roughness created by wrapping wires of different gauges around a vertical tube maintaining a constant pitch to diameter ratio of unity. In addition, an analytical method is proposed to predict free convective heat transfer rates.

FORMULATION OF THE PROBLEM

Recently, it is shown (4) that a monotonic eddy diffusivity expression $\epsilon_M/\nu = 0.4 Y^+ [1 - \exp(-0.0017 Y^{+2})]$ valid for $0 \leq Y^+ \leq \infty$ can be utilized to predict successfully the turbulent free convective heat transfer rates from a vertical plate. A further extension to the problem is done in (5) exhibiting the validity of the similarity approach to the turbulent free convection. Thus, it is thought of that the friction characteristics obtained under forced convective conditions in ducts with discrete roughness can be utilized to the case of free convection from vertical surfaces with similar type of roughness created. For example, the law of friction in fully developed turbulent zone as postulated by Nikuredse(6) is a function of absolute roughness and is independent of Reynolds Number.

Thus, $\quad f = 1/(2 \log \frac{R}{K_s} + 1.74)^2$

This expression can also be put in a simplified form as

$$f = 33 \times 10^{-3} (x/K_s)^{-0.24}$$

Similarly, Hopf(7) and Fromm(8) found out that resistance to flow in rough channels (roughness being caused by wire nets, serrated zinc sheets and corrugated steel) is a function of roughness. Hence, for further analysis of the free convective problem, the friction coefficient f is assumed to be given by

$$f = a (K_s/\delta)^m \qquad \ldots (1)$$

where a and m are constants solely dependent on the geometry of roughness element. Values to a and m cannot be conclusively assigned at this stage. Further, for the problem under consideration by dimensional analysis the governing system of criteria can be expressed as

$$Nu_x = F (Gr_x, Pr, K_s/x, p/x) \qquad \ldots (2)$$

## ANALYSIS

Fig.1a shown as an inset in Fig.1 represents the physical configuration of the model with relevent coordinate system and the governing equations are as follows

$$u_x + v_y = 0 \qquad \ldots (3)$$

$$\rho(u \cdot u_x + v \cdot v_y) = \tau_y - g\beta\rho(T_\infty - T) \qquad \ldots (4)$$

$$u t_x + v t_y = \alpha t_{yy} \qquad \ldots (5)$$

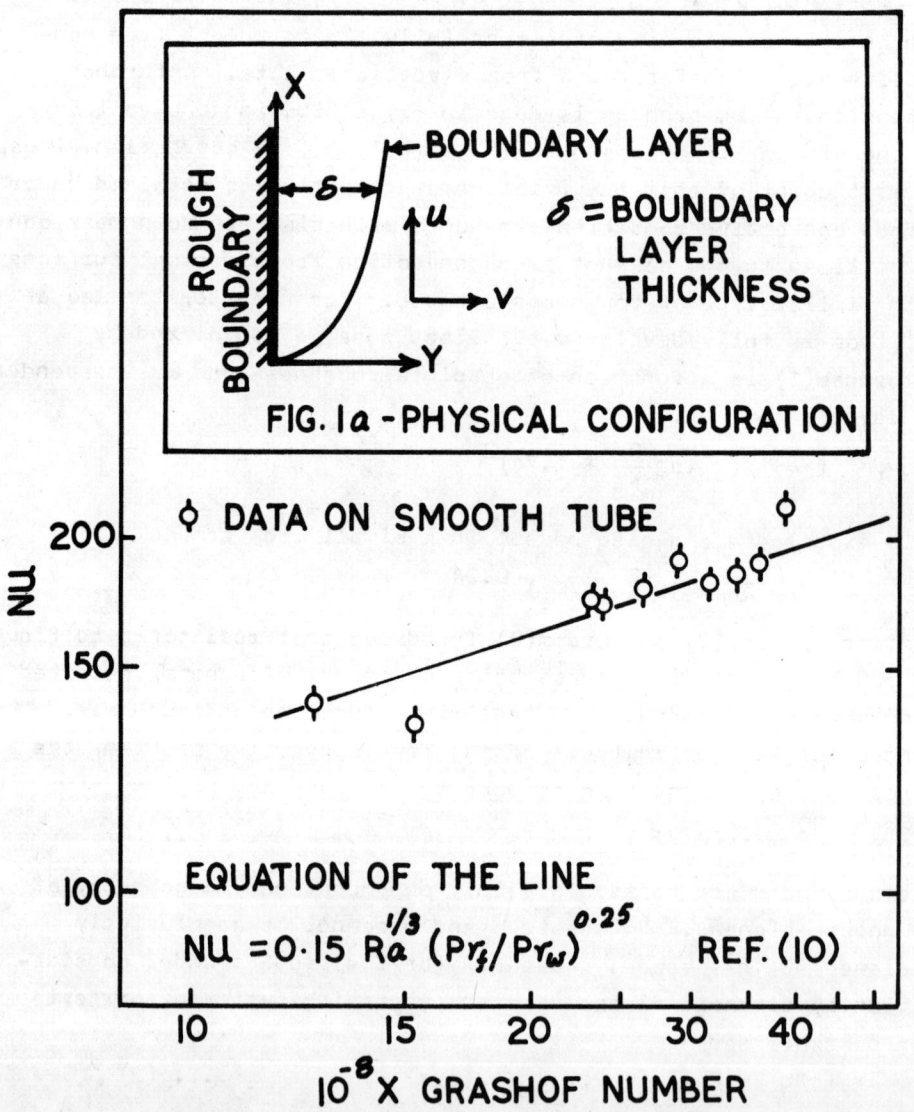

FIG.1 COMPARISON OF DATA WITH EQUATION.

Eqs.(3),(4) and (5) are respectively the laws of conservation of mass, momentum and energy which are to be solved subject to the relevent boundary conditions. Nevertheless, the mechanism of thermal and momentum diffusion near about the solid boundary having certain roughness characteristics are not known. As such, the analysis is further facilitated by assuming the profiles both for velocity and temperature as suggested by Eckert et al(9). Eqs.(4) and (5) with the aid of Eq.(3) can be written as

$$\frac{d}{dx}\int_0^\delta u^2 dy = -\tau_w/\rho + g\beta\Delta t \int_0^\delta \Theta \, dy \qquad \ldots (6)$$

$$\frac{d}{dx}\int_0^\delta u\Theta \, dy = q_w/\rho \Delta t C_p \qquad \ldots (7)$$

Where $\Theta = (T - T_\infty)/(T_w - T_\infty)$

The velocity and temperature profiles are

$$u/u_1 = (y/\delta)^{1/7} (1 - y/\delta)^4 \qquad \ldots (8)$$

$$\Theta = 1 - (y/\delta)^{1/7} \qquad \ldots (9)$$

The shear is given by

$$\tau_w = \tfrac{1}{2}\rho f u_1^2 \qquad \ldots (10)$$

where f is given by Eq.(1). Eqs.(8) and (9) satisfy the boundary conditions $u=0$ at $y=0$ and $\delta$; $\Theta=1$ at $y=0$ and $\Theta=0$ at $y=\delta$. Correcting for Prandtl effects, Reynolds anology gives the following expression

$$q_w = \tfrac{1}{2}(\rho C_p \Delta t \, u_1 f \, Pr^{-2/3}) \qquad \ldots (11)$$

Thus, from Eqs.(6) to (11) we get

$$0.433 \frac{d}{dx}(u_1^2 \delta) = -8\frac{\tau_w}{\rho} + g\beta\Delta t \, \delta \qquad \ldots (12)$$

$$0.0732 \frac{d}{dx}(u_1 \delta) = f u_1 Pr^{-2/3} \qquad \ldots (13)$$

The solution for the Eqs.(12) and (13) are

$$u_1 = C_1 x^n \text{ and } \delta = C_2 x^p \qquad \ldots (13)$$

(n and p are constants)

where $C_1 = \left[ \dfrac{g\beta \Delta t}{0.433 \dfrac{(2+m)}{(1+m)} + \dfrac{4}{13.66} \dfrac{(3+m)}{(1+m)} Pr^{2/3}} \right]^{1/2}$ ... (15)

$C_2 = \left[ 13.66 \dfrac{(1+m)}{(3+m)} a\, K_s^m\, Pr^{-2/3} \right]^{1/(m+1)}$ ... (16)

Thus, in the above equation the unknown values a and m are to be assigned. However, it is clear that the Nusselt number should be proportional to Gr. Hence, from Eqs.(11), (14), (15) and (16) the mean heat transfer coefficient can be evaluated from

$$\overline{Nu} = \int_0^l Nu_x\, d(x/L)$$ ... (17)

Thus Eq.(17) yields

$$\overline{Nu} = A\, (K_s/x)^{m/(1+m)} (Gr)^{\frac{1}{2}}$$

$A = (\dfrac{1+m}{5+m}) \dfrac{a^{1/(1+m)}\, Pr^{(1+3m)/3(1+m)}}{(13.66 \dfrac{1+m}{3+m})^{m/(1+m)}} \dfrac{1}{(0.433 \dfrac{2+m}{1+m} + 0.293 \dfrac{3+m}{1+m} Pr^{2/3})^{\frac{1}{2}}}$

## EXPERIMENTATION

A vertical copper tube $\phi$ = 32.8 mm, L=470 mm is heated indirectly by impressing a voltage across the electrical resistance element kept inside the tube. Six precalibrated iron-constantan thermocouples are silver brazed in flush with the experimental surface. To establish the correctness of all the measurements such as temperature and power, calibration runs are conducted. Taking into the radiation effects, the data are compared with the established equation for turbulent free convection (see Fig.1). The agreement between the data and the equation lends support to the correctness and validity of the data taken. Altogether four wires of different gauges wound around the tube maintaining p/d = 1 are tested and the photographic view of a typical test surface is shown in plate No.1. In all the runs sufficient care is taken to shield the test section from the stray convective currents of air. Steady state conditions are assumed only when the temperatures read by the thermocouples do not vary with time.

## DISCUSSION OF THE RESULTS

The results are shown plotted in Figs.(2) and (3) and it can be observed that the solid lines fared through the data points posses a slope of 0.5 signifying that the data are in confirmity with the theory only for high values of Grashof number.

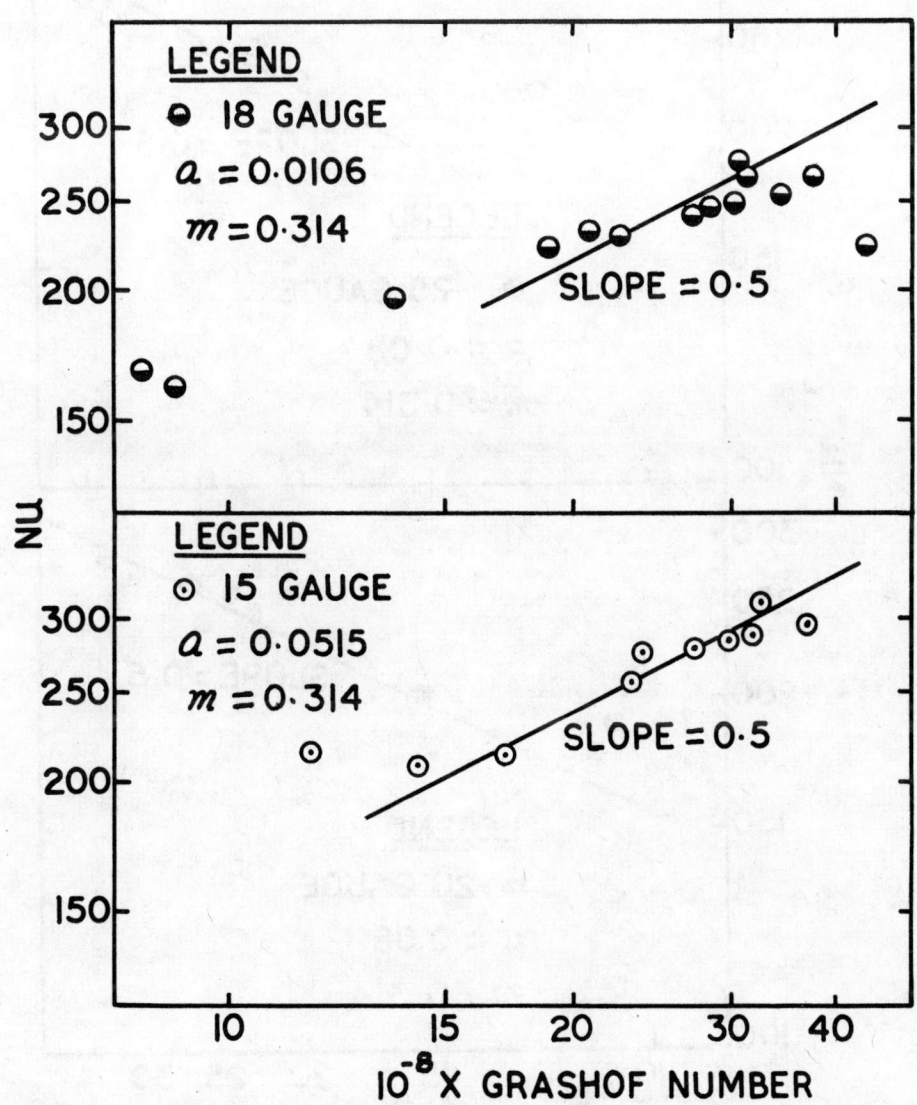

FIG.2    HEAT TRANSFER DATA ON ROUGH SURFACES

The value of m is tentatively fixed at 0.314 based on the data of Hopf(7). Thus, the value of a is estimated and shown in Figs.(2) and (3).

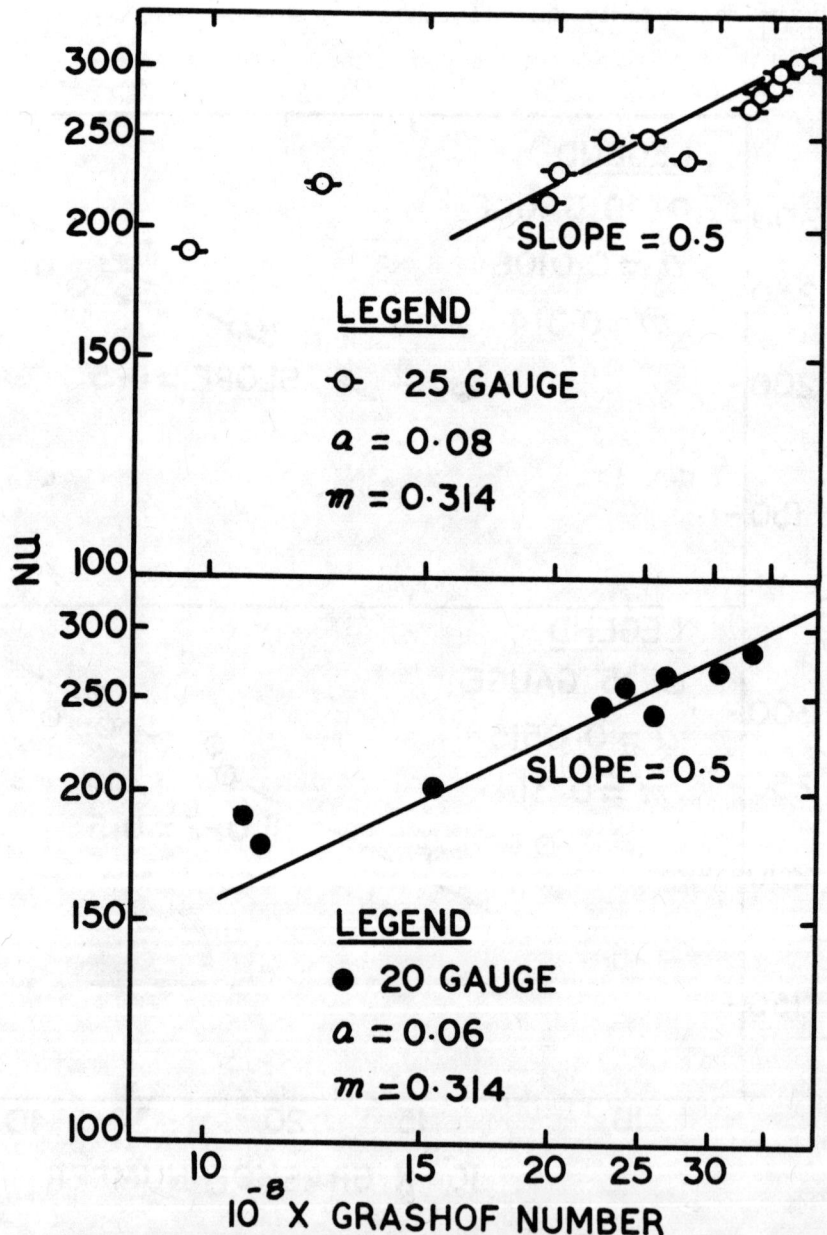

FIG.3 HEAT TRANSFER DATA ON ROUGH SURFACES

However, an attempt is made, to verify whether a single relationship can be obtained for all the runs with in certain limits of accuracy. For this purpose all data are brought on to a single plot in Fig.4. While a single equation with 0.5 as slope is a possible feature only for high values of Gr a line shown in dotted having a slope of 1/3 can correlate all data points with an accuracy of ± 12% from the mean line faired through data points. In all the above calculations, the physical properties are taken at the ambient temperature of air. Thus, the following conclusions can be arrived from both experimental and analytical studies:

1. It is observed that the turbulent free convective heat transfer rates can be considerably increased by providing roughness configuration of the type utilized in the present studies. For example, in the present study

$$h_{rough}/h_{smooth} = 1.5$$

2. The theoretical analysis on the lines outlined by Eckert et al gave that the mean heat transfer coefficient is proportional to $\sqrt{Gr}$ and function of $(K_s/L)$. The values of the unknown constants are evaluated making use of the experimental data.

PLATE 1: TEST SURFACE WITH 15 GAUGE WIRE WOUND AROUND THE TUBE.

3. From the data, the following dimensionless correlation can be obtained

$$Nu = 0.204 \, (Gr.Pr)^{1/3}$$

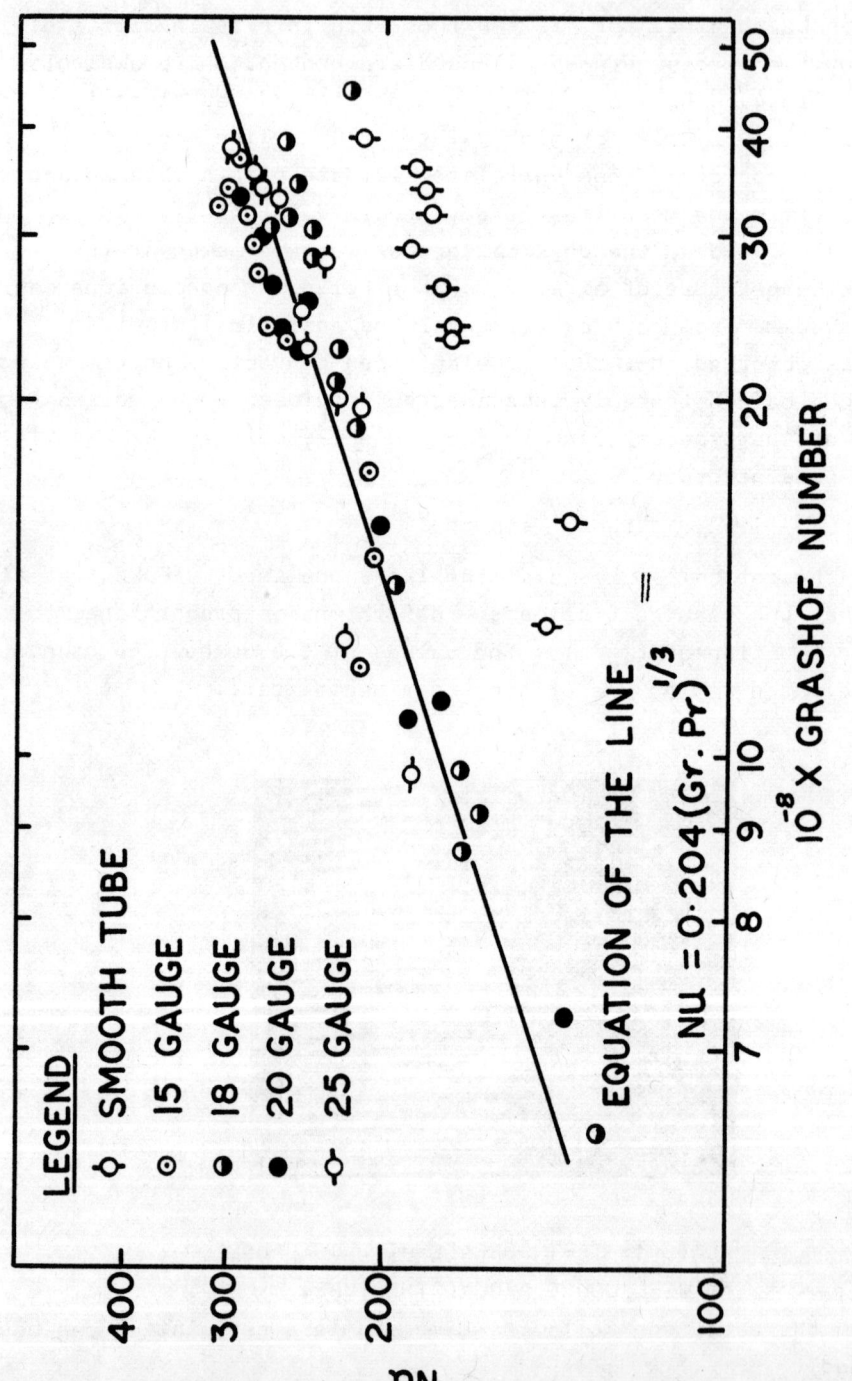

FIG. 4 GENERALIZED PLOT

which is valid for the range of parameters tested. In otherwords the shear relationship valid for turbulent boundary layer can be extended to the present case but for a change in the coefficient.

REFERENCES

1. Ede, A.J., "Advances in free convection", Advances in heat transfer, Academic Press, Vol.4 (1967)

2. Raithby, G.D., and Hollands, K.G.T., "A general method of obtaining approximate solutions to laminar and turbulent free convection problems", Advances in heat transfer, Academic Press, Vol.11 (1975).

3. Rohsenow, W.M., and Hartnett, J.P., Hand book of heat and mass transfer, pp. 7-40 to 7-51 (1973).

4. Kato, H., Nishiwaki, N., and Hiratta, M., Int.J.Heat mass transfer, Vol.11, 1117 (1968).

5. Nato, K., and Matsumoto, K., "Turbulent heat transfer by natural convection along isothermal vertical flat surface" Jl. of heat transfer, Trans. ASME, pp. 621-623 (1975).

6. Schlichting, H., "Boundary layer theory" pp. 525, 4th Edition, McGraw Hill, New York, (1962).

7. Hopf, L.Z., angew.Math.V.Mech., 3:329 (1923).

8. Fromm.K.Z., angew.Math.V.Mech., 3:339 (1923).

9. Eckert, E.R.G., and Drake, M.R., Heat and mass transfer, McGraw Hill (1958).

10. Sorin, S.N., "Heat transfer" (in Russian) Published by "Vishasaya Skola", Moscow (1964).

ACKNOWLEDGEMENTS

The authors sincerely thank the group of under graduate students working with Dr. P.K. Sarma for their kind co-operation in taking all the experimental results.

# ON LAMINAR FREE CONVECTION STAGNATION HEAT TRANSFER FROM A NONISOTHERMAL CYLINDER WITH INTERNAL SOURCES-SINKS

OZER A. ARNAS

*Louisiana State University, Baton Rouge, Louisiana, U.S.A.*

ABSTRACT

An analysis of the laminar free convection stagnation heat transfer from a nonisothermal cylinder with a wall temperature variation of $(x/D)^m$ and with two temperature dependent internal sources-sinks within the fluid is presented. Two cases are discussed: one in which only the x-direction momentum equation is used, and the other in which the two dimensional Navier-Stokes equations are combined into one momentum equation which makes the heat transfer variable $\theta'(0)$ and the temperature profile Grashof number dependent. It is found that the consideration of the two dimensional equations has a distinct effect on the heat transfer results.

NOMENCLATURE

| | |
|---|---|
| $a_1, a_2$ | constants |
| D | diameter of cylinder |
| f | dimensionless stream function, equation (8) |
| g | acceleration due to gravity |
| GR | Grashof number, equation (21) |
| h | convective heat transfer coefficient |
| k | thermal conductivity of fluid |
| m | dimensionless exponent |
| NU | Nusselt number, equation (22) |
| P | dynamic pressure |
| PR | Prandtl number, $[\nu/\alpha]$ |
| R | radius of cylinder |
| T | temperature |
| u | tangential velocity component |
| v | normal velocity component |
| x | coordinate along wall of cylinder |
| y | coordinate normal to wall of cylinder |
| $\alpha$ | thermal diffusivity of fluid |
| $\beta$ | coefficient of thermal expansion of fluid |
| $\delta$ | boundary layer thickness |
| $\zeta$ | dimensionless second source-sink strength, $[\lambda D/\alpha]$ |
| $\epsilon$ | first source-sink strength |
| $\eta$ | dimensionless coordinate, equation (7) |
| $\theta$ | dimensionless temperature, equation (6) |
| $\rho$ | density of fluid |
| $\lambda$ | second source-sink strength |
| $\nu$ | kinematic viscosity of fluid |

ξ            dimensionless first source-sink strength, $[\epsilon D^2/\alpha]$
φ            angle measured from stagnation point
ψ            stream function, equation (5)

Subscripts

∞            ambient conditions
w            wall conditions
$w_o$       conditions at lower stagnation point of cylinder

INTRODUCTION

Laminar free convection heat transfer from nonisothermal bodies has been studied by various researchers. Sparrow and Gregg [1] have presented solutions for the nonisothermal vertical plate. The vertical nonisothermal cone has been analyzed by Hering and Grosh [2]. Koh and Price [3] solved the appropriate boundary layer equations for a nonisothermal horizontal cylinder with the wall temperature varying as $[1 + a_1(x/R)^2 + a_2(x/R)^4]$. Arnas and Valentine [4] applied a method of combination of the two dimensional Navier-Stokes equations with a wall temperature variation of $(x/D)^m$. Low [5] considered heat sources which were explicit functions of the boundary layer similarity variable. Chambre [6] analyzed the problem in which the rate of heat production or absorption was a function of the position in the thermal field. The problem of temperature dependent heat sources or sinks in a stagnation point flow using the conservation of energy equation was considered by Sparrow and Cess [7]. Recently, Arnas [8] analyzed the problem of laminar free convection stagnation heat transfer from an isothermal heated horizontal cylinder with two temperature dependent sources-sinks within the boundary layer which may be due to a combustion process, spraying of a fluid into the medium or after shutdown cooling problems associated with nuclear reactors. In the case of chemical reactions, exothermic and endothermic cases may simulate sources or sinks, respectively.

In the present study, the analysis presented in [8] is extended to a nonisothermal cylinder using the same two source-sink relations: a linear temperature and a gradient of temperature functionals. Single momentum and combined momentum equations cases are considered separately and the heat transfer results for a Prandtl number of 0.72 are compared.

ANALYSIS

The analysis is carried out for a nonisothermal heated horizontal cylinder with a wall temperature variation of $(x/D)^m$, Figure 1. The wall temperature of the cylinder will be $(T - T_\infty) = (T_{w_o} - T_\infty)(x/D)^m$ where $T_{w_o}$ is the specified uniform temperature at the lower stagnation point. For m = 0, the isothermal case, the wall temperature becomes $T_w$, a constant [8].

First Source - Single Momentum Equation

The equations expressing conservation of mass, momentum and energy for steady laminar flow in a boundary layer on a horizontal cylinder with a linear source-sink term in the form $\epsilon(T - T_\infty)$, following [8] are

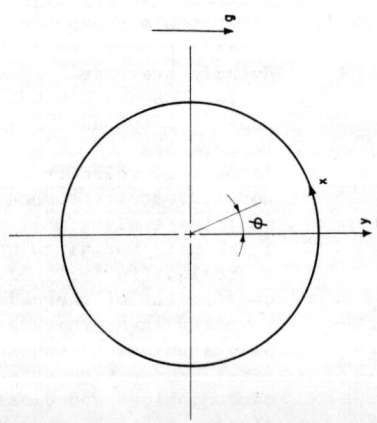

Figure 1    The Coordinate System

# LAMINAR FREE CONVECTION STAGNATION HEAT TRANSFER

$$\frac{\partial u}{\partial x} + \frac{\partial v}{\partial y} = 0 \tag{1}$$

$$u\frac{\partial u}{\partial x} + v\frac{\partial u}{\partial y} = \nu\frac{\partial^2 u}{\partial y^2} + g\beta(T - T_\infty)\sin\varphi \tag{2}$$

$$u\frac{\partial T}{\partial x} + v\frac{\partial T}{\partial y} = \alpha\frac{\partial^2 T}{\partial y^2} + \epsilon(T - T_\infty) \tag{3}$$

The density has been considered a variable only in forming the buoyancy force. Variations of all other properties along with viscous dissipation have been neglected. The appropriate boundary conditions are

$$\begin{aligned} y = 0 \quad & u = v = 0 \quad & T - T_\infty = (T_{w_o} - T_\infty)\left(\frac{x}{D}\right)^m \\ y = \delta \quad & u = \frac{\partial u}{\partial y} = 0 \quad & T - T_\infty = 0 \end{aligned} \tag{4}$$

Equation (1) is automatically satisfied by introducing a stream function such that

$$u = \frac{\partial \Psi}{\partial y} \qquad v = -\frac{\partial \Psi}{\partial x} \tag{5}$$

Defining the nondimensional temperature and similarity variables as

$$\theta(\eta) = (T - T_\infty) / \left[(T_{w_o} - T_\infty)(x/D)^m\right] \tag{6}$$

$$\eta = y(x/D)^{m/4}\left[g\beta(T_{w_o} - T_\infty)/D\nu^2\right]^{\frac{1}{4}} \tag{7}$$

$$\Psi(x,y) = 2x\nu(x/D)^{m/4}\left[g\beta(T_{w_o} - T_\infty)/D\nu^2\right]^{\frac{1}{4}} f(\eta) \tag{8}$$

and with (6), (7) and (8), (2) and (3) may be transformed to give

$$f''' + \tfrac{1}{2}(m + 4) f f'' - (m + 2)(f')^2 + \theta = 0 \tag{9}$$

$$\theta'' + \tfrac{1}{2}(m + 4)\,\mathrm{PR}\, f\,\theta' - (2m\,\mathrm{PR}\, f' - \xi/\mathrm{GR}^{\frac{1}{2}})\theta = 0 \tag{10}$$

where $\xi$, PR and GR are defined in the Nomenclature. The angle $\sin\varphi$ is replaced by $\varphi$ which limits the results to the near stagnation region of the cylinder. Equation (4) transforms into

$$\begin{aligned} \eta = 0 \quad & f' = f = 0 \quad & \theta = 1 \\ \eta = \delta \quad & f' = f'' = 0 \quad & \theta = 0 \end{aligned} \tag{11}$$

## First Source - Combined Momentum Equations

The equations expressing conservation of mass, momentum and energy for this case are [8]

$$\frac{\partial u}{\partial x} + \frac{\partial v}{\partial y} = 0 \tag{1}$$

$$u \frac{\partial u}{\partial x} + v \frac{\partial u}{\partial y} = \nu \left[ \frac{\partial^2 u}{\partial x^2} + \frac{\partial^2 u}{\partial y^2} \right] - \frac{1}{\rho} \frac{\partial P}{\partial x} + g\beta(T - T_\infty) \sin \varphi \qquad (12)$$

$$u \frac{\partial v}{\partial x} + v \frac{\partial v}{\partial y} = \nu \left[ \frac{\partial^2 v}{\partial x^2} + \frac{\partial^2 v}{\partial y^2} \right] - \frac{1}{\rho} \frac{\partial P}{\partial y} - g\beta(T - T_\infty) \cos \varphi \qquad (13)$$

$$u \frac{\partial T}{\partial x} + v \frac{\partial T}{\partial y} = \alpha \frac{\partial^2 T}{\partial y^2} + \epsilon(T - T_\infty) \qquad (3)$$

with (4) as the boundary conditions.

For this case, rather than reducing the two dimensional Navier-Stokes equations by Prandtl's boundary layer approximation, a super-position technique is employed combining the two equations into one [4]. This is accomplished by taking the derivative of (12) with respect to y and (13) with respect to x. After this differentiation, (13) is subtracted from (12) to give

$$u \left[ \frac{\partial^2 u}{\partial y \partial x} - \frac{\partial^2 v}{\partial x^2} \right] + v \left[ \frac{\partial^2 u}{\partial y^2} - \frac{\partial^2 v}{\partial y \partial x} \right] = \nu \left[ \frac{\partial^3 u}{\partial x^2 \partial y} + \frac{\partial^3 u}{\partial y^3} - \frac{\partial^3 v}{\partial x^3} - \frac{\partial^3 v}{\partial x \partial y^2} \right]$$
$$+ g\beta \left[ \frac{\partial T}{\partial y} \sin \varphi - (T - T_\infty) \frac{\partial \varphi}{\partial x} \sin \varphi + \frac{\partial T}{\partial x} \cos \varphi \right] \qquad (14)$$

With (6), (7) and (8), (14) and (3) transform into

$$f^{iv} - \tfrac{1}{2}(3m + 4) f' f'' + \tfrac{1}{2}(m + 4) f f''' + \theta' - (2/GR^{\frac{1}{2}})\theta = 0 \qquad (15)$$

$$\theta'' + \tfrac{1}{2}(m + 4) \text{ PR } f \theta' - (2m \text{ PR } f' - \xi/GR^{\frac{1}{2}})\theta = 0 \qquad (10)$$

with the boundary conditions given in (11). Equation (15) is obtained by neglecting terms of order of the boundary layer thickness $\delta$ or smaller, such as (y/x) and higher order terms [4].

## Second Source - Single Momentum Equation

The form of the second source-sink is assumed to be $\lambda \left[ \dfrac{\partial (T - T_\infty)}{\partial y} \right]$ reducing the transformed momentum and energy equations to

$$f''' + \tfrac{1}{2}(m + 4) f f'' - (m + 2)(f')^2 + \theta = 0 \qquad (9)$$

$$\theta'' + \left[ \tfrac{1}{2}(m + 4) \text{ PR } f + \zeta/GR^{\frac{1}{2}} \right] - 2m \text{ PR } f'\theta = 0 \qquad (16)$$

respectively. The boundary conditions are given in (11).

## Second Source - Combined Momentum Equations

The transformed momentum and energy equations and boundary conditions are those that are given in (15), (16), and (11), respectively.

## SOLUTION OF GOVERNING EQUATIONS

The method of solution used for the various set of transformed momentum-energy equations is an integral method. For single momentum cases, the governing equations are integrated by parts between the limits of zero and $\delta$ and the boundary conditions (11) are used [8]. For combined momentum cases, the momentum equation is first integrated between the limits of $\eta$ and $\delta$ and then the set is once again integrated from zero to $\delta$ [4] utilizing the same methodology. Once these reduced order equations are obtained, series expressions for the velocity and temperature profiles are assumed following Pohlhausen [9]. Using these along with the reduced set of momentum and energy equations and the boundary conditions, the boundary layer thickness for various combinations are determined for a Prandtl number of 0.72.

### First Source - Single Momentum Equation

$$0.000015 \, (3m + 8)\delta^8 - 0.009665 \, \{(3m + 8)\xi/[(5m + 4)GR^{\frac{1}{2}}]\}\delta^6$$

$$+ \{4.438570 \, (3m + 8)\xi^2/[(5m + 4)^2 GR] - 0.038639 \, (3m + 8)/(5m + 4)$$

$$- 0.281876\}\delta^4 + \{35.508557 \, (3m + 8)\xi/[(5m + 4)^2 GR^{\frac{1}{2}}]$$

$$+ 17.156190\xi/[(5m + 4)GR^{\frac{1}{2}}]\}\delta^2$$

$$+ 71.017114 \, (3m + 8)/(5m + 4)^2 + 68.624760/(5m + 4) = 0 \qquad (17)$$

### First Source - Combined Momentum Equations

$$0.000012 \, [(3m + 8)/GR^{\frac{1}{2}}]\delta^{10} + 0.000024 \, [(3m + 8)/GR^{\frac{1}{4}}]\delta^9$$

$$+ 0.000015 \, (3m + 8)\delta^8 - 0.007222 \, \{(3m + 8)\xi/[(5m + 4)GR^{3/4}]\}\delta^7$$

$$- 0.009665 \, \{(3m + 8)\xi/[(5m + 4)GR^{\frac{1}{2}}]\}\delta^6$$

$$- \{0.028884 \, (3m + 8)/[(5m + 4)GR^{\frac{1}{4}}] + 0.130170/GR^{\frac{1}{4}}\}\delta^5$$

$$+ \{4.438570 \, (3m + 8)\xi^2/[(5m + 4)^2 GR] - 0.038639 \, (3m + 8)/(5m + 4)$$

$$- 0.281876\}\delta^4 + \{35.508557 \, (3m + 8)\xi/[(5m + 4)^2 GR^{\frac{1}{2}}]$$

$$+ 17.156190\xi/[(5m + 4)GR^{\frac{1}{2}}]\}\delta^2$$

$$+ 71.017114 \, (3m + 8)/(5m + 4)^2 + 68.624760/(5m + 4) = 0 \qquad (18)$$

Second Source - Single Momentum Equation

$$0.000015\,(3m+8)\delta^8 + 0.025275\,\{(3m+8)\zeta/[(5m+4)GR^{\frac{1}{4}}]\}\delta^5$$

$$- [0.038639\,(3m+8)/(5m+4) + 0.281876]\delta^4$$

$$+ \{31.563162\,(3m+8)\zeta^2/[(5m+4)^2 GR^{\frac{1}{2}}]\}\delta^2$$

$$- \{94.689485\,(3m+8)\zeta/[(5m+4)^2 GR^{\frac{1}{4}}]$$

$$+ 45.749840\zeta/[(5m+4)GR^{\frac{1}{4}}]\}\delta$$

$$+ 71.017114\,(3m+8)/(5m+4)^2 + 68.624760/(5m+4) = 0 \qquad (19)$$

Second Source - Combined Momentum Equations

$$0.000012\,[(3m+8)/GR^{\frac{1}{2}}]\delta^{10} + 0.000024\,[(3m+8)/GR^{\frac{1}{4}}]\delta^9$$

$$+ 0.000015\,(3m+8)\delta^8 + 0.019256\,\{(3m+8)\zeta/[(5m+4)GR^{\frac{1}{2}}]\}\delta^6$$

$$- \{0.028884\,(3m+8)/[(5m+4)GR^{\frac{1}{4}}] + 0.130170/GR^{\frac{1}{4}}$$

$$- 0.025275\,(3m+8)\zeta/[(5m+4)GR^{\frac{1}{4}}]\}\delta^5$$

$$- [0.038639\,(3m+8)/(5m+4) + 0.281876]\delta^4$$

$$+ 31.563162\,\{(3m+8)\zeta^2/[(5m+4)^2 GR^{\frac{1}{2}}]\}\delta^2$$

$$- \{94.689485\,(3m+8)\zeta/[(5m+4)^2 GR^{\frac{1}{4}}]$$

$$+ 45.749840\zeta[(5m+4)GR^{\frac{1}{4}}]\}\delta$$

$$+ 71.017114\,(3m+8)/(5m+4)^2 + 68.624760/(5m+4) = 0 \qquad (20)$$

Equations (17) - (20) were solved on an IBM 7094 computer for Grashof numbers of $10^4$ and $10^9$ and values of nonisothermal parameter m of zero to 4 [10] and are plotted in Figures (2) and (3) for some characteristic values in comparative fashion for the two source-sink functions, respectively. In each case, the asymptotic values for various m at Grashof number of $10^9$ are for the boundary layer thickness of the nonisothermal cylinder single momentum equation.

# LAMINAR FREE CONVECTION STAGNATION HEAT TRANSFER

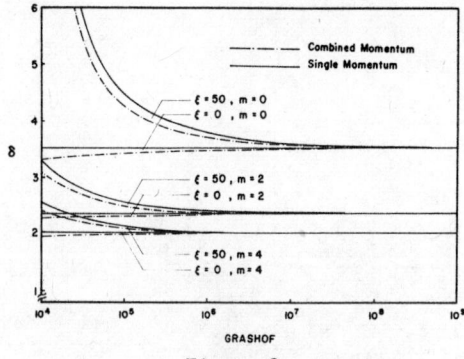

Figure 2
Comparison of Boundary Layer
Thickness for First Source

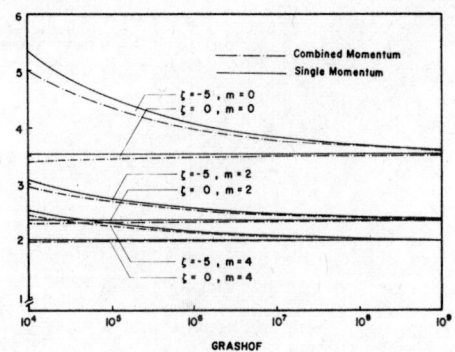

Figure 3
Comparison of Boundary Layer
Thickness for Second Sink

## DISCUSSION OF RESULTS

The velocity and temperature distributions may be easily obtained since the boundary layer thicknesses are known. However, the interest is in the heat transfer, thus the Nusselt number. Therefore, since the cylinder wall temperature depends upon the location, Grashof number becomes a function of x and is given by

$$GR = (z/D)^m \left[ g\beta D^3 \left( T_{w_0} - T_\infty \right)/\nu^2 \right] \tag{21}$$

Note that for x = 0, the absolute stagnation point, the Grashof number becomes zero. Therefore, the results of this study are valid near the stagnation region. The Nusselt number becomes

$$NU = hD/k$$
$$= -\theta'(0) \, GR^{\frac{1}{4}} \tag{22}$$

where $\theta'(0)$ is the dimensionless temperature gradient at the surface of the cylinder and is equal to

$$\theta'(0) = -3/2\delta . \tag{23}$$

The Nusselt number for the first and second source-sink functions for the single and combined momentum equations are given in Figures (4) and (5) for some characteristic values, respectively.

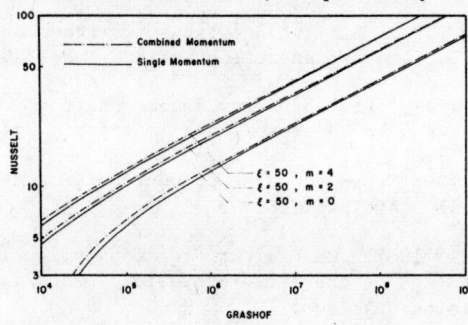

Figure 4
Comparison of Nusselt Numbers
for First Source

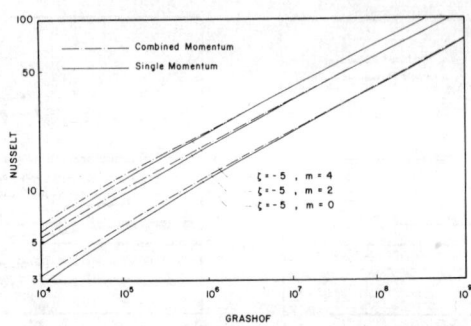

Figure 5
Comparison of Nusselt Numbers
for Second Sink

For m = 0, the results are those of an isothermal cylinder as presented in [8] for all of the equations and graphs. For small values of m, the percent error in Nusselt numbers are distinct and become more appreciable for lower values of the Grashof number in the compared cases of single and combined momentum equations. However as m and Grashof number increase, the deviations become negligible.

For the source-sink function $\xi$ and $\zeta$ zero, the boundary layer thickness equations (18) and (20) reduce to that obtained in [4]. Since previously published experimental data have a better agreement with the theory of [4], the results obtained here, though not verifiable by experiment, are expected to give better results.

CONCLUSIONS

The boundary layer equations for laminar free convection stagnation heat transfer from a nonisothermal cylinder with a wall temperature variation of $(x/D)^m$ and with internal sources-sinks in the boundary layer have been solved. Two functionals are considered: a linear variation with temperature and a gradient of temperature. The deviations in the boundary layer thicknesses and Nusselt numbers with single and combined momentum equations are obtained and presented for various values of m and Grashof number as well as some arbitrary values of the source-sink strengths. It is seen that the effect of the combined momentum equations is distinct and appreciable, though becoming negligible for large values of m and Grashof number.

REFERENCES

1. Sparrow, E. M., and Gregg, J. L. 1958. Similar Solutions for Free Convection from a Nonisothermal Vertical Plate. Transactions of A.S.M.E. XXC:379.

2. Hering, R. G., and Grosh, R. J. 1962. Laminar Free Convection from a Nonisothermal Cone. International Journal of Heat and Mass Transfer. 5:1059.

3. Koh, J. C. Y., and Price, J. F. 1965. Laminar Free Convection from a Nonisothermal Cylinder. Journal of Heat Transfer. 82:1.

4. Arnas, O. A., and Valentine, M. B. 1967. On Laminar Free Convection from a Nonisothermal Cylinder. Proceedings of the Tenth Midwestern Mechanics Conference, Developments in Mechanics. 4:1593.

5. Low, G. M. 1955. Stability of Compressible Laminar Boundary Layer with Internal Heat Sources and Sinks. Journal of Aeronautical Science. 22:329.

6. Chambre, P. L. 1957. The Laminar Boundary Layer with Distributed Heat Sources or Sinks. Applied Scientific Research. A6:393.

7. Sparrow, E. M., and Cess, R. D. 1961. Temperature Dependent Heat Sources or Sinks in a Stagnation Point Flow. Applied Scientific Research. A10:185.

8. Arnas, O. A. 1972. On Laminar Free Convection Stagnation Heat Transfer from an Isothermal Cylinder with Internal Sources-Sinks. Applied Scientific Research. 27:81.

9. Schlichting, H. 1960. <u>Boundary Layer Theory</u>, Fourth Edition, McGraw-Hill Book Company, Inc., New York.

10. Pan American Corporation. 1965. Polynomial Root Extraction. Computer Technical Application Documentation, Program 2-167.0.

# TURBULENT MIXING OF A DUSTSTREAM IN A DUCT

HARALD GROSS

*Institut für Verfahrenstechnik und Dampfkesselwesen*

ABSTRACT

In many duststreams, passing through a duct, concentration resp. precipitation of particles is influenced by the turbulence of the gas in which particles are dispersed. Especially in electric precipitators turbulence is intensified by the "electric wind" which results from the "corona" current. A model for precipitation efficiency was developed, which includes movement of particles by Coulomb forces and remixing by turbulence. The factor $D_{eff}$, representativ for influence of turbulence, was measured by evaluating of mean time spectrums of tracers, passing through the precipitator. Evaluations by mathematical model were compared with measurements.

NOMENCLATURE

| | |
|---|---|
| b | width of the duct (m) |
| i | corona current between precipitator electrodes ($mA/m^2$) |
| j | mass stream ($kg/m^2 s$) |
| L | length of the duct (m) |
| q | portion of dust which passes through the duct, being not precipitated |
| $\rho$ | density of dust in the gasvolume ($kg/m^3$) |
| $\sigma$ | dispersion of mean time spectrum |
| $\tau$ | mean time of the tracer in the precipitator (s) |
| U | voltage on the precipitator electrodes (V) |
| $w_s$ | sinking velocity of a dust particle (m/s) |
| $w_D$ | migration velocity according to "Deutsch" (m/s) |
| $w_T$ | velocity caused by Coulomb forces |
| v | velocity in main stream direction |

Subscripts

| | |
|---|---|
| d | diffusion |
| p | precipitator |

## INTRODUCTION

Many kinds of forces can act upon particles passing with a duststream through a duct. The particles will follow the resulting force, which can cause a concentration of particles for example, near the bottom of the duct. Regions where no particles will be found can also exist. On the other hand turbulence in the gasstream will cause a mixing of the particles in the gasstream. Volume elements with less particles will be transported in regions with high particleconcentration and inverse. By this influence of the turbulence it is almost impossible to clean a turbulent gasstream totaly from particles in any kind of precipitators.

The purpose of this report is to develope and to test a mathematical model, which connects the movement of particles dispersed in a gasstream with the influence of turbulent flow. The model includes a factor, which will be determined by experiments.

## MATHEMATICAL MODEL

Fig. 1 showes a duct with a duststream passing through it. A force eg. gravity is acting upon the particles which accelerates them. But as frictional forces are acting upon them an equilibrium of their velocity will result. By this velocity particles will be concentrated near the bottom of the duct, and it is assumed that particles will be totaly precipitated from the gasstream, if they reach the bottom of the duct. It is further assumed that the flow is a turbulent one. This is important, because in many cases of industrial dust flows, the turbulent flow is dominant. In this special case an electric precipitator was investigated. In this case the turbulence is intensified by an electric gas discharge, the "corona". This turbulence causes a steady mixing of the dust against the concentrating action of Coulomb or gravity forces.

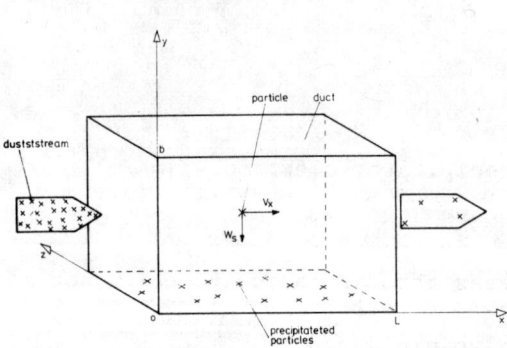

fig. 1 duststream, passing through a duct

The velocity of main flow direction may be $v_x$, the density of the dust in the gas is $\rho$, and is assumed to be distributed homogenously over the cross sectional area of the duct entrance. For simplification only two dimensions will be considered.

The components of the mass stream can be written

$$j_x = \rho v_x$$

$$j_y = \rho w_s$$

# TURBULENT MIXING OF A DUSTSTREAM IN A DUCT

The mass stream caused by turbulence will be approximated by "turbulent diffusion" /1/

$$j_{dx} = -D_{eff} \frac{\partial \rho}{\partial x}$$

$$j_{dy} = -D_{eff} \frac{\partial \rho}{\partial y}$$

In this case turbulence is assumed to be isotropic. For any area of the two dimensional duct the law of the continuity of mass is valid

$$V \frac{\partial \rho}{\partial x} + W_s \frac{\partial \rho}{\partial y} - D_{eff} \frac{\partial^2 \rho}{\partial x^2} - D_{eff} \frac{\partial^2 \rho}{\partial y^2} = 0 \qquad (1)$$

The special boundary conditions for this problem are:

a) $\rho(0, y,) = \rho_o$     dust is mixed homogenously over the cross-sectional area of the duct entrance

b) $\frac{\partial \rho}{\partial x}(L, y) = 0$     after passing the duct no reduction of mass is possible

c) $\frac{\partial \rho}{\partial y}(x, 0) = 0$     if a particle has reached the bottom of the duct, no remixing of turbulence can take place

d) $w \cdot \rho(x, d,) - D_{eff} \cdot \frac{\partial \rho}{\partial y}(x, d,) = 0$     no mass stream is entering through the lead

This two dimensional differential equation can be solved by separation

$$\rho = A\, e^{\alpha x} \cdot e^{\beta y} \qquad (2)$$

It can be shown that, especially by boundary condition d), no analytical solution of differential equation is possible. The solution can be written as a sum of following terms:

$$S = \frac{1}{\rho_o} \int_0^b \sum_{n=0}^{\infty} e^{\delta x} e^{\delta y} (A_1^n e^{\mu_n L} A_2 e^{-\mu_n L})(C_1^n \sin\nu_n b + C_2^n \cos\nu_n b)\, dy$$

Boundary conditions c) and d) lead to a transcendent equation for $\nu_n$

$$\text{tg}\, \nu_n = \frac{2\nu_n}{\delta - \frac{\nu_n^2}{\delta}} \qquad (3)$$

while condition a) and b) will give following result

$$\rho(x,y) = \sum_{n=0}^{\infty} \frac{8e^{\gamma x} e^{\delta y} \delta_o \delta^3 \left[ e^{\mu_n x} \frac{\gamma+\mu}{\gamma-\mu} e^{2\mu_n} e^{-\mu_n x} \right] \left[ \sin\nu_n y \; \frac{\nu_n}{\delta} \cos\nu_n y \right]}{\nu_n \left[ 1 - \frac{\gamma+\mu}{\gamma-\mu} e^{2\mu_n} \right] \left[ e^{2\delta}(5\delta^2 + \nu_n^2) - (\delta^2 + \nu_n^2) \right]} \quad (4)$$

If one is interested in the portion q of the dust which will pass the duct without being precipitated, then following equations should be used:

$$q = \frac{\int_0^b \rho(L,y)dy}{\int_0^b \rho(0,y)dy}$$

$$q = \frac{1}{\delta_o} \cdot \sum_o^n \frac{32 \cdot \mu_n \cdot \left[ \frac{w_T b}{2 D_{eff}} \right]^3 e^{-\left[ \mu_n - \frac{v L}{2 D_{eff}} - \frac{w_T b}{D_{eff}} \right]}}{\left[ \frac{w_T^2 b^2}{4 D_{eff}^2} + \nu_n^2 \right] \cdot \left[ \frac{v L}{2 D_{eff}} + \mu_n \right] \cdot \left[ e^{\frac{w_T b}{D_{eff}}} \left[ 5 \frac{w_T^2 b^2}{4 D_{eff}^2} + \nu_n^2 \right] - \left[ \frac{w_T^2 b^2}{e D_{eff}^-} + \nu_n^2 \right] \right]} \quad (5)$$

$$tg\; \nu_n = -\frac{2 \nu_n \cdot \frac{w b}{D_{eff}}}{\frac{w_T^2 b^2}{4 D_{eff}^2} - \nu_n^2} \quad (6)$$

$$\mu_n = \left[ \frac{v^2 L^2}{4 D_{eff}^2} + \frac{L^2}{b^2} \cdot \left( \nu_n^2 + \frac{w_T^2 b^2}{4 D_{eff}^-} \right) \right]^{\frac{1}{2}} \quad (7)$$

$$\gamma = \frac{v}{2 D_{eff}} \qquad \delta = \frac{w_T}{2 D_{eff}} \quad (8)$$

In most case the first term of equation 5 must be considered only, because further term are very small.

Equation (5) is important for precipitators, e.g. electric precipitators, where Coulomb forces are acting upon the particles and turbulence is intensified by the so called "electrical wind", which is caused by the electrical corona discharge.

The validity of formula (5) was tested in case of an electric precipitator, but the factor $D_{eff}$ had to be determined first by experiments.

EXPERIMENTAL WORKS

Since turbulent mixing is also effective in main flow direction the factor $D_{eff}$ can - assuming isotropic turbulence - be determined by evaluating mean time spectrums of a tracer passing through the precipitator.

Any device (duct, tube, reactor etc.) which is passed through by a stream showes a characteristics response at the exit to a change of a certain condition of the flow at the entrance of the device. One can evaluate the response on the basis of the valid differential equation. In this response equation a characteristic factor like $D_{eff}$ comes into picture. It is possible to vary this factor until the evaluated response equation agrees with the response, which was measured by experiments.

In the case of a one dimensional device, Levenspiel /2/ has evaluated the response equation for a so called "anyone shot" at the entrance of a device By comparison of the dispersion $\sigma$ of the input time function with the dispersion of the output time function it is possible to evaluate the factor $D_{eff}$.

$$\sigma^2 = 2D_{eff} \cdot \frac{1}{v \cdot L} - 2\left(\frac{D_{eff}}{v \cdot L}\right)^2 \cdot \left(1 - e^{-\frac{v \cdot L}{D_{eff}}}\right)$$

This equation is also used in this special case to evaluate the factor $D_{eff}$.

To produce the characteristic signal at the entrance of the precipitator, two methods were used.

1. A short dust impulse was produced and measured by an extinction equipment at the entrance and the exit of the precipitator.
2. The same procedure was used with a gas, $SF_6$ as a tracer. The gas concentration was measured with a special gas probe. The gas tracer method had to be used because with growing precipitation of the dust the signal at the exit becomes too small to be measured. Of course turbulent mixing resp. $D_{eff}$ was expected to be higher in the gas-phase than the mixing of the dust himself. The experiments with $SF_6$ as tracer were done under continuous doting of dust to include the influence of the dust to the turbulence of gas.

MEASURING EQUIPMENT

In fig. 2 the measuring equipment is shown. The lenght of the precipitator is 1,7 m, the height 1,1 m and the width 0,25 m.

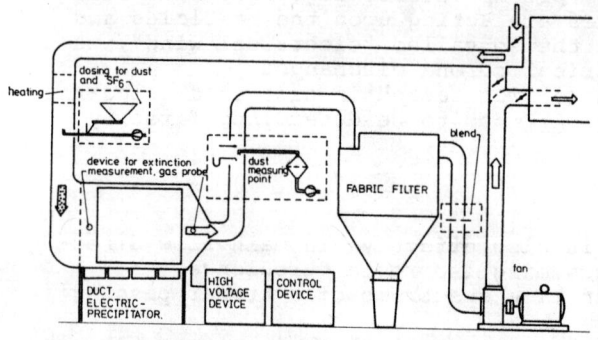

fig. 2 measurement equipment

The gas velocity can be varied up to 1,5 m/s. A device for dust-feeding, a device for extinction measurements and the gas probe at the entrance and the exit of the precipitator are installed. For electrical loading of the dust as well as for electric precipitation by Coulomb forces an equipment for high voltage DC with about 60 kV is installed.

The efficiency of precipitator was evaluated by measuring the dust input and output. The device for detecting the gas tracer was similar to that, used by Colditz (3). A probe, which consists of a flat anode and several steel-needles as a cathode is fed with high voltage DC (fig. 3). At several thousands volt a corona discharge is produced and a current can be measured. If an electro-negativ gas - in this case $SF_6$ - is passing the electrodes, the current decreases. It is possible, to calibrate the probe to a certain concentration of $SF_6$. Calibrating the probe was done with an equipment shown in fig. 4. It was possible to mix definat volume of $SF_6$ gas with air. This mixture was lead to the calibrating tank in which the gas probe was installed. A high voltage supply as well as instruments to record the corona current are installed. Fig. 5 showes some calibrating curves for $SF_6$ at several voltage levels.

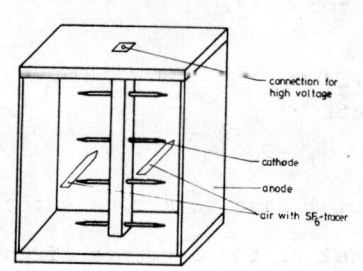

fig. 3 $SF_6$ gas probe

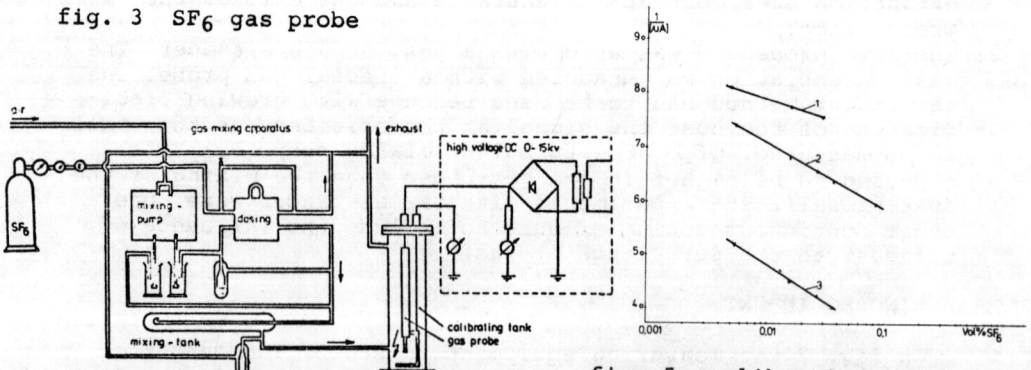

fig. 4 calibrating device for $SF_6$ gas probe

fig. 5 calibrating curves for $SF_6$ gas probe
1. negative corona 10,5 kV
2. positive corona 13,8 kV
3. positive corona 12,0 kV

# TURBULENT MIXING OF A DUSTSTREAM IN A DUCT

By this way it was possible to measure fast changes of gas concentration at the entrace and the exit of the precipitator. Fig. 6 showes an experiment with dust as tracer. In fig. 7 the similar experiment was carried out with $SF_6$. By comparision of the dispersion of the exit and entrance function, $D_{eff}$ is evaluated.

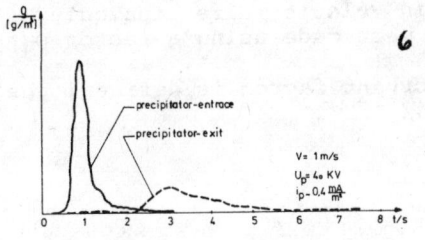

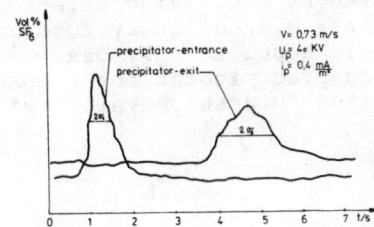

fig. 6  mean time spectrum of dust tracer in the precipitator

fig. 7  mean time spectrum of $SF_6$ gas in the precipitator

The factor $D_{eff}$ was measured mainly as function of the gas velocity and the corona current in the precipitator.

## RESULTS

Some results from $D_{eff}$ measurements are shown in fig. 8 and fig. 9. The factor $D_{eff}$ measured by $SF_6$ gas in higher then $D_{eff}$ measured by dust tracer method. This means that dust can not follow the turbulent motion of the gas at all.

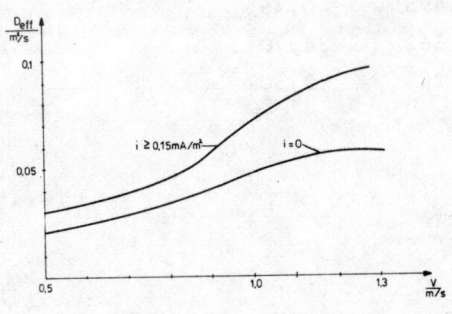

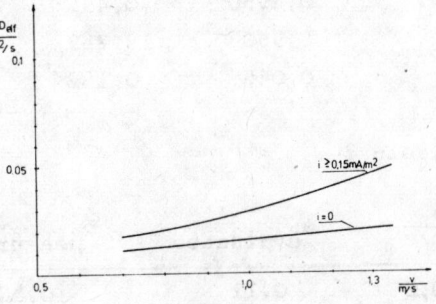

fig. 8  $D_{eff}$ as function of gas velocity and corona current measured by $SF_6$ tracer

fig. 9  $D_{eff}$ as function of gas velocity and corona current measured by dust tracer

The measured factor $D_{eff}$ and the formula (5) were applied to the problem of electric dust precipitation. Under the influence of Coulomb forces the dust particles are carried to the walls of the precipitator with a velocity $w_T$. This migration is disturbed by turbulent motion. This turbulence is intensified by the so called "electric wind" which is blowing transverse to the main velocity. Therefore turbulence will be an anisotropic. So the effectiv acting factor $D_{eff}$ will be higher than the factor which was measured by the mixing effect in main velocity direction only. Therefore evaluation by formula (5) were made using a factor $D_{eff}$ which is equal to $D_{eff}$ gas.

In precipitator theorie an important factor is defined, the so called "migration velocity" $w_D$ /4/

$$w_D = \frac{v \cdot d}{L} \ln q$$

In table 1 the factors which are used for evaluating the efficiency of the model precipitator under certain conditions are given. Also the necessary factors $\rho, \delta, \mu, \nu$ (see formula (6) - (8)) are given. In table 2 the evaluated efficiency resp. the evaluated w-factor are shown and compared with the result of measurements. By evaluating the efficiency with formula (5) it can be shown that the dependence of the factor $w_D$ from velocity v of the gasstream is caused by turbulent mixing.

Table 1
precipitator conditions: $D_{eff}=0.075 \frac{m^2}{s}$ $w_T=0.15 \frac{m}{s}$ L=1,7 m b=0,125 m

| $\frac{v}{m/s}$ | 0,5 | 1,0 | 1,5 | 3 |
|---|---|---|---|---|
| $\gamma$ | 5,666 | 11,333 | 17,0 | 34,0 |
| $\delta$ | 0,125 | 0,125 | 0,125 | 0,125 |
| $\nu_1$ | 0,4950 | 0,495 | 0,495 | 0,495 |
| $\mu_1$ | 8,960 | 13.289 | 18,362 | 34,701 |
| q | 0,045 | 0,152 | 0,265 | 0,501 |

Table 2

| $\frac{v}{m/s}$ | $w_D$ evaluated | $w_D$ measured |
|---|---|---|
| 0,5 | 0,11 | 0,10 |
| 1,0 | 0,138 | 0,135 |
| 1,5 | 0,15 | 0,16 |
| 3 | 0,156 | - |

CONCLUSION

The efficiency of a precipitator under the influence of turbulent was studied. Comparison of measurements with evaluation by a model which was developed, shows that only by considering the influence of turbulence useful results can be obtained.

REFERENCES

/1/ Launder, B., and Spalding, D.B.: Mathematical Models of Turbulence.
Academic Press, London and New York

/2/ Levenspiel, O.: Chemical Reaction Engeneering.
John Wiley and Sons, Inc. New York-London-Sidney

/3/ Colditz, M.: Ein neues Verfahren zur Bestimmung von Verweilzeitspektren in schnell durchströmten Gefäßen.
Chem. Ing. Tech. 44 (1972) Nr. 19, S. 1116-1120

/4/ Deutsch, W.: Ann. der Physik, 68, 335 (1922)

# TURBULENT NATURAL CONVECTION DIFFUSION FLAMES

**LAWRENCE A. KENNEDY**
State University of New York, Buffalo, New York, U.S.A.

ABSTRACT

A boundary layer analysis is performed for the motion of a buoyancy dominated turbulent flow adjacent to a vertical fuel surface and burning in a quiescent atmosphere. No radiation, infinite gas phase reaction rates and unit Lewis numbers are assumed. The flow is numerically examined within the Shvab-Zeldovich approximation and the turbulence is considered to be characterized by the turbulent kinetic energy, dissipation rate and the mean square fluctuation of a scalar.

NOMENCLATURE

| | |
|---|---|
| $a_g$ | gravitational constant |
| B | mass transfer driving force |
| $c_p$ | specific heat |
| C's | model constants |
| $F_1, F_2, F_3, F_\mu$ | wall functions |
| h | enthalpy |
| $h_i^{(o)}$ | enthalpy of formation |
| K | kinetic energy of turbulence (T.K.E.) |
| L | heat of vaporization |
| M | molecular weight |
| Q | net heat release by $\nu'_F$ moles of fuel |
| g | mean square temperature (concentration) fluctuation |
| T | mean temperature |

| | |
|---|---|
| $t$ | fluctuating temperature |
| $U$ | mean vertical velocity |
| $u$ | fluctuating vertical velocity |
| $V$ | mean horizontal velocity |
| $Y_i$ | mass fraction of species i |
| $\beta_1, \beta_2$ | Shvab-Zeldovich variables |
| $\gamma$ | coefficient of thermal expansion |
| $\varepsilon$ | dissipation rate of T.K.E. |
| $\mu$ | molecular viscosity |
| $\mu_t$ | turbulent viscosity |
| $\rho$ | density |
| $\sigma$ | Prandtl and Schmidt numbers |
| $\nu$ | stoichiometric coefficients |
| $\dot{\omega}_i$ | mass source per unit volume of species i |

Subscripts

| | |
|---|---|
| F | fuel |
| o | oxidizer |
| i | $i^{th}$ species |
| t | turbulent |
| $\infty$ | ambient conditions |
| w | condition at wall |

INTRODUCTION

This paper is concerned with developing a flow model to examine the factors influencing the turbulent burning of a fuel surface. The model considers a buoyancy dominated turbulent diffusion flame adjacent to a vertical fuel surface.

The laminar free convection burning of a vertical surface has been studied by a number of authors. Spalding [1] obtained a Pohlhausen solution to this problem. Kosdon, Williams and Berman [2] developed a similarity theory for the combustion of vertical cellulose cylinders. Kim, DeRis and Kroesser [3] formulated this problem similar to that of reference 2 and examined the influence of chemical parameters on the predicted results.

# TURBULENT NATURAL CONVECTION DIFFUSION FLAMES

The investigation of the corresponding turbulent problem has been limited to experimental studies, in particular the recent work at Factory Mutual Research Laboratory (see Ref. 4 and 5). It is important that models be developed to examine the turbulent burning problem in order to access the importance of various fluid mechanical and chemical parameters.

Related turbulent diffusion flame models have been reported by Taminini [6] who examined a buoyancy dominated fuel jet using the K-ε-g turbulence model. Kennedy and Plumb [7] discussed the modeling of a buoyant fuel jet adjacent to a solid surface using a similar approach. The present work considers the more general burning problem of a vertical fuel surface wherein the surface mass transfer is proportional to the energy feedback from the flame.

## FORMULATION

The conservation equations describing the time averaged values for a compressible buoyancy dominated boundary layer flow adjacent to a vertical burning surface are

Continuity: 
$$\frac{\partial(\rho U)}{\partial x} + \frac{\partial(\rho V)}{\partial y} = 0 \tag{1}$$

Momentum:
$$\rho U \frac{\partial U}{\partial x} + \rho V \frac{\partial U}{\partial y} = \frac{\partial}{\partial y}\left[(\mu_{eff}) \frac{\partial U}{\partial y}\right] + a_g (\rho_\infty - \rho) \tag{2}$$

Energy:
$$\rho U \frac{\partial h}{\partial x} + \rho V \frac{\partial h}{\partial y} = \frac{\partial}{\partial y}\left[(\frac{\mu_{eff}}{\sigma}) \frac{\partial h}{\partial y}\right] - \Sigma \dot{\omega}_i h_i^{(o)} \tag{3}$$

Species:
$$\rho U \frac{\partial Y_i}{\partial x} + \rho V \frac{\partial Y_i}{\partial y} = \frac{\partial}{\partial y}\left[(\frac{\mu_{eff}}{\sigma}) \frac{\partial Y_i}{\partial y}\right] + \dot{\omega}_i \tag{4}$$

These differential equations are closed by the following equation:

$$\mu_{eff} = \mu + \mu_t$$
$$= \mu + C_\mu \rho k^2/t \tag{5}$$

where $C_\mu$ is a constant. The length scale is taken to be the

$$\ell = k^{3/2}/t \tag{6}$$

Chemistry: $\nu_F'(F) + \nu_o'(O) \rightarrow \nu_p''(\text{Products}) + Q$ \hfill (7)

Here the $\nu_i$'s are the stoichiometric coefficients and Q is the net heat generated by $\nu_F'$ moles of fuel. Equation (5) implies

$$\Phi_i = (\nu_i'' - \nu_i') M_i \dot{\omega} \tag{8}$$

in which $\dot{\omega}$ is the molar rate.

In the forgoing equations the binary diffusion coefficients were assumed equal; the specific heats and their ratios were assumed constant and the Lewis number, $\rho C_p D/\lambda$, taken to be unity implying that the Schmidt and Prandtl numbers, $\sigma$, are equal. The influence of the hydrostatic pressure upon the density has been neglected and ideal gas laws have been utilized together with an assumption of uniform molecular weights. No radiation has been included in this model eventhough Markstein [8] has shown it to be a non-negligible effect. However, its effect could partially be taken into account by suitably adjusting the heats of vaporization and combustion. The flame sheet approximation is implicitly assumed.

By defining the Shvab-Zeldovich variables [9] the source terms $\Sigma \dot{\omega}_i h_i^{(o)}$ and $\dot{\omega}_i$ can be eliminated from the energy and species equations, thus

$$\beta_1 = \frac{h}{L} + \frac{(Y_o - Y_{o_\infty}) Q}{M_o \nu_o' L} \tag{9}$$

$$\beta_2 = \frac{h}{L} + \frac{Y_F Q}{M_F \nu_F' L} \tag{10}$$

where L is the effective heat of vaporization of the fuel. Equations (3) and (4) then become

$$\rho U \frac{\partial \beta_1}{\partial x} + \rho V \frac{\partial \beta_1}{\partial y} = \frac{\partial}{\partial y} \left[ \left(\frac{\mu_{eff}}{\sigma}\right) \frac{\partial \beta_1}{\partial y} \right] \tag{11}$$

$$\rho U \frac{\partial \beta_2}{\partial x} + \rho V \frac{\partial \beta_2}{\partial y} = \frac{\partial}{\partial y} \left[ \left(\frac{\mu_{eff}}{\sigma}\right) \frac{\partial \beta_2}{\partial y} \right] \tag{12}$$

which will yield the distributions for h, $Y_F$ and $Y_o$.

The boundary conditions of interest are that the vertical velocity, U, must vanish at both the wall and infinity (ambient conditions)

$$U(x,o) = U(x,\infty) = 0 \tag{13}$$

At infinity the temperature and mass fraction of the oxidizer are known constants, while the mass fraction of the fuel vanishes. This implies that

$$\beta_1(x,\infty) = \beta_2(x,\infty) = 0 \tag{14}$$

At the wall, the temperature is a constant and the oxidizer mass fraction vanishes, hence $\beta_1(x,0)$ is a constant

$$\beta_1(x,o) = \frac{hw}{L} - \frac{Y_{o_1\infty} Q}{M_o \nu'_o L} = -B \qquad (15)$$

Here the first term is the ratio of the specific enthalpy of the gas at the fuel surface to the heat of vaporization of the fuel and the second is the ratio of chemical heat release per unit mass of oxidizer to the heat of vaporization of the fuel. The parameter B is the mass transfer driving force.

The surface mass transfer rate $\dot{m}$ is related to the surface heat transfer by

$$\dot{m} = \rho_w v_w = \left.\frac{k}{C_p}\right)_w \frac{1}{L} \left.\frac{\partial h}{\partial y}\right) = \left.\frac{\mu}{\sigma}\right)_w \left.\frac{\partial \beta_1}{\partial y}\right) \qquad (16)$$

The conservation of mass for fuel at the wall states

$$\dot{m} = \dot{m} Y_{F,w} - \left.(\rho D)\right)_w \left.\frac{\partial Y_F}{\partial y}\right)_w \qquad (17)$$

Assuming laminar and turbulent unit Lewis numbers together with equation (16) and the definitions of $\beta_1$ and $\beta_2$ yields

$$\left.\frac{\partial \beta_2}{\partial y}\right)_w = \left[1 - \frac{Q}{M_F \nu'_F L} + \beta_2(x,o) - \frac{h_w}{L}\right] \left.\frac{\partial \beta_1}{\partial y}\right)_w \qquad (18)$$

## TURBULENCE MODEL

Numerous authors have proposed closure models for turbulent flow in an attempt to accurately predict the turbulent shear stresses. The model utilized in the present work is the K-$\epsilon$-g model [10]. This approach writes transport equations for the turbulent kinetic energy, dissipation rate and the mean square fluctuation of a scalar and uses a length scale given by equation (8). Including the contribution of buoyancy to these quantities, the three governing turbulent equations for isotropic conditions are

$$\rho \frac{U \partial K}{\partial x} + \rho \frac{V \partial K}{\partial y} = \frac{\partial}{\partial y}\left[\frac{\mu_t}{\sigma_K} \frac{\partial K}{\partial y}\right] + \mu_T \left(\frac{\partial u}{\partial y}\right)^2 - \rho\epsilon + \rho a_g \beta \overline{ut} \qquad (19)$$

$$\rho \frac{U \partial \epsilon}{\partial x} + \rho \frac{V \partial \epsilon}{\partial y} = \frac{\partial}{\partial y}\left[\frac{\mu_t}{\sigma_\epsilon} \frac{\partial \epsilon}{\partial y}\right] + C_1 \frac{\epsilon}{K} \mu_t \left(\frac{\partial U}{\partial y}\right)^2 - C_2 \frac{\rho \epsilon^2}{K}$$

$$+ C_3 \rho a_g \gamma \frac{\epsilon}{K} (\overline{ut}) \qquad (20)$$

$$\rho \frac{U \partial g}{\partial x} + \rho \frac{V \partial g}{\partial y} = \frac{\partial}{\partial y}\left[\frac{\mu_t}{\sigma_g} \frac{\partial g}{\partial y}\right] + C_{g1} \mu_t \left(\frac{\partial \beta_i}{\partial y}\right)^2 - C_{g2} \rho \frac{\epsilon}{K} g \qquad (21)$$

The velocity-temperature fluctuation correlation is taken to be

$$\overline{ut} = C_4 (gK)^{\frac{1}{2}} \qquad (22)$$

Equation (15)-(18) together with equations (1), (2), (7), (11) and (12) describe the turbulent buoyancy driven flow. The model constants are given in Table 1.

TABLE I. MODEL CONSTANTS FOR TURBULENCE MODEL

| $C_\mu$ | $C_1$ | $C_2$ | $C_3$ | $C_4$ | $C_{g1}$ | $C_{g2}$ | $\sigma_k$ | $\sigma_\epsilon$ | $\sigma_g$ |
|---|---|---|---|---|---|---|---|---|---|
| .09 | 1.44 | 1.92 | 1.44 | .5 | 2.8 | 1.7 | 1.0 | 1.3 | .9 |

To account for the nonisotropic behavior of the flow near the surface wall functions and wall terms [11] were added to the K-ε-5 equations yielding

$$\rho\frac{U\partial K}{\partial x} + \rho\frac{V\partial K}{\partial y} = \frac{\partial}{\partial y}\left[\left(\mu + \frac{\mu_t}{\sigma_K}\right)\frac{\partial K}{\partial y}\right] + \mu_T\left(\frac{\partial U}{\partial y}\right)^2 - \rho\epsilon + \rho a_g \gamma\, C_4 (gK)^{1/2}$$
$$- 2\mu\left(\frac{\partial K^{1/2}}{\partial y}\right)^2 \tag{23}$$

$$\rho\frac{U\partial \epsilon}{\partial x} + \rho\frac{V\partial \epsilon}{\partial y} = \frac{\partial}{\partial y}\left[\left(\mu + \frac{\mu_t}{\sigma_\epsilon}\right)\frac{\partial \epsilon}{\partial y}\right] + C_1 F_1 \frac{\epsilon}{K}\mu_T\left(\frac{\partial U}{\partial y}\right)^2 - C_2 F_2 \rho\frac{\epsilon^2}{K}$$
$$+ C_3 F_3 C_4 \rho a_g \gamma\, \epsilon \left(\frac{g}{K}\right)^{1/2} \tag{24}$$

$$\rho\frac{U\partial g}{\partial x} + \rho\frac{V\partial g}{\partial y} = \frac{\partial}{\partial y}\left[\left(\frac{\mu}{\sigma} + \frac{\mu_t}{\sigma_q}\right)\frac{\partial g}{\partial y}\right] + C_{g1}\mu_T\left(\frac{\partial \beta_i}{\partial y}\right)^2 - C_{g2}\rho\frac{\epsilon}{K}g$$
$$- 2k\left(\frac{\partial g^{1/2}}{\partial y}\right)^2 \tag{25}$$

$$\mu_t = C_\mu F_\mu \frac{\rho K^2}{\epsilon} \tag{26}$$

The form of wall functions $F_1$, $F_2$ and $F_\mu$ used in this study are those of Jones and Launder [11] and are given in table II.

TABLE II. WALL FUNCTIONS

$F_1 = 1.0$ $\qquad\qquad F_\mu = \exp[-2.5/(1 + Re_t/50)]$

$F_2 = 1.0 - .3 \exp(-Re_t^2)$ $\qquad F_3 = 1.0$

Here the turbulent Reynolds number is defined $Re_t = \rho K^2/\mu\epsilon$. The function $F_3$ was taken to be unity. It should be noted that future work may reveal that these are not the optimum values.

For non-reacting flow the work of Blom et al discussed in Ref. 12 indicates that $\sigma_T$ should be higher inside the velocity maximum relative to its value outside. In the related heat transfer problem the $\sigma_T$ which gave the best agreement with heat transfer data [13] was

$$\sigma_T = 2.5 - 2.0 \, y/\delta$$

This value was used in the present calculations.

## COMBUSTION MODEL

The flame sheet approximation is invoked so that fuel and oxidant will combine to yield a single product whenever they simultaneously exist at a point. Since no fuel or oxidizer then exists, this location is determined by the relation $\beta_2 - \beta_1 = Y_{O,\infty} Q/(M_O \nu'_O L)$. However, in order to obtain the average concentrations of $Y_F$ and $Y_O$, account should be taken of their fluctuations. In the case of fires these may be large compared to their mean values. Since we only have the time-averaged and the mean square fluctuations of $\beta_i$, from the governing equations an additional constraint on the variation of the instantaneous value $\beta_i$ with time must be assumed.

A variety of different forms for this variation have been used by different authors e.g., ref. 14-17. For the present study the "clipped" Gaussian distribution discussed by Elghobashi and Pun [17] was adopted. This approach may be summarized as follows. The probability density function $p(\tilde{\beta}_i)$ is given by

$$p(\tilde{\beta}_i) = \exp[-(\tilde{\beta}_i - \bar{\beta})^2/2\bar{\sigma}^2][H(\tilde{\beta}_i - 0) - H(\tilde{\beta}_i - 1)]/\bar{\sigma}(2\pi)^{\frac{1}{2}} + 2A_1 \delta(\tilde{\beta}_i - 0) + 2A_2 \delta(\tilde{\beta}_i - 1) \quad (27)$$

Here $H(\xi)$ is the unit step function and $\delta(\xi)$ denotes the Dirac delta function. $\bar{\beta}$ and $\bar{\sigma}$ are respectively the mean and standard deviation for the complete Gaussian distribution which forms a part of $p(\tilde{\beta}_i)$. $A_1$ and $A_2$ are given by

$$A_1 = \int_{-\infty}^{0} \exp[-(\tilde{\beta}_i - \bar{\beta})^2/2\bar{\sigma}^2] d\tilde{\beta}_i / \bar{\sigma}(2\pi)^{\frac{1}{2}}$$

$$A_2 = \int_{1}^{\infty} \exp[-(\tilde{\beta}_i - \bar{\beta})^2/2\bar{\sigma}^2] d\tilde{\beta}_i / \bar{\sigma}(2\pi)^{\frac{1}{2}}$$

Taking the first moment of $p(\tilde{\beta})$ about $\tilde{\beta} = 0$ and the second about the mean value of the are $p(\tilde{\beta})$ gives

$$\beta_i = A_2 + \int_0^1 \tilde{\beta} \exp[-(\tilde{\beta} - \bar{\beta})^2/2\bar{\sigma}^2] d\tilde{\beta}/\bar{\sigma}(2\pi)^{\frac{1}{2}} \quad (28a)$$

$$g = A_2 - \beta_1^2 + \int_0^1 \tilde{\beta}_1^2 \exp[-(\tilde{\beta}_i - \bar{\beta})^2/2\bar{\sigma}^2] d\tilde{\beta}/\bar{\sigma}(2\pi)^{\frac{1}{2}} \quad (28b)$$

Therefore when both $\beta_i$ and $g$ are known from the initial solution of the governing equations, the probability density function, $p(\tilde{\beta}_i)$ may be obtained from Eq. (28a, b) and Eq. (27). Then the mean value $\varphi$ of a fluctuating quantity $\varphi(\tilde{\beta}_i)$ can be evaluated from

$$\varphi = \int_0^1 \tilde{\varphi}(\tilde{\beta}_i) p(\tilde{\beta}_i) d\tilde{\beta}_i$$

## DISCUSSION AND RESULTS

The Patankar-Spalding procedure was used to solve the governing set of equations. The advantage of this technique is that it allows the grid to expand with the boundary layer and the resulting tridiagonal system is rapidly inverted.

The calculations reported were initiated in the laminar region and a small amount of turbulent kinetic energy introduced at a point which yielded a numerical transition coincident with the experimental transition to turbulence. The model was also checked by calculating the heat transfer to a vertical wall for a non-reacting buoyant flow which agreed with experimental data [13].

The wall terms and wall functions act to damp the turbulent kinetic energy in the viscous sublayer where the molecular viscosity dominates the turbulent transport. In inert forced flows, the importance of these wall terms decrease as one moves towards the free stream values. However, in buoyant flames, the maximum velocity occurs within the boundary layer and these wall terms will also influence the damping at the outer edge of the boundary layer. Thus in order to overcome this difficulty the wall functions and wall terms were only applied out to the velocity maximum.

The calculations utilized 80 cross stream grids which were not evenly distributed across the boundary layer. Roughly half of the grid points were inside of the velocity maximum where steep gradients of all of the flow variables occur. In addition it was essential to maintain several grids within the viscous sublayer in order to obtain satisfactory results. Forward step sizes of 4 or 5% of the total boundary layer thickness proved to be effective.

The entrainment rate at the outer edge of the boundary layer was calculated from a momentum balance at this location. Typical computing times required to produce a solution was about 10 minutes on a CDC 6400.

In the present study, the burning surface was taken to be cellulose. The pyrolysis of this material is complex but formaldehyde represents an average pyrolysis product [2]. Thus the gas phase combustion reaction was taken as

$$H_2CO + O_2 \rightarrow H_2O + CO_2$$

with a heat release of 4.18 kcal/gm of formaldehyde.

Figure 1 shows the vertical velocity and mass fraction profiles at a height of 2 meters. Also shown are the mean velocity profiles in a non-reacting turbulent boundary layer for mass addition through a heated surface and for an impermeable heated surface [18].

Figure 2 shows the mean temperature profiles at a height of 2 meters for the case of a burning surface and for a non-reacting flow with and without mass addition. The maximum velocity is seen to occur inside the maximum temperature (flame front). This has also been observed for the case of laminar flames [3].

Figure 3 shows a typical predicted mean temperature profile T, and the RMS temperature fluctuations, $< t^2 >^{1/2}$. It is interesting to note that the maximum value of the fluctuations is about 1200°C while the mean temperature peak is about 2000°C.

Figure 4 gives the predicted burning rate at various distances along the fuel surface. The results gives a fluctuating rate which could be due to either numerical problems or to physical fluctuations in the turbulent flows. The burning rate is very sensitive to the concentration profiles and both (or either) numerical or physical anomalies have a strong influence. Regardless it is seen that the burning rate increases approximately as $x^{0.4}$.

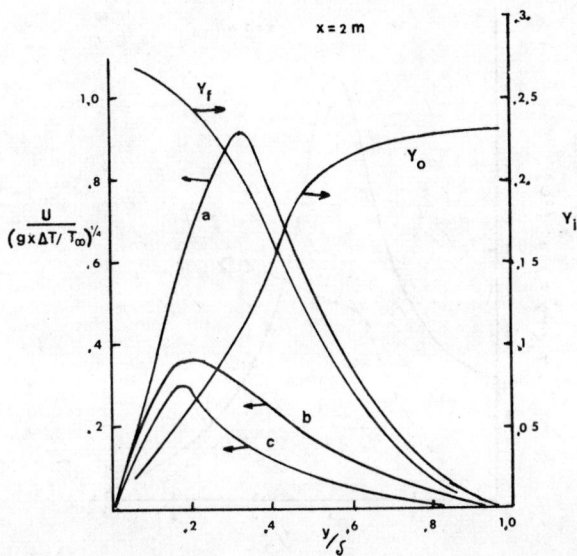

Fig. 1 Mean vertical velocity and mass fraction profiles adjacent to a vertical fuel surface. Curve (a): velocity profile with combustion, $\dot{m}_w = -5.1 \times 10^{-4}$ g/cm$^2$/sec, $T_w = 600°C$. Curve (b): velocity profile in a non-reacting flow with mass addition, $\dot{m}_w = -5.1 \times 10^{-4}$ g/cm$^2$/sec, $T_w = 600°C$. Curve (c): velocity profile over heated impermeable surface $T_w = 600°C$.

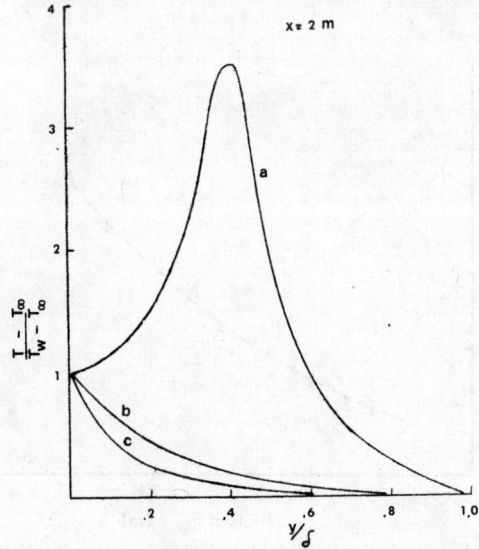

Fig. 2 Mean temperature profiles adjacent to a vertical fuel surface. Curve (a): temperature profile with combustion and mass addition $\dot{m}_w = -5.1 \times 10^{-4}$ g/cm$^2$/sec, $T_w = 600°C$. Curve (b) temperature profile in a non-reacting flow with mass addition $\dot{m}_w = -5.1 \times 10^{-4}$ g/cm$^2$/sec, $T_w = 600°C$. Curve (c) temperature profile over heated impermeable surface $T_w = 600°C$.

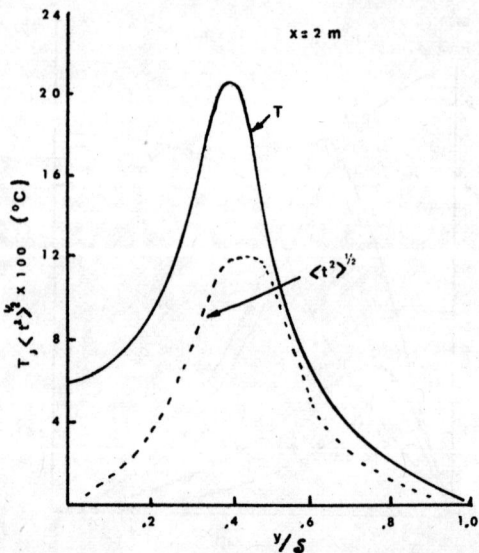

Fig. 3 Mean temperature T and rms temperature fluctuations $<t^2>^{1/2}$ profiles for reacting flow adjacent to vertical fuel surface.

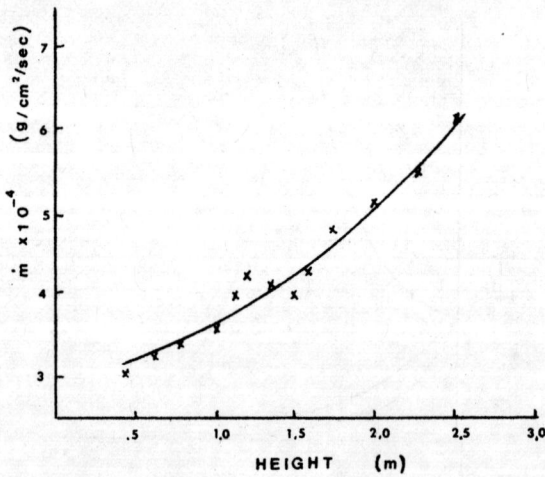

Fig. 4 Burning rate variation with distance along the fuel surface.

## CONCLUSIONS

It has been shown that buoyant diffusion flames can be modelled using modifications of the techniques developed for forced flows. In particular the influence of buoyancy on turbulence model can easily be included. The fluctuating scalar quantities can be predicted and their effects on the mean flow be taken into account.

Further studies are required to develop an approach to realistically account for radiative transfer.

## REFERENCES

1. Spalding, D. B., "Mass Transfer in Laminar Flow", Pro. Roy. Soc. A221, 78 (1954).
2. Kosdon, F. J.; Williams, F. A.; Buman, C., "Combustion of Vertical Cellulosica Cylinders in Air", 12th Symposium (International) on Combustion, The Combustion Institute (1969).
3. Kim, J. S.; DeRis, J.; Kroesser, F. W., "Laminar Free Convection Burning of Fuel Surfaces", 13th Symposium (International) on Combustion, The Combustion Institute (1971).
4. DeRis, J. and Orloff, L., "The Role of Buoyancy Direction and Radiation in Turbulent Diffusion Flames on Surfaces, 15th Symposium (International) on Combustion, The Combustion Institute (1974).
5. Orloff, L., Modak, A. T. and Alpert, R. L., "The Effect of Radiative Heat Transfer from Flames on Burning Rate of Large-Scale Vertical Plastic Surfaces", Presented at the Eastern Section of the Combustion Institute, Upton Long Island (1975).
6. Taminini, F., "On the Numerical Prediction of Turbulent Diffusion Flames", paper presented at the Central and Western State Section of the Combustion Institute, San Antonio, Texas, April 1975.
7. Kennedy, L. A. and Plumb, O. A., "Prediction of Buoyancy Controlled Turbulent Wall Diffusion Flames", to be presented at the 16th Symposium (International) on Combustion (1976).
8. Markestein, G. H., "Radiative Energy, Transfer from Gaseous Diffusion Flames", 15th Symposium (International) on Combustion, The Combustion Institute, (1974).
9. Williams, F. A., Combustion Theory, Addison Wesley Press, Reading, Mass. 1965.
10. Launder, B. E. and Spalding, D. B., "Mathematical Models of Turbulence", Academic Press, 1972.
11. Jones, W. P. and Launder, B. E., "The Prediction of Laminarization with a 2 Equation Model of Turbulence", Intern. J. Heat Mass Transfer, 15, 301 (1972).
12. Rodi, W., "A Note on the Empirical Constants in the Kolmogorov-Prandtl Eddy Viscosity Expression", Trans. Am. Soc. Mech. Engr., Series I, 97, 386 (1975).
13. Plumb, O. A., "An Experimental and Numerical Examination of Buoyancy Driven Wall Boundary Layers", Ph.D. Dissertation, State University of New York at Buffalo (1976).
14. Hawthorne, W. R., Weddel, D. S. and Hottel, H. C., "Mixing and Combustion in Turbulent Gas Jets", Third Symposium (International) on Combustion, p. 266, Williams and Wilkins (1949).
15. Richardson, J. M., Howard, Jr., H. C. and Smith Jr., R. W., "The Relation Between Sampling Tube Measurements and Concentration Fluctuations in a Turbulent Gas Jet", Fourth Symposium (International) on Combustion, p. 814, Williams and Wilkins (1953).

16. Gosman, A. D. and Lockwood, F. C., "Prediction of the Influence of Turbulent Fluctuations of Flow and Heat Transfer in Furnaces", Imperial College Heat Transfer Report HTS/73/52.
17. Elghobashi, S. E. and Pun, W. M., "A Theoretical and Experimental Study of Turbulent Diffusion Flames in Cylindrical Furnaces", Fifteenth Symposium (International) on Combustion, The Combustion Institute, p. 1353 (1974).
18. Plumb, O. A. and Kennedy, L. A., "Application of a K-$\epsilon$ Turbulence Model to Natural Convection from a Vertical Isothermal Surface", to be presented at the 1976 National Heat Transfer Conference, St. Louis, Missouri.

# EXPERIMENTAL STUDY OF HEAT AND MASS TRANSFER IN A VERTICAL AIR LAYER WITH BLOWING FOREIGN GASES

G. V. TSIKLAURI, V. G. PUZACH, and L. M. SOLOVIEV
*Institute of High Temperatures, Moscow, USSR*

## ABSTRACT

The experimental study of the natural convection in vertical air cell with injection of admixture of the foreign gases. Grashof number changes between $8 \cdot 10^4 \div 10^6$. Three gases were used during experiments: He, Ne and $CO_2$. The experiment were made with a constant geometric configuration of the cell.

## NOMENCLATURE

| | |
|---|---|
| c | concentration |
| S | displacement of interferometric bands |
| $K = \frac{n-1}{\rho}$ | Gladson-Dal constant |
| g | acceleration of gravity |
| h | layer height |
| L | layer width |
| n | refraction coefficient |
| p | pressure |
| T | temperature |
| $\lambda$ | wavelength |
| $\rho$ | density |
| R | gas constant |
| $Gr = \frac{g \beta_T (T_w - T_0) \delta^3}{\nu^2}$ | Grashof thermal criterion |
| $Gr_D = \frac{g \beta_D (m_w - m_0) \delta^3}{\nu^2}$ | Grashof diffusion criterion |
| $Ra = Gr \, Pr$ | Rayleigh number |
| $Le = \frac{a}{D}$ | Lewis number |

Subscripts

w    wall

a    air

g    gas

INTRODUCTION

    The study of the mechanism of free-convective motion in closed layers and spaces is of interest for cleaning, ventilation and thermal insulation systems. In such systems the heating of air and mixing it with foreign gases can generally result in stratification of the considered volume in a vertical direction. The available publications on this topic deal primarily with the questions associated with gas pollution of large air volumes /1/.

    The paper presents the results of the experimental study of temperature and concentration fields in a naturally convective gas flow in a vertical air layer formed by one heated wall and one cooled wall when blowing in foreign gases. The experiments were conducted at a constant geometric configuration of the cell.

EXPERIMENTS

    A schematic diagram of the experimental facility is shown in Fig.1. The vertical boundaries of the layer sized 600 x 100 x 200 mm are formed by adiabatic walls (one wall is heated by incandescent lamps, the other is cooled by cold water). The lower horizontal boundary is made of porous material through which various gases were supplied. In the process of experiments the temperature field in the layer was measured by means of prob with nine copper-constantan thermocouples located at five levels along the layer height and also using a great number of thermocouples built-in the wall.

    The concentration field of the blown gases was determined by the interferometric method at the same levels at which the temperature levels were measured. The Mach-Zehnder interferometer was used for this purpose. The preliminary experiments at the experimental facility have shown that optical glasses do not disturb the intergerogram of

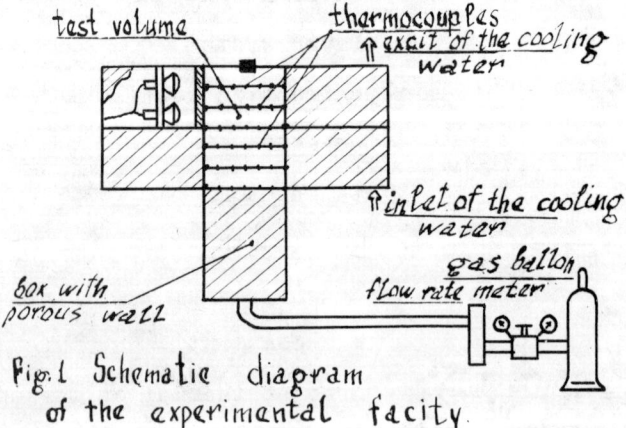

Fig.1 Schematic diagram of the experimental facility.

the object to be investigated. Owing to the fact that the interferograms carry the information of the change in the density of the medium through which passes the working beam, it is important that the field is two-dimensional in the plane normal to the beam axis with negligible changes along the beam. In the process of adjustment of the equipment it has been found that the temperature variation across the air layer is small.

The concentration of admixture in the air was found by the formula /2/

$$C = \frac{\frac{\lambda s t}{\rho \rho} + \frac{K_a}{R_a}\left(\frac{T}{T_\infty} - 1\right)}{\frac{K_g}{R_g} - \frac{K_a}{R_a}} \qquad (1)$$

The fringe spacing inclination in the photographic films of the interferogram was measured using a microscope; the interferograms was taken before injecting of the gas and after that upon obtaining the stationary operating conditions. It should be noted that with the gas concentrations up to 10% the temperature fields in the layer volume did not change until the stationary conditions were reached, i.e. they were maintained the same as prior to blowing the gas. Fig.2 illustrates typical interferograms taken before and after injection the gas.

## THE RESULTS OF THE INVESTIGATIONS

Shown in Fig.3 are temperature profiles across the cells in the upper ($\bar{h}$ = 4.62) and lower ($\bar{h}$ = 1.05) cross sections of the model for the range of change in the Grashof number $8 \cdot 10^4 <$ Gr $< 10^6$. As shown in the temperature curves at low Grashof number, the temperature changes are practically linear, i.e. here we have conditions analogous "thermal conductivity" operating conditions. Of course, a buoyant convective flow takes place near the hot wall and a descending flow occurs near the cold wall; however, this process is scarcely noticeable. As the Grashof number increases, the positive temperature gradient near the hot wall and the negative temperature gradient near the cold wall rise up; there is no heat transfer in the medium part of the air layer, i.e. $\partial T/\partial x$ = 0. Further increase in the Grashof number results in formation of a stabilized flow of the "boundary layer" type with low temperature gradients in the medium part of the layer and high temperature gradients near the wall.

In the upper part of the layer (Fig.3b), at numbers $Gr > 7 \cdot 10^5$ the phenomenon of "boundary layer inversion" takes place when the temperature of the gas layers in the medium part of the layer near the hot wall is lower than that near the cold wall. In this case the picture of the flow is very complex: a secondary naturaly convective cirvulating stream is generated in the "core" of the flow. The direction of this stream is opposite to that of the basic circulation in the cell.

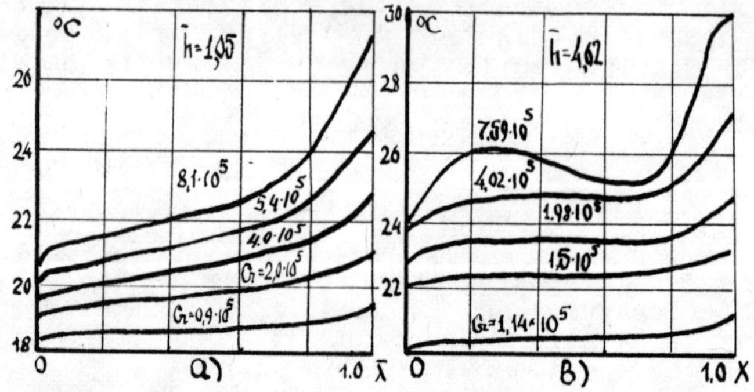

Fig.3. Temperature profiles

The profiles of the concentration of the gases in air are shown in Fig.4. Three gases were used during the experiments: He, Ne and $CO_2$. The analysis of the curves in Fig.4 shows that the concentration of the light gas He follows the isotherms, namely: near the hot wall the concentration of He is lower than near the cold wall. the maximum concentration of He being in the lower part of the layer. On the contrary, for Ne and $CO_2$ the highest value of concentration is observed near the hot wall. The concentration of heavy gases increases along the height of the layer.

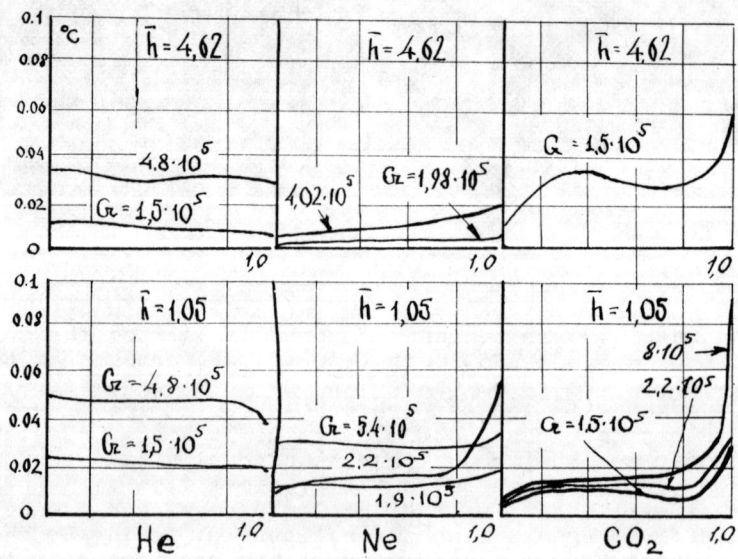

Fig. 4 Experimental distribution of gas concentrations along the width of sell.

Physically this phenomenon is explained as follows. The molecules of Ne or $CO_2$ supplied from the lower part of the gap with no heating (temperature field) are distributed uniformly across the layer, while along the same it is distributed in accordance with the diffusion process, the intensity of which is determined by the diffusion Grashof number $Gr_D = \frac{g \beta m (m_w - m_o) \delta^3}{\nu^2}$. When one of the side walls is heated, convection motion appears, the heavy gas molecules are entrapped by this flow thus increasing the gas molecules are entrapped by this flow thus increasing the gas concentration in the upper part of the layer near the hot surface and reducing the same near the cold wall. In other words, the character of distribution of concentrations of heavy gases is determined by the fact that the convection prevails over the diffusion at $Gr > Gr_D$.

The dimensional analysis of the above case shows that the heat transfer coefficient can be found from the following relation

$$Nu = a Ra^m (h/L)^n \qquad (2)$$

For the "boundary layer" conditions the interpretation of the experimental data is given in Fig.5. It is seen that the dependence

$$Nu = 0.18 Ra^{0.25} (h/L)^{-0.2} \qquad (3)$$

is in good correspondence with the experience.

Fig.6 gives experimental points for the dimensionless complex $Nu_D (h/L)^{-0.46}$ of the product of the criteria $(Gr^* Le)$ $Gr^* = Gr + Gr_D$ is taken as a predominant criterion; the thermal Grashof number was found from the temperature drop between the hot wall and the centre of the layer, while the diffusion Grashof number was found by the difference of the gas concentrations near the wall and in the centre.

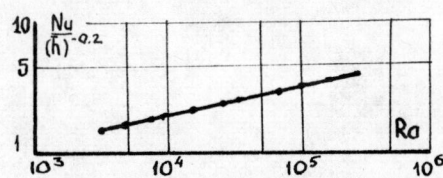

Fig. 5 Heat transfer results

As shown in Fig.6, the experimental points for the three investigated gases are well described by the following dependence

$$Nu_D = 0.06 (Gr^* Le)^{0.33} (h/L)^{-0.46} \qquad (4)$$

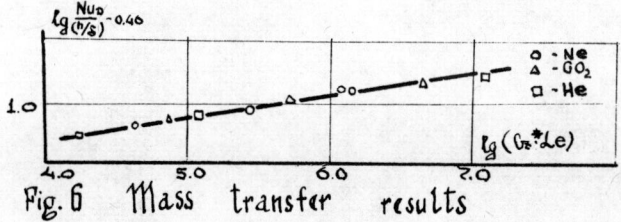

Fig. 6 Mass transfer results

REFERENCES

1. Averyanov A.G., Gurvich B.I. Principles of Ventilation for Removing Gases. "Heating and Ventilation" No.5 (1954).

2. Brdlik P.M., Dubovik V.N., Molchansky I.S. Heat and Mass Transfer with Natural Convection on Vertical Porous Surface when Blowing Carbon Dioxide in Air" "PMTF" No.6, 1971.

# TURBULENT FLOW AND HEAT TRANSFER IN PIPES UNDER CONSIDERABLE EFFECT OF THERMOGRAVITATIONAL FORCES

B. S. PETUKHOV

*Institute of High Temperatures, USSR Academy of Sciences, Moscow, USSR*

## ABSTRACT

The paper deals with a turbulent flow and heat transfer in pipes with combined forced and free convection. The effect of thermogravitation on the averaged flow and turbulent transfer is noted. The results of an approximate theoretical analysis are given and the boundaries of influence of the thermogravitational forces on the heat transfer are determined. Discussion is presented dealing with the results of measurement of the velocity and temperature profiles, heat transfer and friction with ascenting and descending flows in heated vertical and horizontal pipes. The problems of further investigations are formulated.

## NOMENCLATURE

- $T$ — temperature
- $T_b$ — bulk temperature
- $T_w$ — wall temperature
- $p$ — pressure
- $W_i, W_\kappa$ — velocity vector components
- $\overline{W}$ — average velocity of liquid across the pipe section
- $X_i, X_\kappa$ — Cartesian coordinates
- $\rho$ — density
- $C_p$ — specific heat
- $\lambda$ — thermal conductivity coefficient
- $a$ — temperature diffusivity coefficient
- $\mu, \nu$ — dynamic and kinematic viscosity coefficients
- $\beta$ — thermal expansion coefficient

$g, g_i$ — modulus of gravity force vector and its component
$d$ — pipe diameter ($d = 2r_0$)
$\varphi$ — angle read from pipe upper generatrix
$q_w$ — heat flux per unit area on the wall
$\alpha$ — heat transfer coefficients
$Nu = \alpha \cdot d / \lambda$ — Nusselt number
$Nu_0$ — Nusselt number with purely turbulent forced flow, see equation (7)
$Re = \overline{w} \cdot d / \nu$ — Reynolds number
$Pr = \nu / a$ — Prandtl number
$Gr_q = \dfrac{g \beta q_w \cdot d^4}{\nu^2 \cdot \lambda}$, $Gr_A = \dfrac{g \beta d^4 \cdot A}{\nu^2}$ — Grashof numbers
$Pe = \overline{w} \cdot d / a$ — Peclet number
$Ra_A = Gr_A \cdot Pr$ — Rayleigh number
$\xi$ — friction factor
$\xi_0$ — friction factor at purely turbulent forced flow, see equation (8)
$A = dT_b/dx$ — longitudinal temperature gradient

## INTRODUCTION

At the present time great interest is being shown in the study of turbulent flows and heat transfer under conditions of essential influence of thermogravitational forces. The interest in this problem, including the case of such flows in pipes, is due both to scientific and practical considerations. Until recently, it was considered that with a developed turbulent flow of liquid in tubes ($Re > 10^4$) the influence of the thermogravitational forces on the heat transfer and resistance is insignificant. However, the latest experimental data have shown that this influence can be both significant and determining under certain conditions. It has been found that sufficiently high Grashof numbers (or Rayleigh numbers) and moderate Reynolds numbers result in a change in the characteristics of the turbulent transfer under the action of thermogravitational forces, deformation of the velocity and temperature profiles, and in a change in the heat transfer and resistance. The character of these changes depends on the mutual direction of the forced flow and the gravity force vector. As applied to pipes, the following three typical situations should be distinguished:

(1) a flow in vertical pipes upward when the liquid is heated and downward when the liquid is cooled;
(2) a flow in vertical pipes upward when the liquid is cooled and downward when the liquid is heated;
(3) a flow in horizontal pipes with heating and cooling of the liquid.

In each of these cases the character of the flow appearing

Fig.4 shows the Nu number as a function of $Ra_A/Re$ with an ascending flow of mercury in a heated pipe /10/. At first, the Nu number decreases with an increase in $Ra_A/Re$ and then rises up*. The decrease in the Nu number may be explained by a decrease in the turbulent transfer under the effect of thermogravitation, whereas its subsequent rise is explained by the influence of thermogravitation on the averaged flow (an increase in the velocity near the wall).

The effect of thermogr vitation has been clearly exposed in our experiments made together with B.K.Strigin /13/ on heat transfer and friction with ascending and descending flows of water in heated pipes (constant density of the heat flux was maintained on the wall). The results of measurements in the case of the ascending flow are given in Fig.5. At low values of $Gr_A$ and Re > 5000 the Nu numbers are independent of $Gr_A$ and are in agreement with the known dependencies for purely forced convection (Nu = $Nu_o$). With further increase in $Gr_A$ (at the same Re number) the Nu number drops down due to a decrease in the turbulent transfer under the conditions of stable density stratification. The flow in this regions characterized by the presence of low-frequency fluctuations which result in corresponding pulsations of the wall temperature. With subsequency increase in the $Gr_A$ number over a certain value (the higher, the grater Re) the character of the dependence changes: the Nu number increases with $Gr_A$ and becomes weakly dependent on Re. This section of the curve correspond to the dominating role of free convection in the formation of the averaged flow. In this case the $Gr_A$ number is substantially "responsible" for the heat transfer instead of the Re number. The role of forced convection is reduced assentially to maintenance of a specified flow rate. This, in particular, is confirmed by the fact that the points for low Re values (about 500) also lie on this section of the curve.

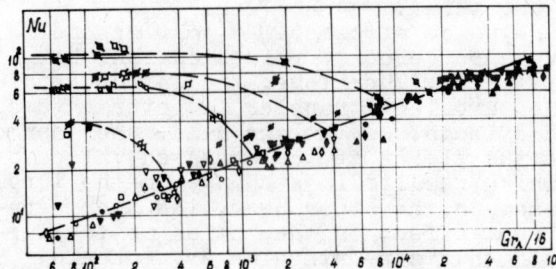

Fig.5. The heat transfer with an ascending water flow in a heated vertical pipe.

It should be noted that the scatter of the experimental points in Fig.5 is not connected to the accuracy of the experimental data but is caused by the fact that the dependence of the Nu number on the Pr and Re numbers is not taken into account in given system of coordinates.

---

* In this connection, it should be noted that the discrepancy of the avilable experimental data on heat transfer to liqued metals, particularly at low values of the Re number, is probably due to a different degree of the influence of the free convection, which was not taken into account during the analysis.

due to the interaction of forced and free convection is identical. Therefore, the dependencies for the heat transfer and resistance are also identical (provided that the properties of the liquid, except for density, are practically constant).

## THE RESULTS OF A THEORETICAL ANALYSIS

The system of equations of conservation of energy, momentum and mass for a stationary (on the average) turbulent flow of incompressible fluid with constant physical properties (with the exception of the density in the expression for a thermogravitational force) in the absence of internal heat sources and energy dissipation is written as

$$\left.\begin{array}{l} \rho C_p \overline{W}_K \dfrac{\partial \overline{T}}{\partial X_K} = \dfrac{\partial}{\partial X_K}\left(\lambda \dfrac{\partial \overline{T}}{\partial X_K} - \rho C_p \overline{W'_K T'}\right), \quad K=1,2,3 \\[6pt] \rho \overline{W}_K \dfrac{\partial \overline{W}_i}{\partial X_K} = -\dfrac{\partial \overline{p}}{\partial X_i} + \rho g_i + \dfrac{\partial}{\partial X_K}\left(\mu \dfrac{\partial \overline{W}_i}{\partial X_K} - \rho \overline{W'_i W'_K}\right), \quad i=1,2,3 \\[6pt] \partial \overline{W}_K / \partial X_K = 0 \end{array}\right\} \quad (1)$$

The nonuniform density distribution in the flow results in appearance of thermogravitational forces acting both on the averaged flow and on the turbulent transfer. The latter can easily be seen from the equation of the energy balance of pulse motion, which is written in a simplified form as

$$\rho \overline{W'_i W'_K}\,(\partial \overline{W}_i/\partial X_K) - \overline{\rho' W'_i}\, g_i + \rho D = 0 \qquad (2)$$

Here we take into account only the generation of turbulent energy (first term), its dissipation (third term) and the work of thermogravitational forces (second term).

As seen from (2), the Reynolds stresses $\rho \overline{W'_i W'_K}$ depend both on the field of the averaged velocity values and on the work of thermogravitational forces. Therefore, the influence of the thermogravitation on the turbulent transfer is determined by two mutually interconnected effects. On the one hand, the thermogravitational forces change the fields of the averaged values of velocity and temperature, which results in a change in the turbulent transfer characteristics. On the other hand, the thermogravitational forces have a direct effect on the motion of the turbulent elements of the liquid thus strengthening or weakening the intensity of the turbulent transfer.

In the case of stable density stratification the vertical displacement of the turbulent elements is accompanied by consumption of energy for the work of Archimedean forces and this leads to attenuation of turbulence and, therefore, turbulent transfer. In the case of unstable density stratification the Archimedean forces perform work of vertical motion of the turbulent elements and this rises the turbulence energy. The first case is observed, for example, in an escending flow and the second - in a descending flow in a heated vertical pipe. The thermogravitational forces have different effects on not only the turbulent pulse transfer but on the heat transfer and, therefore, on the turbulent Prandtl number /1/. The influence of thermogravitational forces on the turbulent transfer is not low compared to their influence on the averaged flow. Depending on the conditions, both effects are commensurable or one of them is predominaut.

The problem of a turbulent flow and heat transfer in tubes with combined forced and free convection is analyzed in a number of papers. These can be subdividied into two groups. The first group includes papers /2,3,4/, in which the influence of thermogravitational forces is taken into account only when they act on an averaged flow; as for the turbulent transfer characteristics, they are accepted by the data purely forced convection. Such an approach does not allow the character of the number Nu as a function of the Gr number to be correctly described either quantitatively or even qualitatively. Thus, in the case of an ascending flow in a heated vertical pipe the calculations have shown that the number of Nu rises up monotonically with an increase in the Gr number at Re = idem; meanwhile, the experimental data indicate to the fact that the Nu number first decreases with an increase in the Gr number and only after that it starts increasing.

The second group includes papers /5,6,7,8,9/, in which the influence of thermograviational forces on the turbulent transfer is somewhat taken into account. Thus, in the L.E.Ber papers /5,6/ an attempt is made to take into account the effect of thermogra- viational forces on the turbulent transfer of a pulse in a flow in a vertical pipe. The corrections to the turbulent transfer coefficient for purely forced convection taking into account the influence of thermogravitation were selected on the basis of experimental data on heat transfer. This fact did not allow the corrections to be adequately grounded and accurate. The Ber calculations are quite approximate and actually are not brought to numerical results, and this causes trobles in their analysis.

The A.F.Polyakov papers /7,8,9/ are interesting since they present an approximate analysis of a turbulent flow and heat transfer in vertical and horizontal pipes at a distance from the input with weak influence of the thermogravitational forces on the forced flow and the heat transfer. The influence of thermogravitation is taken into account both on the averaged flow and on the turbulent transfer. The latter effect is taken into account by the equation of balance of the turbulent energy.

The analysis given in /7/ has shown that in the case of an ascending flow in heated vertical pipes at $Pr > 0.5$ and low Gr numbers the effect of thermogravitation on the turbulent transfer in predominant*. Therefore, the influence of thermogravitation on the averaged flow is not taken into account. Owing to a decrease in the intensilty of turbulent transfer under the conditions of stable stratification the Nu number decreases with an increase in the Gr number and this is in qualitative correspondence with the experimental data (see Fig.1). After the Gr number has reached a certain value, the effect of thermogravitation becomes predominant and the Nu number increases with the Gr number. However, the latest situation is not quantitatively analyzed in the paper /7/.

In the case of a descending flow in heated vertical pipes the influence of thermogravitation on the turbulent transfer considerably exceeds its effect on the averaged flow. The intensity of turbulent transfer under conditions of unstable stratification increases, and the Nu number rises up monotonically with an

---

\* At $Pr \ll 1$ the predominance of any effect depends on the relation between the Re and Pr numbers.

increase in the Gr number in complete agreement with the experimental data (see Fig.1).

The approximate analysis of a flow in heated horizontal pipes /8,9/ has shown that at low values of the Gr number the thermogravitational forces act, first of all, on the averaged flow stimulating appearance of secondary flows in a plane normal to the tube axis. The action of thermogravitational forces on the turbulent transfer apparently manifests itself at higher Gr numbers. The secondary flows break the axial symmetry of the field of velocity and temperature, and this results in nonuniform distribution of the heat transfer over the pipe perimeter. As the number increases, the heat transfer increases near the lower generatrix and decreases near the upper generatrix.

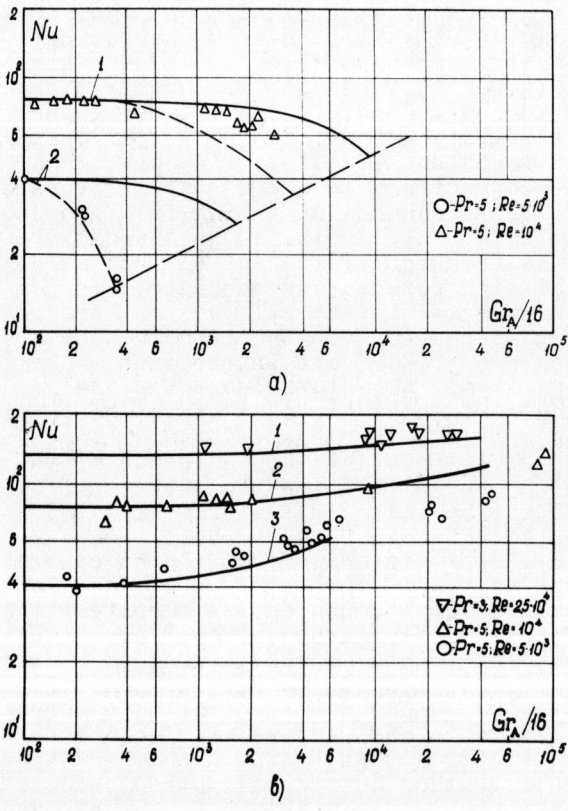

Fig.1 The Nu number with ascending (a) and descending (b) flows in a heated vertical pipe. Continuous lines - theoretical calculations; dashed lines - empirical equations (5) and (6); botted lines - experimental data.

The approximate calculation of the flow and heat transfer at very low influence of thermogravitation conducted in /7,9/ made it possible to obtain equations descibing the boundaries of the beginning of the influence of thermogravitation on the heat transfer. 1% deviation of the Nu number under the effect of thermogravitation from its value with purely forced convection is observed at values of the number determined from the relations: for vertical pipes

$$Gr_q = \frac{1{,}3 \cdot 10^{-4} \cdot Re^{2{,}75} \cdot Pr\, [Re^{1/8} + 2{,}4(Pr^{2/3}-1)]}{lg\, Re + 1{,}15\, lg\,(5Pr+1) + 0{,}5 Pr - 1{,}8} \qquad (3)$$

for horizontal pipes

$$Gr_q = 3 \cdot 10^{-5} \cdot Re^{2.75} \cdot Pr^{0.5}[1+2.4(Pr^{2/3}-1)Re^{-1/8}] \qquad (4)$$

Equations (3) and (3) are valid for the region remote from the inlet at $q_w$ = cost and at Pr 0.5. The limiting curves constructed by equations (3) and (4) are in agreement with the results of the measurements. Fig.2 shows the limiting curves for vertical tubes.

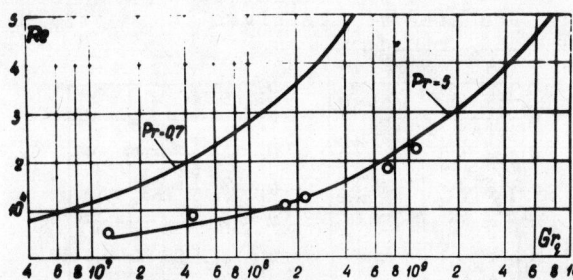

Fig.2. Boundaries of the effect of thermogravitation on the heat transfer with a turbulent flow in vertical pipes.

In conclusion, it should be recognized that at the present time we have no adequately complete analysis of problems of turbulent flows and heat transfer in pipes under considerable influence of the thermogravitational forces. In order to fill this gap, it is necessary to have experimental data of the structure of the flow and heat transfer under the given conditions.

EXPERIMENTAL DATA FOR VERTICAL PIPES

When a liquid moves through a vertical pipes in a gravitational field, the ascending flow with heating and the ascending flow with cooling are practically equivalent; in a similar way, there are equivalent an ascending flow with cooling and a descending flow with heating. In accordance with the experimental data available, we will further consider the ascending and descending flows when the liquid is heated.

The ascending flow with heating of the liquid
─────────────────────────────────────────────

In the near-wall region of such a flow the thermogravitational forces councide in direction with the velocity of the averaged flow. Therefore, the velocity of the liquid increases near the wall and decreases near the axis with an increase in the Gr number. At a certain value of the Gr number minimum velocity occurs at the tube axis, while maximum velocity is observed between the axis and the wall. The maximum velocity approaches the wall with an increase in the Gr number. Such a character of the change in the velocity profiles under the effect of thermogravitation has been found experimentally in flows of air and mercury /10,11,12/. As an example, Fig.3 presents the velocity and temperature profiles measured by Buhr, Horsten and Carr for mercury for $Re \simeq 2 \cdot 10^4$, $3 \cdot 10^4$ and $6 \cdot 10^4$.

There is a few data of the influence of thermogravitation on the turbulent transfer in vertical pipes. This problem was not studied systematically. Paper /11/ gives the results of a few measurements of Reynolds stresses when air moved upwards in a heated pipe, i.e. under the conditions of stable stratification of the density in height. It has been found that at $Re \simeq 5000$ the Reynolds stresses decrease with an increase in the Gr number and at $Gr_\Delta \simeq 2.5 \cdot 10^4$ they approach zero through the entire cross section of the pipe. The coefficients of turbulent transfer of the momentum and heat, calculated in paper /10/ by measuring the velocity and temperature profiles in a mercury flow, depend not only on the Reynolds number but also to a great extent, on the Rayleigh number. An idea of the influence of the thermograviation on a turbulent transfer may be given indirectly by studying the data of the heat transfer.

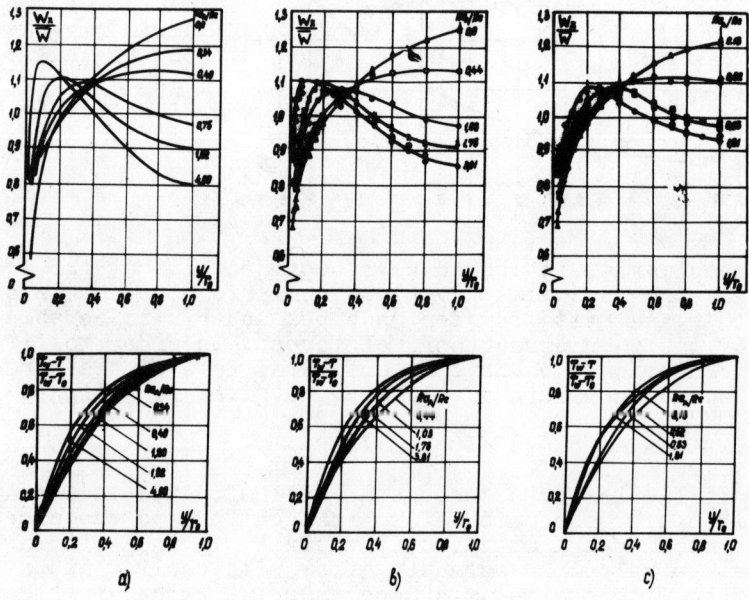

Fig.3. The velocity and temperature profiles with an ascending flow in a heated vertical pipe: a) $Re \simeq 2 \cdot 10^4$, b) $Re \simeq 3 \cdot 10^4$, c) $Re \simeq 6 \cdot 10^4$.

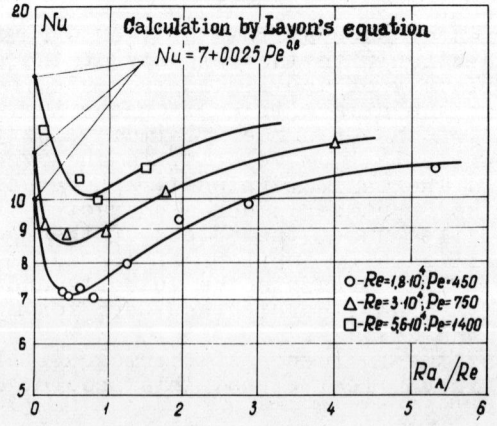

Fig.4. The heat transfer with an ascending mercury flow in a heated vertical pipe.

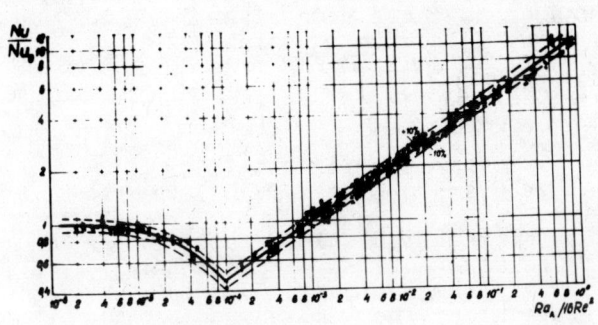

Fig.6. Generalization of the experimental data on the heat transfer with an ascending water flow in a heated vertical pipe. Continuous lines - equations (5) and (6), dotted lines - experimental data.

Generalization of experimental data on heat transfer with an ascending flow of water in heated vertical pipes (partially these data are given in Fig.5) is obtained in /13/ and is given in Fig.6. For the region of the developed flow and heat transfer ($x/d > 40$) the experimental data are described with ±10% accuracy by the equations:

for $Ra_A/Re^2 \leq 10^{-4}$

$$Nu/Nu_0 = [1 + 720 (Ra_A/Re^2)]^{-1}, \qquad (5)$$

for $Ra_A/Re^2 > 10^{-4}$

$$Nu/Nu_0 = 3.97 (Ra_A/Re^2)^{1/3} \qquad (6)$$

Here $Nu_0$ is calculated by the formula /14/

$$Nu_0 = (\xi/8) Re \cdot Pr / K + 12.7 \sqrt{\xi/8} \; (Pr^{2/3} - 1), \qquad (7)$$

where

$$\xi = (1.82 \, lg \, Re - 1.64)^{-2}, \qquad (8)$$

$$K = 1 + 900/Re \qquad (9)$$

Equations (5) and (6) describe the experimental data obtained in the range of values $300 \leq Re \leq 3 \cdot 10^4$; $5 \cdot 10^3 \leq Ra_A \leq 8 \cdot 10^6$ and $2 \leq Pr \leq 6$.

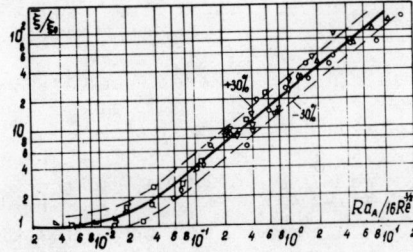

Fig.7. Friction factor with an ascending water flow in a heated vertical pipe. Continuous line - equation (10), dotted line - experimental data.

In paper /13/ the friction factor, which is average along the pipe length, is also measured. These data given in Fig.7 are satisfactorily described by the following equation

$$\xi / \xi_0 = [1 + 3.5 (Ra_A / Re^{3/2})]^{0.4}, \qquad (10)$$

where $\xi_0$ is the value of $\xi$ found from formula (8).

Equation (10) holds at values $300 \leq Re \leq 3 \cdot 10^4$, $8 \cdot 10^4 \leq Ra_A \leq 6.5 \cdot 10^6$ and $2 \leq Pr \leq 6$.

The physical properties of the liquid in equations (5) to (9) are chosen at a bulk temperature of the liquid in the given cross-section of the pipe, while in equation (10) they are selected with a temperature average along the length.

The above experimental data are obtained at low Reynolds numbers. Of course, the effect of thermogravitation manifests itself the stronger, the lower Re. However, if the Gr numbers are sufficiently large, the effect of thermogravitation will be significant at high Reynolds numbers as well.

This is confirmed by the data of heat transfer for single-phase heat carriers at near-critical parameters of state. In this case the Gr numbers may attain high values due to pronounced dependence of the density on the temperature near the pseudocritical point. Thus, according to the data obtained in paper /15/, the heat transfer for carbon dioxide at near-critical parameters increases under the effect of thermogravitation 1.5 times at a value $Re \simeq 2.5 \cdot 10^5$ and 2 times at a value $Re \simeq 10^5$. The near-critical parameters are characterized by specific conditions of flows and heat transfer which are accompanied by sharp drop of heat transfer coefficients (up to five times) at some sections of the pipe. According to modern concepts the appearance of such conditions is also associated with the action of the thermogravitation and thermal acceleration of the flow.

## A descending flow with heating of the liquid.

In the near-wall region of such a flow the thermogravitational forces act in the direction opposite to the direction of the averaged flow velocity. In this case we may expect some decrease of the velocity near the wall at small Gr/Re ratios, as it is observed in the case of a laminar flow. Unfortunately, the flow velocity near the wall was not measured; therefore, the presence of such an effect may only be contemplated on the basis of some heat transfer data (see below). At higher values of Gr/Re we have favourable conditions for intensive generation of turbulent energy due to the opposite directions of the forced and free convection near the wall and due to the unstable density distribution. In this case the velocity profiles become filled, which, in particular, is seen in Fig.8, where the results of a few measurements are given. However, the given flow can essentially be characterized by the heat transfer data.

Fig.9 presents the results of the heat transfer measurements with an ascending and descending flow of water in a heated pipe /12/. The measurements have been conducted at a distance from the heated section inlet (at $x/d \simeq 50$). The Gr/Re ratios for each experimental point are given in the drawing. Attention is immediately called to the striking difference in the character of the dependence of the Nu number on the Re number at a descending flow (upper curve) and at an ascending flow (lower curve).

# TURBULENT FLOW AND HEAT TRANSFER IN PIPES

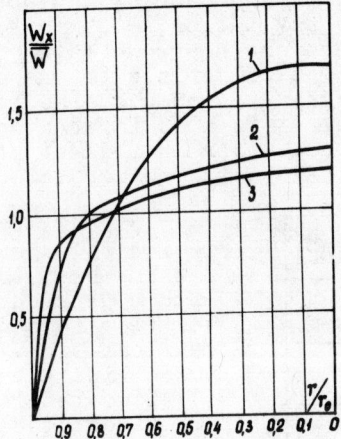

Fig.8. The velocity profiles with a descending air flow in a heated vertical pipe.
1 - Re = 2650; $Gr_q/Re$ = 178;
2 - Re = 5780, $Gr_q/Re$ = 206;
3 - Re = 16890, $Gr_q/Re$ = 15500.

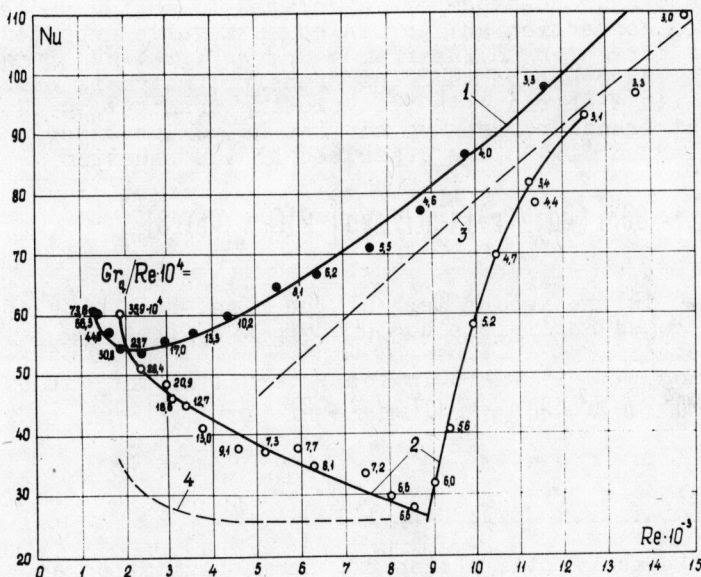

Fig.9. The heat transfer with a turbulent water flow in a heated vertical pipe. 1 - descending flow with combined forced and free convection; 2 - ascending flow with combined convection; 3 - purely forced turbulent flow; 4 - ascending laminar motion with combined convection.

The character of the dependence for the ascending flow confirms the considerations expressed above. In the case of a descending flow, owing to the action of thermogravitation, much more intensive heat transfer is observed, therefore, here we have more intensive turbulent transfer not only in comparison to the ascending flow in the presence of thermogravitation but also in comparison to purely forced convection. It is also interesting to note that intensive mixing in the flow under the effect of thermogravitation is observed already at Reynolds numbers of $Re \simeq 10^3$.

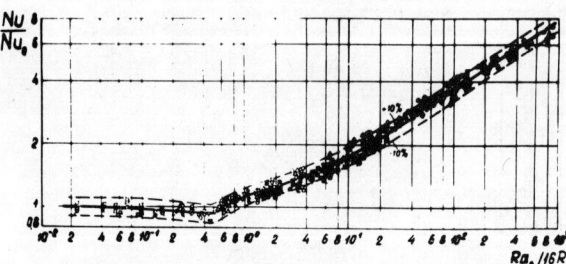

Fig.10. Generalization of experimental data on heat transfer with a descending water flow in a heated vertical pipe. Continuous line - equation (11), dotted line - experimental data.

Fig.10 presents the experimental data of local heat transfer obtained in paper /13/ for the case of a descending flow of water in heated pipes. The date relate to a developed flow pipe section. The measured Nu number is related to the Nu number for purely forced convection and presented as a value dependent on $Ra_A/Re$. With parameter $Ra_A/Re$ rising up, we first observe some insignificant decrease on the heat transfer, which is probably associated with a decrease in the velocity gradient on the wall; then the heat transfer rises up continuously. The experimental data presented in Fig.10 are described by the equation

$$Nu/Nu_0 = [1 + 0.031(Ra_A/Re)]^{1/3} - 0.15 \exp\{-2[(Ra_A/Re)-8]^2\} \qquad (11)$$

At $Ra_A/Re \geq 16$ the second term at the right-hand side of equation (11) becomes negligible compared to the first term and it can be rejected.

Equation (11) is valid in the interval of the values $300 \leq Re \leq 2.5 \cdot 10^4$, $5 \cdot 10^3 \leq Ra_A \leq 13 \cdot 10^6$ and $2 \leq Pr \leq 6$.

EXPERIMENTAL DATA FOR HORIZONTAL PIPES

As noted above, the action of thermogravitation on the averaged flow in horizontal pipes manifests itself in forming secondary flows in a plane normal to the pipe axis. When the liquid is heated, ascending flows appear near the side walls and descending flows appear in the central part of the flow. Due to the interaction of these free-convective flows with the forced flow along the axis a complex flow appears, which may chematically be presented as a flow ascending along two vertical lines, one line featuring clockwise rotation of the liquid and the other line providing counterclockwise rotation of the same.

Such a character of the averaged flow is rather convincely confirmed by the results of direct measurements of the profiles

# TURBULENT FLOW AND HEAT TRANSFER IN PIPES

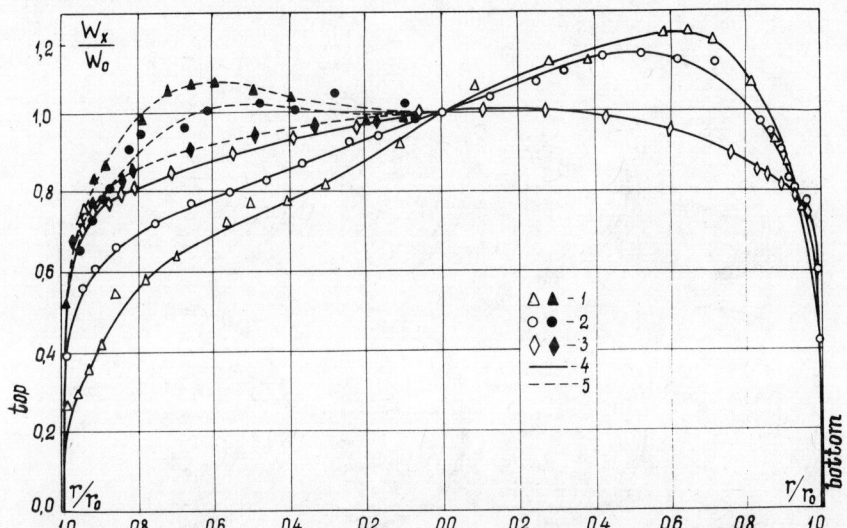

Fig.11. The velocity profiles. 1 - $Re = 1.3 \cdot 10^4$, $Gr_q = 4.2 \cdot 10^8$; 2 - $Re = 2.6 \cdot 10^4$; $Gr_q = 7.7 \cdot 10^8$; 3 - $Re = 5.2 \cdot 10^4$; $Gr_q = 1.03 \cdot 10^9$; 4 - in vertical diametral palne; 5 - in horizontal diametral plane.

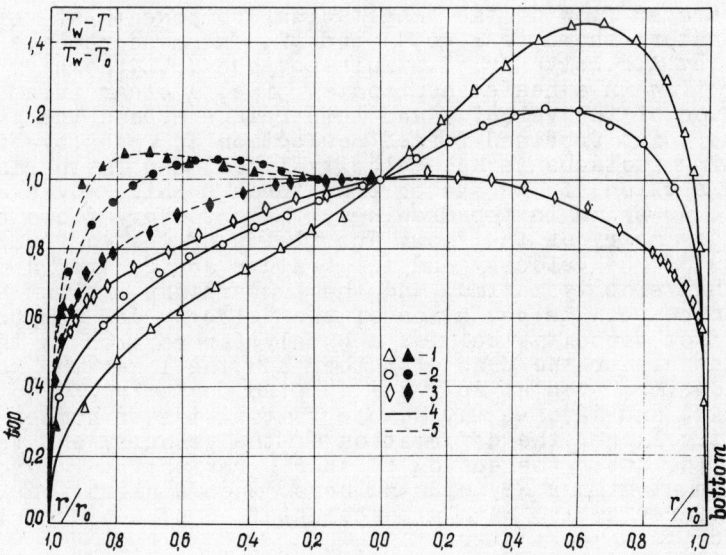

Fig.12. The temperature profiles with an air flow in a heated horizontal pipe. The designations are same as in Fig.12.

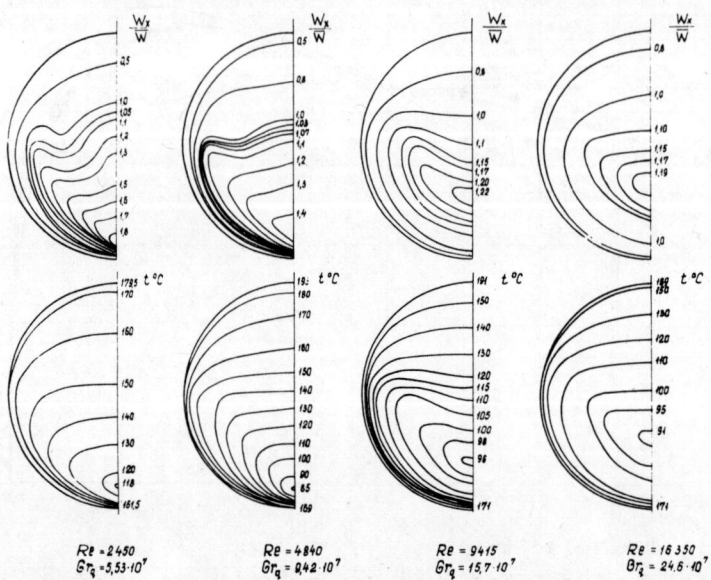

Fig.13. The isotachs and isotherms in an air flow in a heated horizontal pipe.

of the averaged flow of the longitudinal components of velocity and temperature shown in Figs.11 and 12. The measurements were conducted by A.F.Polyakov, V.A.Kuleshov and Yu.L.Shekhter /17/ in an air flow in a heated horizontal pipe. A clear idea of the distribution of the velocity and temperature across the pipe during the joint free and forced convection is given by Fig.13 illustrating isotachs (equal-velocity lines) and isotherms in the cross section of the heated pipe through which moves an air flow /16/. As shown in the drawings, the secondary flows break the axial symmetry of the flow. The fields of the logitudinal components of the velocity and temperature are considerably deformed. The velocity maximum and the temperature minimum are shifted downward. The deviation of the velocity and temperature profiles from those typical for a purely forced flow is the larger, the higher the Grashof number and the lower the Reynolds number. However, even at $Re \simeq 5 \cdot 10^4$ the deviation is noticeable (see Figs.11 and 12). We may assume that if the Gr number is sufficiently large, the deformation of the velocity and temperature profiles under the action of thermogravitational forces takes place at higher Reynolds numbers. Such a situation is observed, for example, with a turbulent flow of heat carriers near the critical parameters.

The action of thermogravitational forces on the turbulence in the case of a flow in heated horizontal pipes manifests itself in a different way in different regions of the flow. The gradients of the averaged velocity are reduced near the upper generatrix and this leads to a decrease in the generation of energy of pulsatory motion. Furthermore, in this case the density stratification is stable and this also stimulates a decrease in the turbulence energy. Therefore considerable reduction of the intensity of the pulsatory motion is to be expected near the upper generatrix. On the contrary, the velocity gradients rise up near the lower generatrix under the effect of thermogravitation while the density stratification is unstable. Both these factors increase the turbulence energy; therefore, the intensity of pulsatory motion should be high near the lower generatrix. This is confirmed by the experimental data /17/ about the turbulence intensity distribution in an air flow at $Re = 1.2 \cdot 10^4$ and $Gr_q = 4 \cdot 10^8$ shown in Fig.14. The intensity is maximum near the lower generatrix ($\varphi = 180°$, $r/r_o = 0.95$), drops down with a decrease in the angle $\varphi$ and near the upper generatrix ($\varphi = 0$, $r/r_o = 0.95$) is about ~1/20 of the value near the lower generatrix.

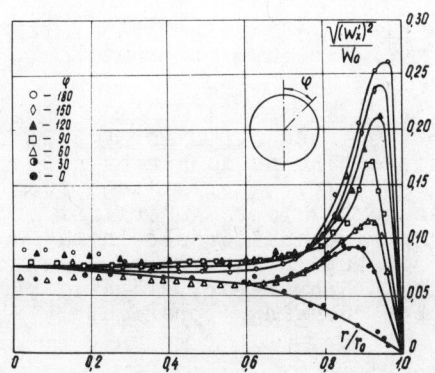

Fig.14. The distribution of the turbulence intensity across a heated horizontal pipe at the Reynolds numbers of $Re = 1.2 \cdot 10^4$ and $Gr_q = 4 \cdot 10^8$.

The above-mentioned characteristic features of the flow determine also the nature of the heat transfer. During a combined forced and free convection in horizontal pipes the heat transfer varies over the pipe perimeter the stronger, the higher the $Gr_q$ number. On the upper generatrix the Nu number is minimum and decreases with an increase in the number $Gr_q$; near the lower generatrix the Nu number is maximum and increases with $Gr_q$.

Only secondary flows develop near the input to the heated part of the pipe; therefore, the Nu number is practically independent of the $Gr_q$ number along a certain section of the pipe having a length of $X < X_{lim}$, and the Nu number remains constant over the perimeter. Under $X_{lim}$ there is understood such a distance from the input on which $(Nu_{\varphi=\pi} - Nu_{\varphi=0})/Nu_{\varphi=\pi} = 0.05$, and this may be regarded as indication of a start of the influence of the thermogravitation on the heat transfer. Let $Gr_{q\,lim}$ stand for the number $Gr_q$ at which we observe the beginning of such an effect over the distance $X = X_{lim}$. According to the analysis of the experimental data, there is unambiguous connection between the ratios $Gr_{q\,lim}^{\infty}/Gr_{q\,lim}$ and $X_{lim}/d$. As $X_{lim}/d$ increases, the number $Gr_{q\,lim}$ decreases and assumes a constant value equal to $Gr_{q\,lim}^{\infty}$ at $X_{lim}/d \geqslant 45$. The values of $Gr_{q\,lim}^{\infty}$ are found using equation (4).

Fig.15 shows $Nu/Nu_o$ as a function of $Gr_q/Gr_{q\,lim}$ on the upper and lower generatrices of the pipe. The experimental data given in this drawing are obtained for motion of air /17/ and water /18/ through heated pipes and cover the range of values

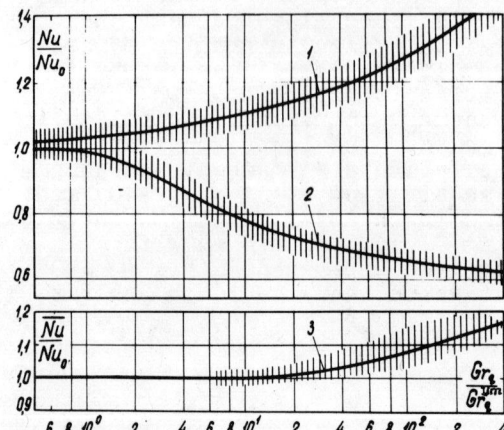

Fig.15. The Nusselt number on the lower generatrix (1), upper generatrix (2) and average-over-perimenter (3) with air and water flows in heated horizontal pipes.

$8 \cdot 10^3 \leq Re \leq 10^5$, $10^7 \leq Gr_q \leq 10^{10}$ and $0.7 \leq Pr \leq 3.3$. Shown in the same drawing is dependence of the Nu number average over the perimeter (related to $Nu_0$) on $Gr_q/Gr_{q\,lim}$. As seen from the drawing, the Nu number increases on the lower generatrix and decreases on the upper generatrix with an increase in the number $Gr_q$. Deviation to either side from $Nu_0$ is as high as 40%. The $Nu_q$ number, which is average over the perimeter, slightly rises with an increase in the $Gr_q$ number changing approximately for 15%.

CONCLUSIONS

Further progress in the study of turbulent flows and heat transfer with combined forced and free convection requires systematic experimental study of the structure of stratified flows: velocity and temperature fields, turbulent stresses, turbulent heat fluxes, and other correlation functions. In addition to the study of the flow structure, it is necessary to obtain more detailed experimental data on heat transfer and friction in a wide range of Grashof, Reynolds and Prandtl numbers with different orientation of the system in the gravitation field. General methods of calculation of turbulent stratified flows and heat transfer in pipes can be developed from the experimental data and methods used in the semi-empirical turbulence theory.

REFERENCES

1. Monin A.S., Yaglom A.S., Statistical Hydromechanics, Part I, "Nauka" Publishers, 1965.

2. Ber L.E. Bulletin of the USSR Academy of Sciences, No.11, 1957.

3. Ber L.E. Bulletin of the USSR Academy of Sciences, No.6, 1962.

4. Ojalvo M.S., Anand D.K., Dunbar R.P., Journal of Heat Transfer, vol.89, No.4, 1967.

5. Ber L.E. PMTF, Journal of Applied Mechanics and Technical Physics, No.4, 1967.

6. Ber L.E. Bulletin of the USSR Academy of Sciences, Fluid and Gas Mechanics, No.3, 1974.

7. Polyakov A.F. Teplofizika vysokikh temperatur (High Temperatures), vol.11, No.1, 1973.

8. Polyakov A.F. In collected paper on "Teploobmen i fizicheskaia gazodinamika" (Heat Transfer and Physical Gas Denamics", Nauka Publishers, Moscow, 1974.

9. Polyakov A.F. Journal of Applied Mechanics and Technical Physics, PMTF, No.5, 1974.

10. Buhr H.O., Horsten E.A., Carr A.D. Journal of Heat Transfer, vol.96, No.2, 1974.

11. Carr A.D., Connor M.A., Buhr H.O. Journal of Heat Transfer, vol.95, No.4, 1973.

12. Sarabi A.R. Theses de docteur-ingenier, A L'Universite Paris, 1971.

13. Petukhov B.S., Strigin B.K. Teplofizika Vysokikh Temperatur, (High Temperatures), vol.6, No.5, 1968.

14. Petukhov B.S., Genin L.G., Kovalev S.A., Heat Transfer in Nuclear Power Plants. "Atomizdat" Publishers, 1974.

15. Ikryannikov N.P., Petukhov B.S., Protopopov V.S. Teplofizika Vysikikh Temperatur (High Temperature) vol.10, No.1, 1972.

16. Mreiden A. Theses de docteur-ingenier, A L'Universite Paris, 1968.

17. Petukhov B.S., Polyakov A.F., Shekhter Yu.L., Kuleshov V.A. Report at the International Seminar "Heat Transfer at Turbulent Free Convection", Yugoslavia, 1976.

18. Petukhov B.S., Polyakov A.F. 4th International Heat Transfer Conference, Versailles, September, 1970.

# EXPERIMENTAL STUDY OF THE EFFECT OF THERMOGRAVITATION UPON TURBULENT FLOW AND HEAT TRANSFER IN HORIZONTAL PIPES

B. S. PETUKHOV, A. F. POLIAKOV, Yu. L. SHEKHTER, and V. A. KULESHOV

*Institute of High Temperatures, USSR Academy of Sciences, Moscow, USSR*

## ABSTRACT

The paper contains the results of experimental studies of local heat transfer, fields of averaged velocities and temperatures, as well as the distributions of fluctuation intensities of temperature and velocity in turbulent flow of air within a horizontal pipe under conditions of marked effect of thermogravitational forces.

## NOMENCLATURE

| | |
|---|---|
| $r_0 = d_0/2$ | — pipe radius |
| $x$ | — distance from the beginning of heating |
| $R = r/r_0$ | — running relative radius |
| $\varphi$ | — angle read from the top generatrix |
| $u$ | — axial component of velocity |
| $\lambda, \nu, \rho$ | — coefficients of thermal conductivity, viscosity and density, respectively |
| $T, t$ | — temperature |
| $\vartheta = (T_w - T)/(T_w - T_0)$ | — relative temperature |
| $q$ | — heat flux per unit area |
| $T_x = \lambda_B (T_w - T_B)/\bar{q}_w \cdot d_0$ | — dimensionless temperature |
| $Nu = q_w \cdot d_0 / \lambda_B (T_w - T_B)$ | — Nusselt number |
| $Re = \bar{u} \cdot d_0 / \nu_B$ | — Reynolds number |
| $Gr = g\beta q_w \cdot d_0^4 / \lambda_B \cdot \nu_B^2$ | — Grashof number |
| $Pr$ | — Prandtl number |

$\sqrt{\overline{T'^2}}$ and $\sqrt{\overline{u'^2}}$ —root-mean-square values of temperature and velocity fluctuations, respectively

Subscripts

o   — value on the axis

b   — found at bulk temperature of liquid

w   — value on the wall

n   — relates to the coordinate of the beginning of separation of Nu numbers on the top and bottom generatrices

lim — limiting value

t   — theoretical value

The line over a symbol denotes averaging over the perimeter.

In monographs and textbooks on heat transfer the effect of thermogravitational forces on friction and heat transfer is considered in application to the inner problem only for the case of the laminar flow of heat-transmission medium. None of those books contain data on the effect of thermogravitation upon turbulent wall flows. However, analytical results /1, 2/ and experimental studies /4, 7/ show that, in the case of turbulent flow in horizontal pipes, thermogravitational forces may have a marked effect both on the velocity and temperature and on the friction and heat transfer.

Presented in this paper are the results of measurements upon turbulent flow of air in a horizontal pipe. The basic parameters vary within the following ranges: $Re_b = (1.2-21) \cdot 10^4$, $Gr = (0.2-9) \cdot 10^9$ and $T_w/T_b = 1.07-2.1$.

The experimental study has been conducted in an air loop described in detail in Ref./5/. The test section is a stainless steel (Grade 1X18H9T) pipe with an internal diameter of 144 mm and wall thickness of 4 mm. The length of the preliminary isothermal section is equal to about 16 gages, while the heat transfer measurement section is somewhat less than 52 diameters. Heating is effected by way of passing alternating current directly through the pipe wall. The use of radial and end-face guarding heaters helps reduce heat losses to (5-10) per cent of the heat flux per unit area on the internal surface of the pipe.

In the couse of experiments, there have been measured local (over the length and perimeter) heat transfer, averaged temperatures and velocities, as well as radial distributions of temperature and velocity fluctuation intensities. The section in which probe measurements have been taken is at a distance of $x/d_o = 49.2$ from the beginning of heating.

The measurements of averaged wall and air temperatures, including the radial distribution, have been carried out with the aid of Chromel-Alumel thermocouples whose readings were recorded by an LM-1604 digital voltmeter using a DTU data acquisition system and a Clary printer.

The velocity profile at high Reynolds numbers ($Re \geq 5 \cdot 10^4$) was found with the aid of a Pitot tube whose signal had been supplied to a DISA Electronik capacitive low differential pressure

limiting $Gr_{lim}$ for the given $x/d_o$.

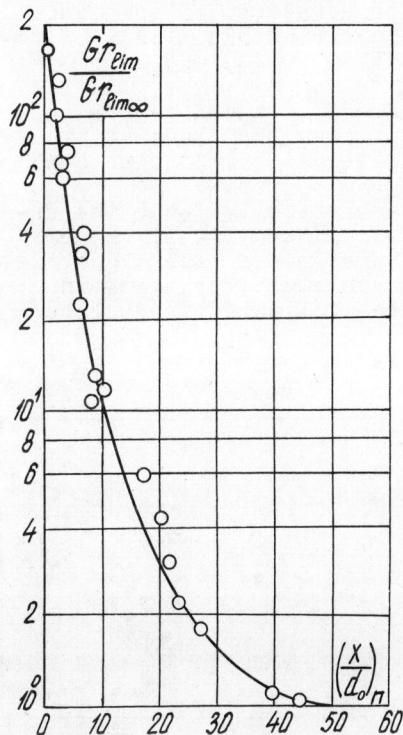

Fig.2. Variation of the relative limiting Grashof number in the entry region.

The present paper studies the variation of heat transfer over both the length and perimeter of the pipe.

The distribution of Nusselt numbers over the angle $\varphi$ is calculated with due regard for the variation of the heat flux per unit area over the perimeter due to losses of heat over the wall. The latter is found from the distribution of the wall temperature, $T_w$, measured experimentally in the cross-section, with the aid of the equation of thermal conductivity for a round thin ring with internal heat sources:

$$q_w/\bar{q}_w = 1 + 2(\lambda_w/\lambda_\delta)\cdot(d^2T_x/d\varphi^2)\ln(\tau_2/\tau_1) \quad (2)$$

where

$$T_x = \lambda_\delta(T_w - T_\delta)/\bar{q}_w \cdot d_o \quad (3)$$

When calculating $q_w/\bar{q}_w$, the $T_w$ distribution over the pipe perimeter is approximated by the dependence of the form

$$(T_w - T_{w,\pi})/(T_{w,0} - T_{w,\pi}) = \cos^k(\varphi/2) \quad (4)$$

Under the investigated conditions corresponding to a considerable effect of thermogravitational forces on local heat transfer, the $Gr/Gr_{lim}$ ratios have values of from 8 to 280. While so doing, the exponent k in Eq.(4) varies from 2.3 to 3.5. It is seen from Fig.3 that the dependence (4) describes well the measured temperature distributions over the pipe perimeter for these limiting $Gr/Gr_{lim}$ values. One should bear in mind, however, that the calculation of the heat flux distribution over the pipe perimeter with the aid of Eq.(2) is approximate. The accuracy of $d^2T_x/d\varphi^2$ depends on the reliability of description of the wall temperature distribution by the interpolation equation. A more accurate result may be obtained by using the solution of the two-dimensional problem of thermal conductivity in a ring with internal heat sources.

Presented in Fig.4a is the variation of heat transfer on the lower generatrix depending on $Gr/Gr_{lim}$. The $Gr_{lim}$ values were calculated from the curve presented in Fig.2 with due regard for the expression of (1). It is obvious that, at a certain ratio of the conditional parameters, $Nu_\tau$ may exceed by 40 per cent $Nu_t$ calculated in accordance with the recommendations contained in Ref./6/ for turbulent flow in a pipe of gases with variable physical properties without the effect of thermogravitational forces. The ratio of heat transfer coefficients on the bottom and top generatrices as a function of $Gr/Gr_{lim}$ is shown in Fig.4b. As seen from the Figure, at high $Gr/Gr_{lim}$ ratios the Nusselt number

transducer. The average and fluctuation velocities at low Reynolds numbers (Re ≤ 1.2·10⁴) were measured with the aid of an ETAM-5A VEI hot-wire anemometer with temperature compensation.

Temperature fluctuations were measured with a DISA Electronik d.c. hot-wire anemometer using a 55M20 bridge and a 55D35 root-mean-square voltmeter.

The diameter of the sensitive element of the hot-wire anemometer probes was 5 microns. When measuring the fluctuation velocity and temperature components, the signal from the probes was integrated by the 55D35 voltmeter for 30 sec.

The authors of Refs./2, 4/ have analytically found the dependences for Nu near the top and bottom generatrices, which helped find the boundaries of 1% of the effect of thermogravitational forces on local heat transfer. For stabilized flow in a horizontal pipe this is described by the following dependence:

$$Gr_{lim_\infty} = 3 \cdot 10^{-5} Re^{2.75} \cdot Pr^{0.5} \left[ 1 + 2.4 Re^{-1/8}(Pr^{2/3} - 1) \right] \qquad (1)$$

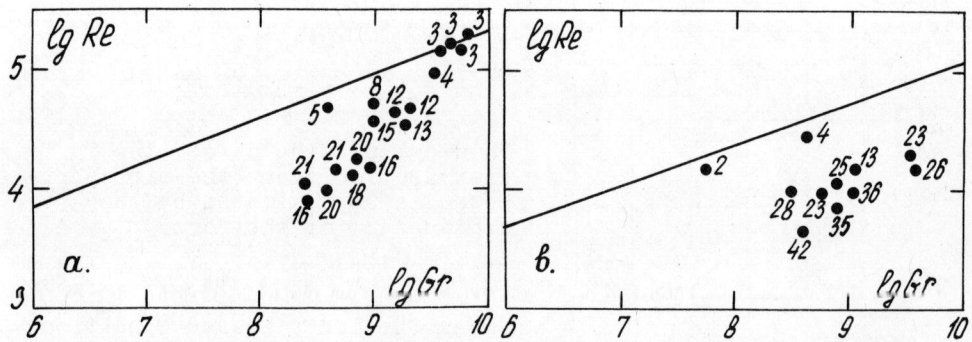

Fig.1. Comparison of the results of calculation (lines) from Eq.(1) with the experimental data for $x/d_o > 40$: a - air, $Pr = 0.71$; b - water, $Pr = 2.8-6.5$ /3/, calculation at $Pr = 3$.

The comparison of calculation results on the basis of Eq.(1) with the experimental data for $x/d_o \geq 40$ presented in Fig.1 shows their satisfactory agreement. Shown at each point is the per cent deviation of $(Nu_\pi - Nu_o)/2$ relative $Nu_\pi$.

It follows from the Figure that, even at adequately high Re (Re ≥ 2.10⁵), there is possible a marked effect of mass forces on local heat transfer.

The effect of thermogravitational forces develops over the pipe length (cf., Ref./4/). Near the beginning of heating, the Nusselt numbers on the top and bottom generatrices are equal. With an increase of $x/d_o$, there occurs a difference between the heat transfer coefficients on the top and bottom generatrices, increasing towards the outlet section of the pipe. At $x/d_o > 40$, stabilized conditions heat transfer set in.

Fig.2 shows the variation of the relative Grashof number in the entry region corresponding to the coordinate of the beginning of separation of Nu numbers on the top and bottom generatrices. The abscissa $(x/d_o)_n$ is defined as a cross-section in which occurs a 5% deviation of $Nu_\pi$ (or $Nu_o$) value from the 1/2 ($Nu_\pi$ + $Nu_o$) value. The Grashof number calculated from the flow parameters in the $(x/d_o)_n$ cross-section can be interpreted as the

# EFFECT OF THERMOGRAVITATION UPON TURBULENT FLOW

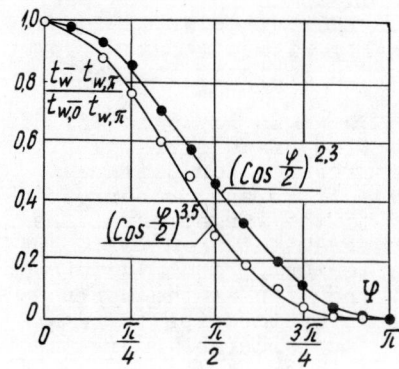

Fig. 3. Distribution of the relative wall temperature over the pipe perimeter.

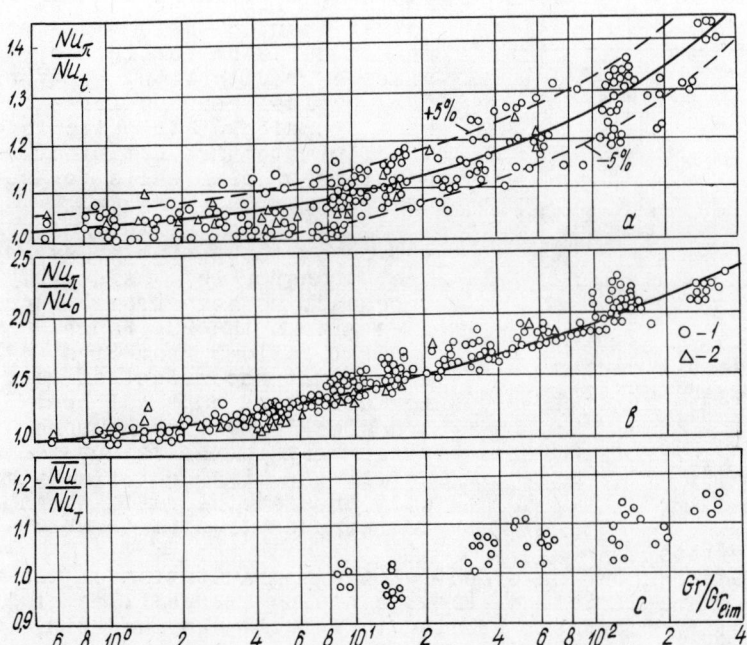

Fig. 4. Variation, as a function of the relative Grashof number: a) of the relative Nusselt number on the bottom generatrix; b) of the ratio between Nusselt numbers on the bottom and top generatrices; and c) of the circumference-average Nusselt number: 1 - air; 2 - water, Pr = 2.7-3.3

on the top generatrix may be almost 2.5 times less than that on the bottom generatrix. Also plotted on these graphs are experimental data from Ref./3/ obtained upon a flow water in a pipe 18.8 mm in dia., which are in satisfactory agreement with the result of air experiments. The obtained divergence between heat transfer on the top and bottom generatrices leads to a considerable excess of the wall temperature on the top generatrix over the wall temperature in the lower portion of the pipe and perimeter-average wall temperature. This is indicative of the need for taking into account the non-uniformity of heat transfer over the perimeter when designing heat-transfer systems with horizontal heating surfaces.

Shown in Fig.5 is the distribution of Nu over the pipe circumference for different values of Re and Gr corresponding to the variation of the thermogravitation effect parameter $Gr/Gr_{lim}$ from 14 to 250. Heat transfer in the upper portion of the pipe ($0<\varphi<\pi/2$) varies considerably more than in the lower portion ($\pi/2<\varphi<\pi$). It is worthy of note that the maximum heat transfer occurs in the $\varphi$ region near $3\pi/4$ rather than on the bottom generatrix as found earlier for viscous-gravity flow. However, the physical reasons for such heat transfer distribution over the perimeter are not clear, and this problem calls for special analysis.

The effect of thermogravitation on the perimeter-average heat transfer is considerably less and, as seen from Fig.4c, lies within the 20% range.

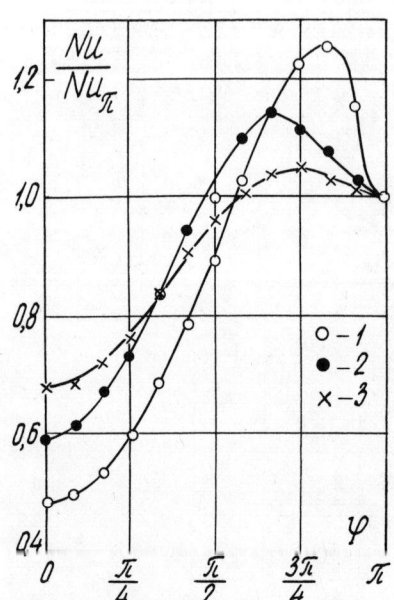

Fig.5. Nusselt number distribution over the pipe circumference:
1 - $Gr/Gr_{lim}$ = 255;
2 - $Gr/Gr_{lim}$ = 63;
3 - $Gr/Gr_{lim}$ = 14.

It follows from Figs. 4 and 5 that the effect of thermogravitation on local heat transfer is most pronounced at relatively low Re values and relatively high heat flux per unit area on the wall, which corresponds to considerably high $Gr/Gr_{lim}$ ratios.

Figs. 6 and 7 show the effect of thermogravitation on the averaged flow characteristics. Fig.6 contains the profiles of the axial velocity component related to velocity on the axis, measured over 12 radii. The same Figure shows the isotach distribution over the pipe cross-section, plotted on the measured profiles. Fig.7 shows the relative temperature profiles measured over 13 radii, and the corresponding isotherm distribution. It is seen from the Figures that in the wall region (R = 0.9) the velocities and temperatures in the vertical diametral plane near the top and bottom generatrices differ strongly. An analysis of isotherms and isotachs leads one to assume that, by analogy with viscous-gravity flow (cf., Ref./3/), in turbulent flows there also forms a system of whirls parallel to the channel axis.

Upon forced turbulent flow, thermogravitational forces affect both the averaged flow and turbulent heat- and momentum transfer. One can judge the effect of thermogravitation on turbulence by the measured fields of root-mean-square values of fluctuations of temperature and axial velocity component.

EFFECT OF THERMOGRAVITATION UPON TURBULENT FLOW 725

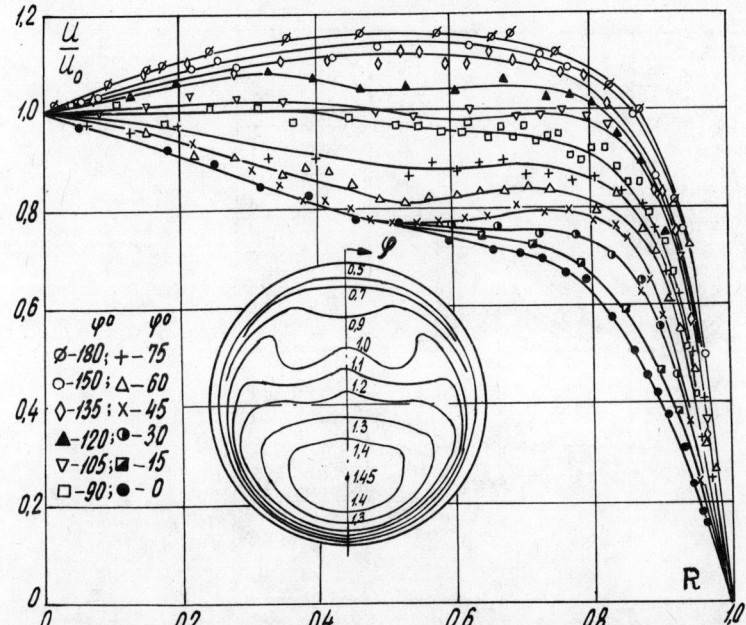

Fig.6. Distribution of relative velocity over the flow cross-section at Pr = 0.71, Re = $1.2 \cdot 10^4$ and Gr = $5 \cdot 10^8$.

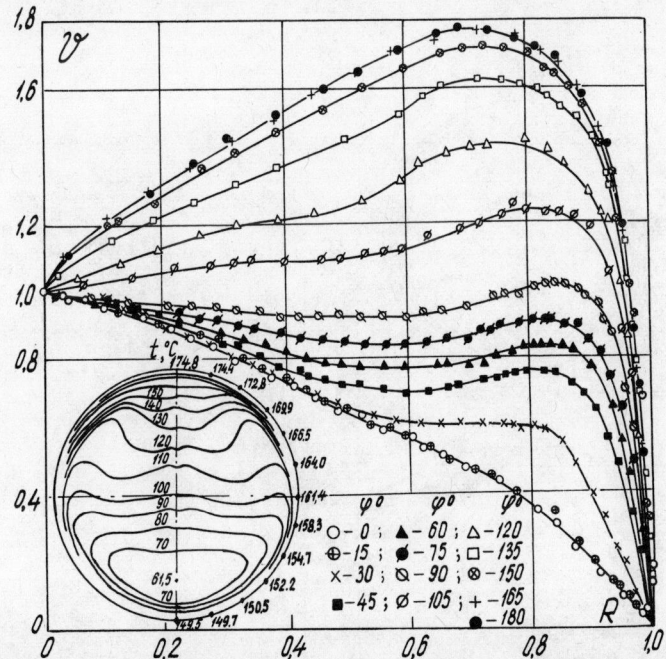

Fig.7. Distribution of relative temperature over the flow cross-section at Pr = 0.71, Re = $1.2 \cdot 10^4$ and Gr = $5 \cdot 10^8$.

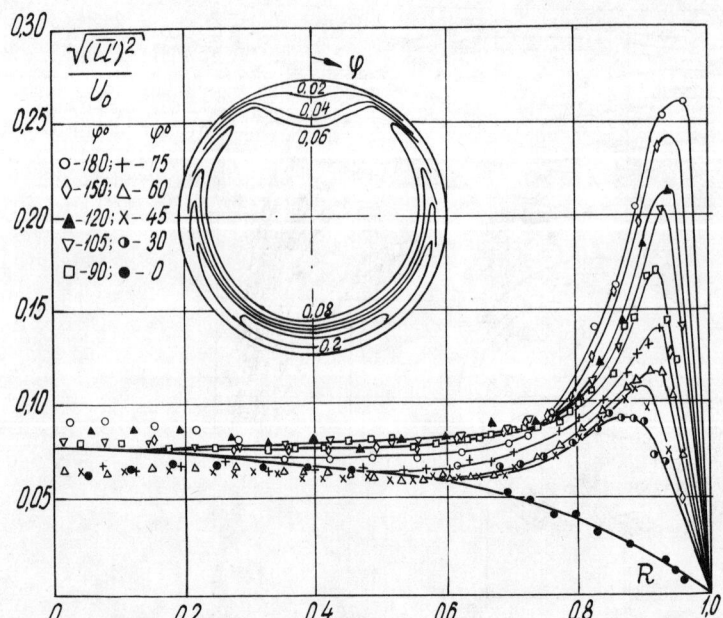

Fig.8. Distribution of velocity fluctuation intensities over the flow cross-section at Pr = 0.71, Re = 1.2 · 10$^4$ and Gr = 5·10$^8$.

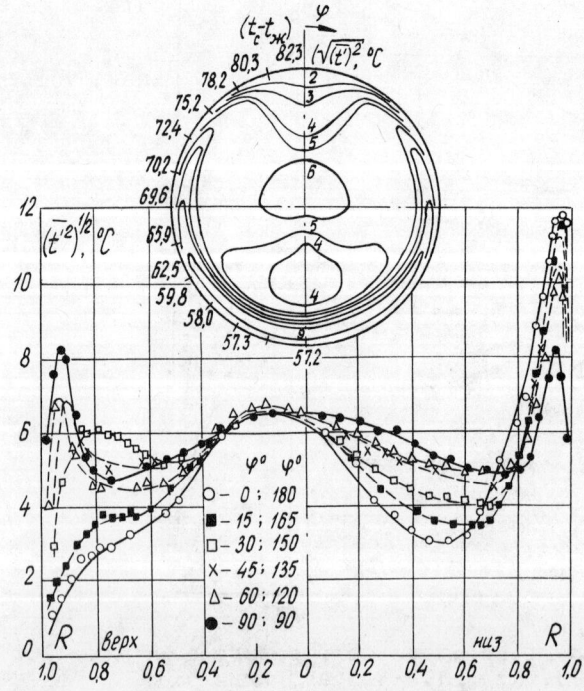

Fig.9. Distribution of temperature fluctuation intensities over the flow cross-section at Pr=0.71, Re=1.2·10$^4$ and Gr = 5·10$^8$.

Presented in Figs. 8 and 9 are distributions of temperature and velocity fluctuation intensities. As seen from the Figures, temperature and velocity fluctuations in the lower portion of the pipe are considerably higher than in the upper one. This can probably be attributed to two reasons. First, there takes place in the lower portion of the pipe an unsteady distribution of density over the radius, which provides an additional source of turbulence generation under the effect of thermogravitational forces; at the same time, in the upper portion of the pipe the distribution of density over the radius is steady, which leads to the attenuation of turbulence. Second, intensive secondary flows likewise lead to a considerable change in the nature of turbulence.

REFERENCES

1. Poliakov, A.F. In collected papers on "Teploobmen i fizicheskaia gazodinamika" (Heat Transfer and Physical Gas Dynamics), Nauka Publishers, Moscow, 1974.

2. Poliakov, A.F. PMTF (Journal of Applied Mechanics and Technical Physics), No.5, 1974.

3. Petukhov, B.S., Poliakov, A.F. Teplofizika vysokikh temperatur (High Temperatures), No.2, 1967.

4. Petukhov, B.S., Poliakov, A.F., Kuleshov, V.A., Shekhter, Yu.L. 5th International Heat Transfer Conference, Tokyo, 1974.

5. Poliakov, A.F., Kuleshov, B.A., Shekhter, Yu.L. IFZh (Journal of Engineering Physics), vol. XXVII, No.5, 1974.

6. Petukhov, B.S., Kurganov, V.A., Gladuntsov, A.I. In collected papers on "Teploperenos" (Heat Transfer), Minsk, vol.1, 1972.

7. Amine Mreiden, DT.h., L'universite de Paris, 1968.

# THE EFFECT OF THERMAL INSTABILITY ON LAMINAR CHANNEL FLOW

**SIMON OSTRACH and YASUHIRO KAMOTANI**

*Case Western Reserve University,
Cleveland, Ohio*

ABSTRACT

Results are presented of experimental investigations of flow and heat transfer in horizontal parallel-plate channels with the lower plate heated and the upper one cooled. The experiments covered a range of Rayleigh numbers between 100 and 13,500 and Reynolds numbers less than 100 for fully-developed air flows and Rayleigh numbers between $10^3$ and $3.1 \times 10^4$ and Reynolds numbers ranging from 30 to 1100 for flows in the thermal entrance region.

For fully-developed flows the heat transfer measurements indicate an increase in the Nusselt number when longitudinal vortex rolls appear in the channel. This indicates the considerable heat transfer augmentation obtained in this way. No appreciable difference was found in the heat transfer coefficients for fully-developed flows and for confined natural convection at the same Rayleigh number. Periodic spanwise temperature distributions created by the vortex rolls are distorted by a second type of vortex rolls as the Rayleigh number is increased. Because of this interaction the regular vortex structure disappears around a Rayleigh number of 8000.

In the thermal entrance region experimentally determined critical Rayleigh numbers were much higher than theoretically predicted vales. Beyond critical Rayleigh numbers the second-type vortex rolls are predominant and the local Nusselt number increases gradually with Rayleigh number. The thermal-entrance length does not change appreciable with Rayleigh number.

NOMENCLATURE

D     height of test section

Gr     Grashof number, $g\beta\Delta TD^3/\nu^2$

g     acceleration of gravity

h     heat transfer coefficient

Nu     Nusselt number, $hD/k$

Pe     Peclet number, PrRe

$Q_{NET}$     net energy input

Ra     Rayleigh number, $g\beta\Delta TD^3/\nu\kappa$

Re     Reynolds number, $\bar{U}D/\nu$

T     temperature

$\bar{T}$     average temperature at a given z

T'     temperature fluctuation

$T_m$     mean temperature in the passage, $1/2\,(T_1 + T_2)$

$T_o$     temperature of main stream and the cooled plate

$T_1$     surface temperature of the heated plate

$T_2$     surface temperature of the cooled plate

$\Delta T$     temperature difference, $T_1-T_0$ or $T_1-T_2$

$\bar{U}$     volume averaged speed of the main flow

u     velocity fluctuation

x     coordinate parallel to the main flow direction measured from the leading edge

x'     dimensionless axial distance, x/DPe

y     coordinate in spanwise direction

z     coordinate normal to horizontal plates

Greek Symbols

$\beta$     coefficient of volumetric expansion

$\varepsilon$     emissivity

$\kappa$     thermal diffusivity of air at $T_m$

$\lambda$     wavelength (pitch) of vortex rolls

$\nu$     kinematic viscosity of air at $T_m$

$\sigma$     Stefan-Boltzman constant

INTRODUCTION

    It is well known that a horizontal layer of fluid, when heated from below and cooled above, becomes unstable when the temperature difference is increased above a critical value. With a superposed unbounded flow above the critical value several forms of fluid motion have been observed [1]. In the case of a superposed fully-developed flow between two horizontal flat plates where the lower plate is heated and the upper one is cooled, longitudinal vortex rolls appear in the passage when the temperature difference becomes larger than a critical value [2]. In either case, below the critical point heat is transported from the lower plate to the upper one by conduction through the fluid, and the temperature changes linearly in the fluid. Above the critical point the heat transfer rate is increased by the thermal instability and the temperature field is strongly influenced by the motion of vortex rolls.

There are several works available concerning the change of heat transfer rate due to the thermal instability for a confined horizontal layer of fluid. Mull and Reiher (reported by Jacob [3]) measured the heat transfer rate in the range of Ra = 1,500 to 6.3 x $10^6$ in air. Malkus [4] conducted experiments using distilled water and acetone in the range of Ra = 10 to $10^{10}$. The work of Silveston [5] was done on several fluids and covered Ra = 350 to 3.8 x $10^5$. Globe and Dropkin [6] used several fluids to investigate the effect of Prandtl number on heat transfer rate in the range of Ra between 1.51 x $10^5$ and 6.76 x $10^8$. Plows [7] studied the problem numerically, and calculated the change of heat transfer rate due to two-dimensional vortex rolls.

As for heat transfer with a superposed fully-developed flow on a thermal instability between two horizontal plates only a few papers are available. Mori and Uchida [2] solved the governing non-linear equations of the problem approximately using the energy integral method. They also reported experimental data taken mainly at Ra $\doteq$ 8,000, Re $\doteq$ 520 to compare with their theoretical results.. Ogura and Yagihashi [8], and Hwang and Cheng [9] solved the problem by numerical methods. As noted by Hwang and Cheng [9], the governing equations for the fully-developed case are identical to those for the confined case with two-dimensional vortex rolls studied by Plows [7]. Nakayama et al [10] studied theoretically the onset of instability when both lower and upper plates were subjected to identical uniform axial temperature gradients, and Akiyama et al [11] verified the results experimentally.

In the present experiments the heat transfer rate was measured for a fully-developed flow of air in the range of Ra = 100 to 13,500. Temperature distributions were investigated more extensively and for a wider range of Rayleigh number in the present work than in Mori and Uchida's experiments in order to study the effects of vortex rolls on the temperature field in more detail.

In practice fully-developed conditions are usually not met because of the limited length of the flow passage. Thus the study of thermal instability in the entrance region is very useful. However, very little work has been done on this problem. Hwang and Cheng [12] determined theoretically the conditions marking the onset of thermal instability in a hydrodynamically fully-developed but thermally-developing region. They found that for Pr $\geq$ 0.7, the flow is more stable in the thermal entrance region than in the fully-developed region, but the opposite is true for Pr $\leq$ 0.2. They calculated critical Rayleigh numbers at different locations in the thermal boundary layer.

In the present experiments the heat transfer rate was measured in the thermal entrance region in the range Ra = $10^3$ to 3.1 x $10^4$ and Re = 30 to 1,100. The onset of thermal instability was determined experimentally and was compared to the theoretical prediction of Hwang and Cheng. In addition, temperature distributions were investigated to study the effects of longitudinal vortex rolls on the temperature field in the thermal entrance region and, hence, on the augmentation of the heat transfer.

EXPERIMENTAL APPARATUS AND PROCEDURE

The test apparatus for each experiment is shown in Fig. 1. The test section length was 26.7 cm and for the fully-developed tests (Fig. 1a) the test section width was 26.7 cm whereas for the thermal-entrance region tests it was 31.8 cm wide (Fig. 1b). The lower plate of the test section was made of a 6.35 mm thick aluminum plate whose surface was polished and the flatness was carefully checked. Electrical heating mats were bonded to the back of the aluminum plate. Guard heaters were used to compensate for the heat loss to both sides of the plate. The input to each heater was regulated individually by a voltage controller. The upper plate of the test section was made of a 6.35mm thick plexiglass plate which was reinforced by aluminum beams to keep it flat. The

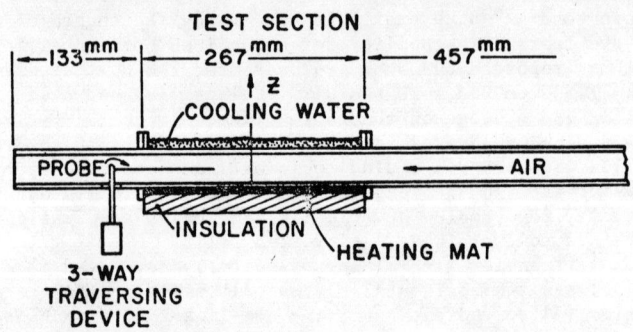
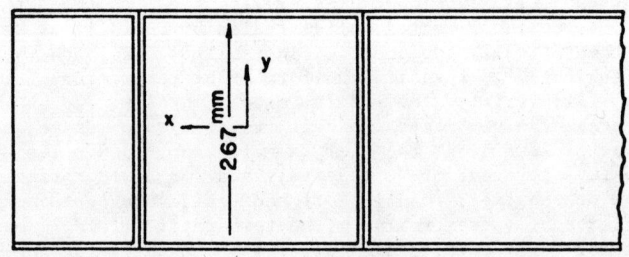
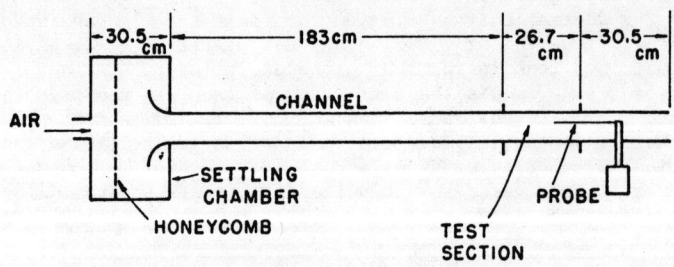
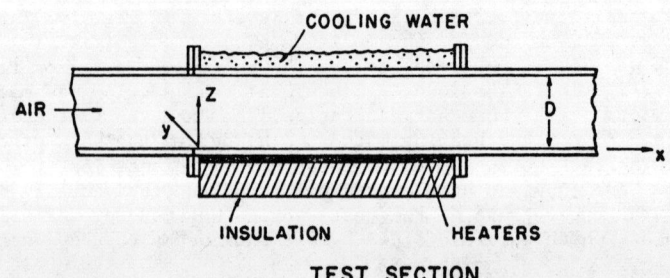

Figure 1. Sketch of Test Apparatus

upper plate was cooled by water flowing over it. The cooling water temperature was maintained to $\pm 0.01°C$ by a constant temperature circulator. The surface temperatures of the lower and upper plates were measured by several copper-constantan thermocouples embedded in the plates. The surface temperatures were kept uniform (within $\pm 1\%$) throughout the experiments. For the thermal entrance region tests the upper plate temperature was kept at the main stream temperature to avoid a thermal boundary layer.

The side walls of the test section were made of 3.2 mm thick insulating walls to minimize heat conduction through them. The test section height was adjusted by the height of the side walls.

The air was supplied from a compressor. For the fully-developed tests (Fig. 1a) the air passed through a channel diffuser and a 45.7 cm long rectangular channel before entering the test section. For the thermal entrance region tests (Fig. 1b) the air passed through a settling chamber and a 183 cm long channel before entering the test section. The volume flow rate was measured by a flow meter. The flow rate was regulated by a flowrate controller. The mean air speed in the test section was calculated from the total flow rate. To obtain fully-developed velocity and temperature profiles over a large part of the test section the Reynolds number (based on mean speed and test section height) was kept below 100 for those tests.

The heat transfer rate was calculated from the net heat input to the air in the test section. The calculation of the net heat input is explained in detail in Ostrach and Kamotani [13]. The heat transfer coefficient and Nusselt number are defined as

$$\frac{Q_{NET}}{area} = h\,(T_1 - T_0)$$

$$Nu = \frac{hD}{k}$$

The experimental error in the value of Nu was estimated to be $\pm 10\%$.

To measure velocity and temperature distributions in the test section a probe was inserted from the downstream end of the test section. The probe was supported by a 3-way traversing device. Temperature was measured by a copper-constantan thermocouple probe. Velocity was measured by a hot-wire probe which was operated by a constant current anemometer.

## EXPERIMENTAL RESULTS - FULLY DEVELOPED FLOW

### Preliminary Investigation of Flow Field

To observe the flow field it was visualized by slowly introducing cigarette smoke into the test section. Photographs were taken through the transparent uppper wall for both the confined (no through flow) cases and the fully developed cases. For the confined cases both upstream and downstream ends of the test section were closed. Sample photographs are shown in Fig. 2.

For the confined cases the flow structure above the critical Rayleigh number changes with the gap between two plates as observed by Chandra [1]. At $D \doteq 9$ mm so called 'Benard cells' were clearly observed. For $D > 10$ mm hexagonal cells appeared only sporadically, and the flow field was dominated by longitudinal cells. Such variation of the flow structure with D is probably due to the variation of physical parameters (e.g. viscosity) with temperature (Schlüter et al [14], Segel and Stuart [15]). As D becomes smaller, the critical temperature difference rapidly increases (proportionally to $D^{-3}$) for a fixed Ra so that the non-uniformity of physical parameters in the flow field become important.

When the fluid in the test section is moving at a constant speed, longitudinal vortex rolls aligned to the flow direction are predominant. However, when the speed is very small, the flow structure resembles that of the enclosed

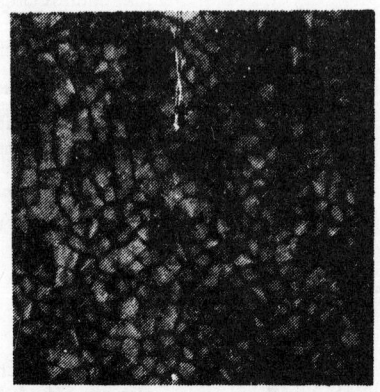

D = 9 mm, No Flow
Ra = 2,850

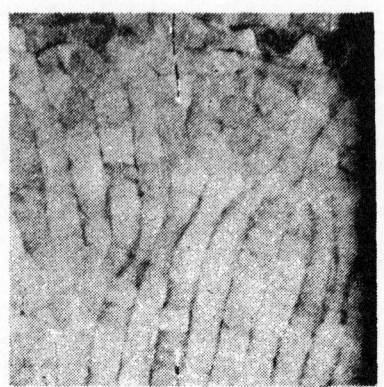

D = 11 mm, No Flow
Ra = 2,850

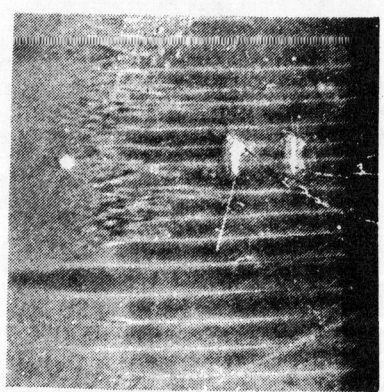

D = 12 mm, Re = 58
Ra = 3,200

Figure 2. Flow Patterns

cases. For Re > 10 (in all cases studied herein) we observed well defined longitudinal vortex rolls in the range of temperature difference and passage height studied herein.

Transverse (z-direction) velocity distributions were measured at the location half way from the beginning of the test section when both walls were at equal temperature. The data are presented in Fig. 3 together with the parabolic profile for fully developed channel flow. Since hot wires cannot measure small speeds, the speed $\bar{U}$ was set higher (20cm/sec) for this measurement than the value used for the present experiments. Figure 3 confirms that the velocity profile is fully developed at the measuring location.

Similarly, transverse temperature distributions were measured at the same location with through flow when there was a temperature difference which did not exceed the critical value. The data given in Fig. 4 show that the temperature distribution is linear as in a fully developed flow.

Thus, in the range of Reynolds number studied here, both velocity and temperature distributions were found to be fully developed over the downstream half of the test section where the heat transfer measurements were made.

Heat Transfer Rate

In Fig. 5 are plotted Nusselt numbers versus Rayleigh numbers for both the fully developed case (7.5 mm < D < 15 mm, 10 < Re < 100) and the confined case. The present data are compared with the empirical curve obtained by Silveston [5] for the confined case, and with the experimental data and the theoretical curve given by Mori and Uchida for the fully-developed case which is written as

$$Nu = 1 + 1.413 \left(1 - \frac{Ra_{cr}}{Ra}\right) \tag{1}$$

The present results are also compared with the numerical results obtained by Hwang and Cheng [9].

As can be seen in Fig. 5 there is a sharp increase of Nusselt number across the critical Rayleigh number, and there is no appreciable difference in heat transfer rate between the fully-developed case and the confined case. Both sets of the present data agree very well with Silveston's curve. Nu calculated from Eq. (1) is smaller than the present data, and the difference increases as Ra increases. Considering the approximation method adopted by Mori and Uchida, their results are expected to be good for Rayleigh number not very much larger than the critical value. In their experiments Mori and Uchida calculated the Nusselt number from the temperature gradient at the surface of the heated plate. Such a procedure is not very reliable when the temperature field is complicated like in the present problem. Moreover, as explained in the next section, for Ra > 8,000 regular vortex rolls disappear, and the fluid motion becomes irregular, which makes it almost impossible to calculate with reasonable accuracy the over-all heat transfer rate from temperature distributions. For those reasons, the data taken by Mori and Uchida differ from the present data by as much as 20%, and the difference is largest for Ra > 8,000. Numerical results of Hwang and Cheng show slightly larger Nu than the present data. The difference is about 10%, which is within the error of the present experiments.

For large Ra (> 8,000) Nu changes proportionally to $Ra^{1/3}$. Noting the definition of Ra and Nu, this means that the heat transfer rate becomes independent of D.

The data also showed that as long as the flow in the test section was fully developed the heat transfer rate was independent of Reynolds number in the range of Reynolds number studied herein (Re < 100), which agrees with Eq. (1).

The parameter which shows the importance of buoyancy forces in a forced flow problem is called the Froude number ($\equiv Re^2/Gr$). Body forces become

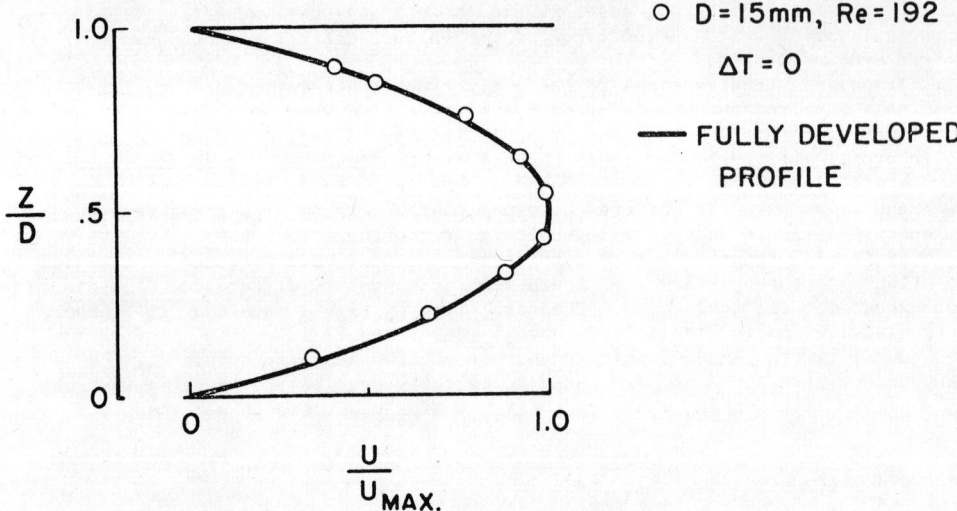

Figure 3. Fully Developed Velocity Distribution

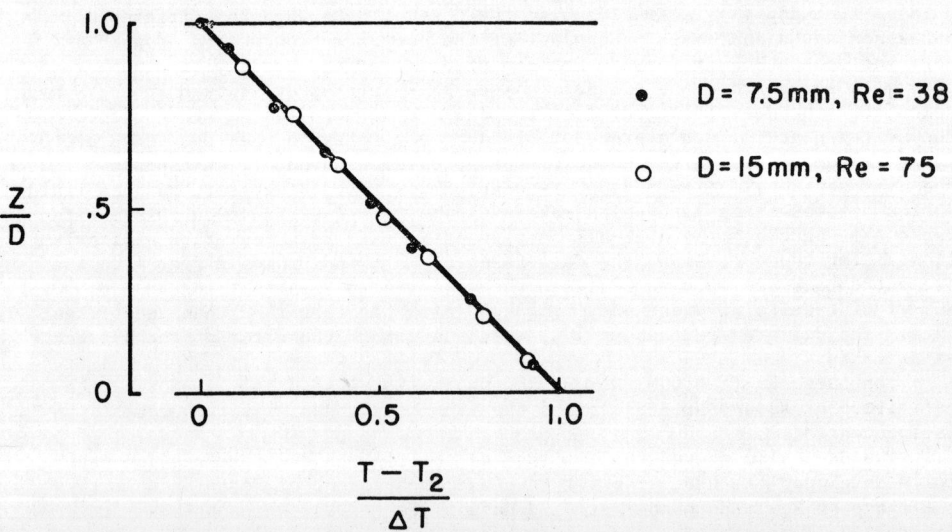

Figure 4. Fully Developed Temperature Distribution

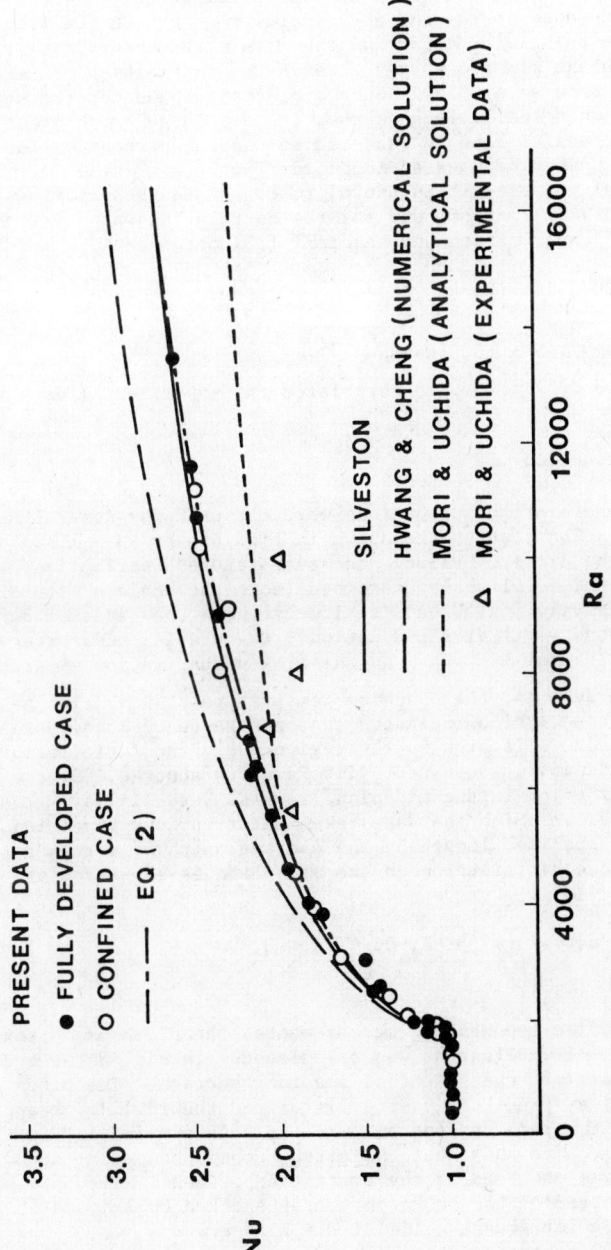

Figure 5. Nusselt Number versus Rayleigh Number

significant when the Froude number is of unit order of magnitude or smaller. In the present experiments the Froude number ranged from .04 to 10. However, if longitudinal vortex rolls are fully developed, the temperature distribution is independent of the axial distance and the axial velocity. Consequently, the heat transfer rate does not depend on Reynolds number. On the other hand, when the flow field is turbulent, a part of the main stream kinetic energy is converted into turbulent kinetic energy through the interaction of Reynolds stresses and the main shear flow. Such a process depends on the Reynolds number. Therefore, when the Rayleigh number is very large (much greater than 8,000) the heat transfer rate is expected to depend on the Reynolds number and, thus, the Froude number becomes an important factor.

For practical purposes it is useful to correlate the experimental data by a single equation. The method used for the correlation was developed by Churchill and Usagi [16] and Ostrach and Kamotani [13] obtained:

$$Nu = 1 + \frac{8.30 \times 10^{-4}(Ra-Ra_{cr})}{\sqrt{1 + 1.45 \times 10^{-4}(Ra-Ra_{cr})^{4/3}}} \qquad (2)$$

Eq. (2) is plotted in Fig. 5. It correlated the experimental data very well.

Temperature Distribution

Spanwise temperature distributions were measured for several Rayleigh numbers at $z/D = .5$. The results are shown in Fig. 6. At $Ra = 1,530$ and $Re = 14$ (slightly below the critical value), the temperature distribution, which is uniform at lower Ra, is slightly disturbed (note the scale of the ordinate). The disturbance is very small, but it already shows the trace of periodicity, which indicates weak vertical fluid motion. Chandra [1] also observed the vertical motion of fluid at $Ra < Ra_{cr}$ in the flow visualization experiments for confined cases.
At $Ra = 1,960$ and $Re = 14$, the spanwise temperature distribution has very regular sinusoidal shape, which indicates the appearance of longitudinal vortex rolls. The temperature is high in the region where the fluid is moving upward and low where it is moving downward. For Ra up to about 8,000, the spanwise temperature distribution in the mid-plane remains very similar. The similarity is shown in Fig. 7, in which the data taken under various conditions are plotted. Scaling the spanwise length by the average pitch of vortex rolls the spanwise temperature distribution in the mid-plane is expressed as

$$\frac{T - T_m}{\Delta T} = .20 \cos(2\frac{y}{\lambda}) \text{ for } 1,900 < Ra < 8,000$$

where $T_m = 1/2(T_1 + T_2)$.

From the spanwise temperature measurements, the pitch of vortex rolls averaged over several wavelengths was calculated. In Fig. 8 the average pitch is plotted against the height of the test section. The pitch was found to be equal to 2D, as predicted by the linearized theory [2]. Measurements showed that the pitch remains constant with Ra. As plotted in Fig. 8, the data taken by Mori and Uchida show that the pitch becomes constant for $D > 15$ mm. No such behavior was observed in the present experiment. The data by Mori and Uchida seem to contradict the smoke photographs taken by Akiyama et al [11] which show very regular square cells at $D = 25.4$ mm.

For $Ra > 8,000$, the spanwise temperature distribution becomes irregular (Fig. 6), which suggests the disappearance of regular vortex rolls. At the same time fluctuations of the temperature begins to be noticeable. Fig. 9 shows the temperature fluctuation level measured by a hot wire located at the mid-plane. The temperature fluctuation remains nearly zero up to $Ra = 8,000$,

# THE EFFECT OF THERMAL INSTABILITY ON LAMINAR CHANNEL FLOW

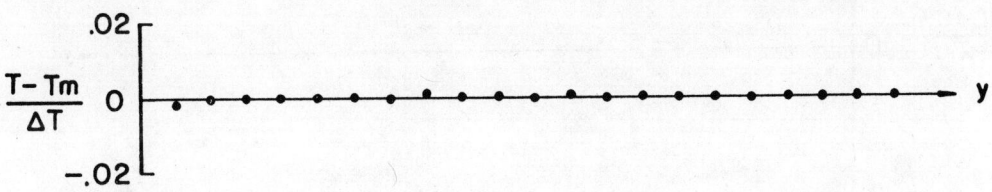

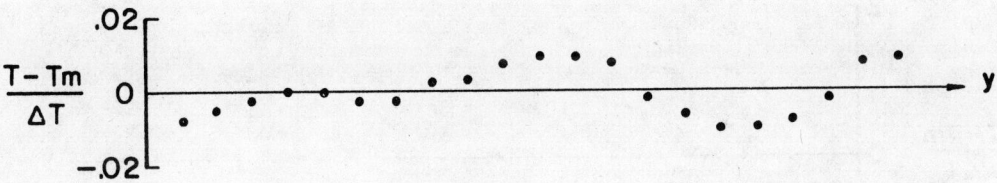

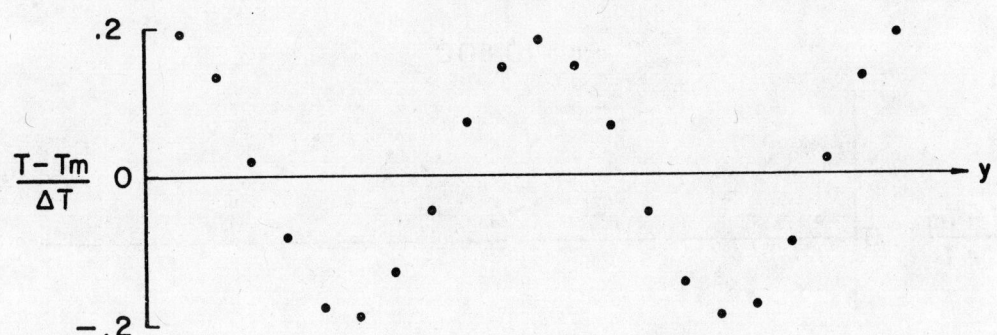

Figure 6a. Spanwise Temperature Distribution

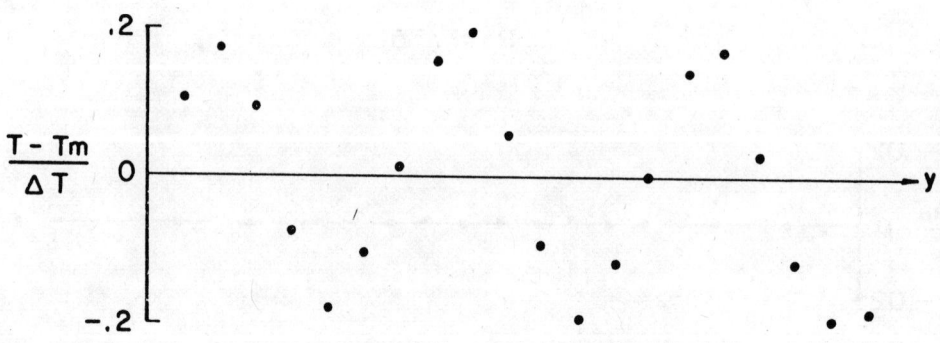

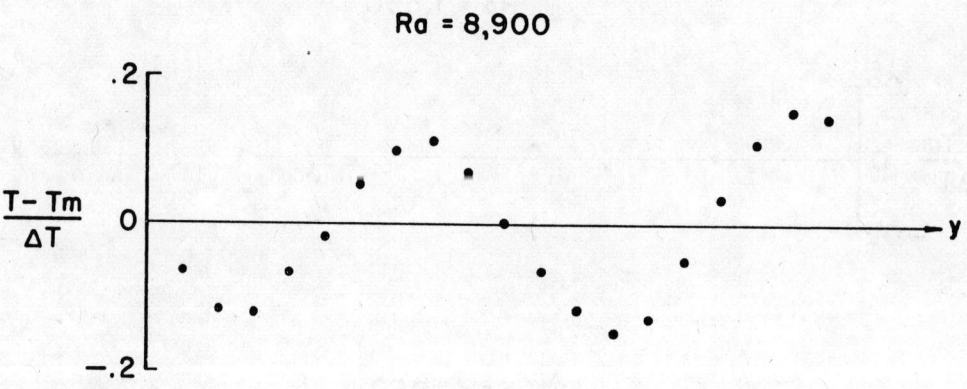

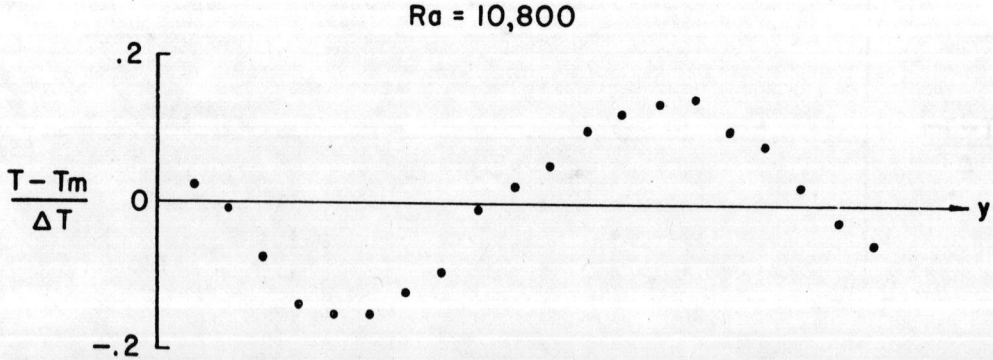

Figure 6b. Spanwise Temperature Distribution

# THE EFFECT OF THERMAL INSTABILITY ON LAMINAR CHANNEL FLOW

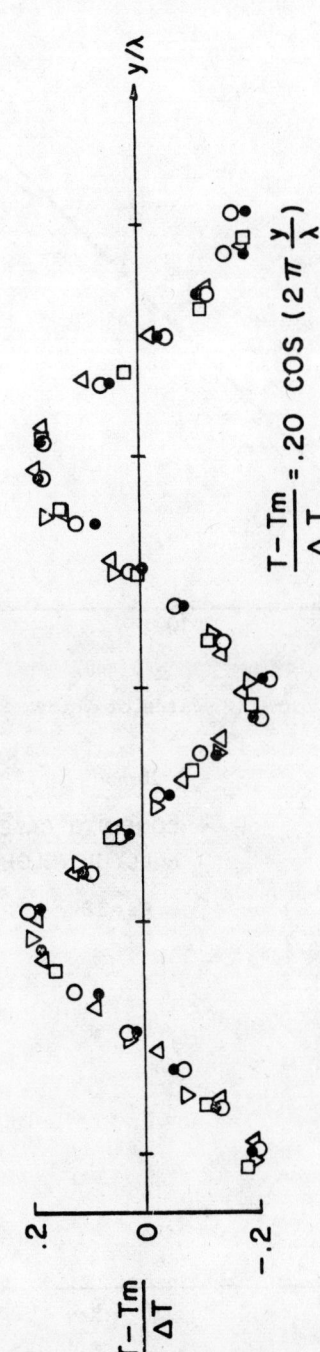

Figure 7. Similarity of Spanwise Temperature Distribution

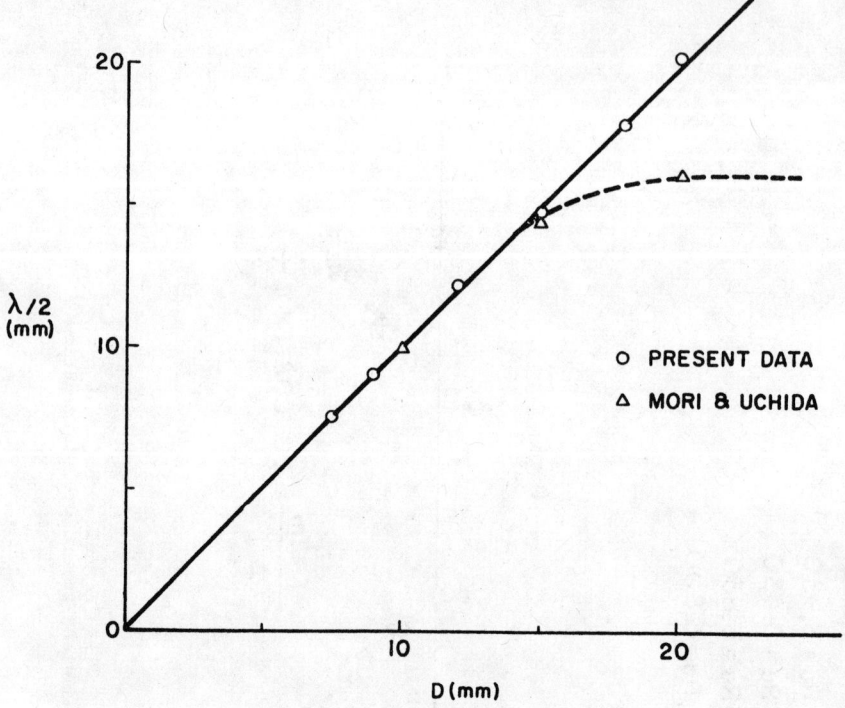

Figure 8. Pitch of Vortex Rolls

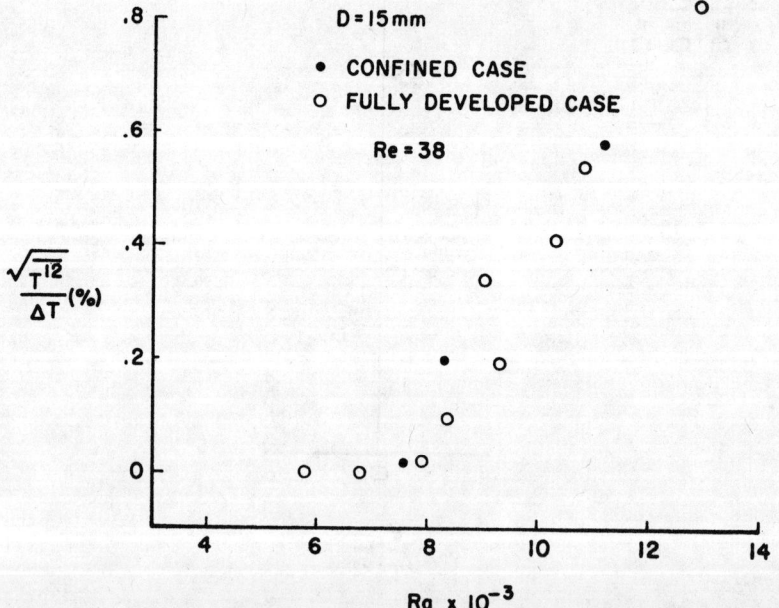

Figure 9. Temperature Fluctuation Level

# THE EFFECT OF THERMAL INSTABILITY ON LAMINAR CHANNEL FLOW 743

but for Ra > 8,000 it increases sharply with increasing Ra. This behavior was observed for both the fully-developed case and the confined case. The frequency of the fluctuation was rather low, mostly under 150 cycles/sec. No jump of heat transfer rate was observed in this range of Ra. As mentioned in the previous section, Nu becomes proportional to $Ra^{1/3}$ from about Ra = 8,000.

These fluctuations are caused by what Mori and Uchida called "second type vortex rolls". These rolls have half the size of "first type vortex rolls" (Fig. 10). They are associated with a second instability which is predicted to occur at Ra = 18,352 by the linearized theory [2] assuming the theory holds at such high Ra. However, as Mori and Uchida noted, the second type of vortex rolls can exist below the critical value deriving their energy not from buoyancy but from Reynolds stress. As the second type of vortex rolls grow stronger, they destroy the orderly vortex structure of first type vortex rolls. Whether this eventually leads to turbulent flow or to another stable state was not studied in the present experiment.

Fig. 11 shows temperature distributions in z-direction. Each point represents the average value over several wavelengths at a given z, that is

$$\overline{T}(z) = \frac{1}{L} \int_{y_o}^{y_o+L} T(y,z) dy$$

The vortex motion associated with the thermal instability and superposed flow induces the mixing of hot and cold fluid. Consequently, as Ra increases the temperature becomes uniform in the region near the mid-plane. At Ra =7,500 the temperature gradient changes sign near the mid-plane, as was predicted by Mori and Uchida's analysis. In such a region the direction of heat flux is opposite to that of conduction, which means that the amount of heat convected is larger than the amount of heat transferred from the lower plate to the upper one. Beyond Ra = 8,000 the temperature distribution becomes uniform again near the mid-plane because of random mixing.

Whereas the mean temperature distribution reflects the overall mixing of cold and hot fluid, another aspect of the effect of vortex rolls on the temperature distribution can be seen from the distribution of disturbance temperature which is defined as the difference between the local temperature and the mean temperature. Fig. 12 shows contours of constant dimensionless disturbance temperature $\theta$. $\theta$ is positive in the region where the fluid is moving upward, and negative where it is moving down. As can be seen in Fig. 12, contours of $\theta$ are not symmetric with respect to the mid-plane (z/D = .5). The centers of contours are below the mid-plane in the positive $\theta$ region, and above in the negative $\theta$ region. Such asymmetry was found to become more noticeable as Ra increases. This asymmetry is due to non-sinusoidal spanwise temperature distributions off the mid-plane caused by second type vortex rolls. The temperature distribution is distorted by second type vortex rolls in such a way that $\theta$ decreases in the upper half of the passage (.5 < z/D < 1) and increased in the lower half (0 < z/D < .5), resulting in the asymmetry of the contours.

EXPERIMENTAL RESULTS - THERMAL ENTRANCE REGION

Preliminary Investigation of Flow Field

In order to confirm the fully-developed laminar velocity profile at the test section, cross-sectional velocity distributions were measured near the entrance of the test section using a hot wire. Fig. 13 shows that in the range of Reynolds numbers covered in the present experiments the velocity profile is fully-developed at the test section. The turbulence level of the main stream was small but increased with the Reynolds number. Since most of the present data were taken for Re < 200, the effect of the free stream turbulence was negligible. To check the two-dimensionality of the main stream, spanwise

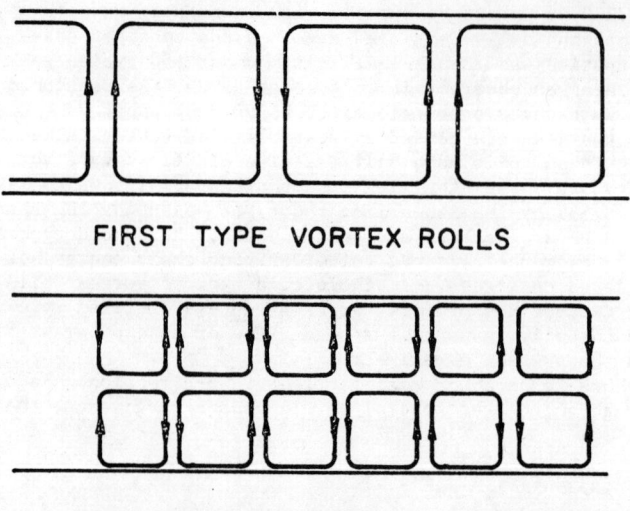

Figure 10. First and Second Type Vortex Rolls

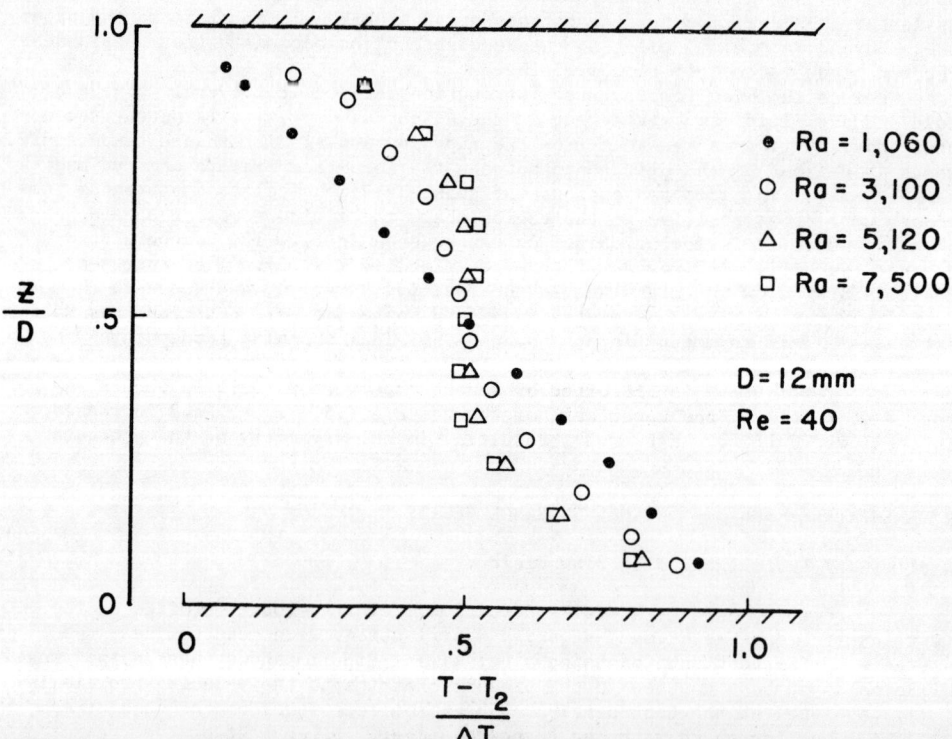

Figure 11. Mean Transverse Temperature Distribution

# THE EFFECT OF THERMAL INSTABILITY ON LAMINAR CHANNEL FLOW

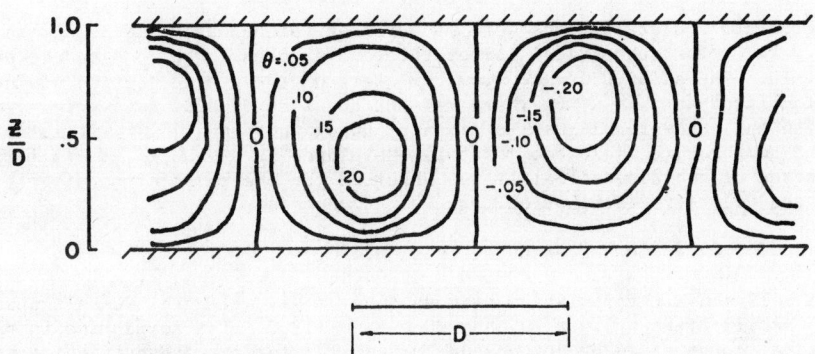

Figure 12. Contours of Constant Disturbance Temperature

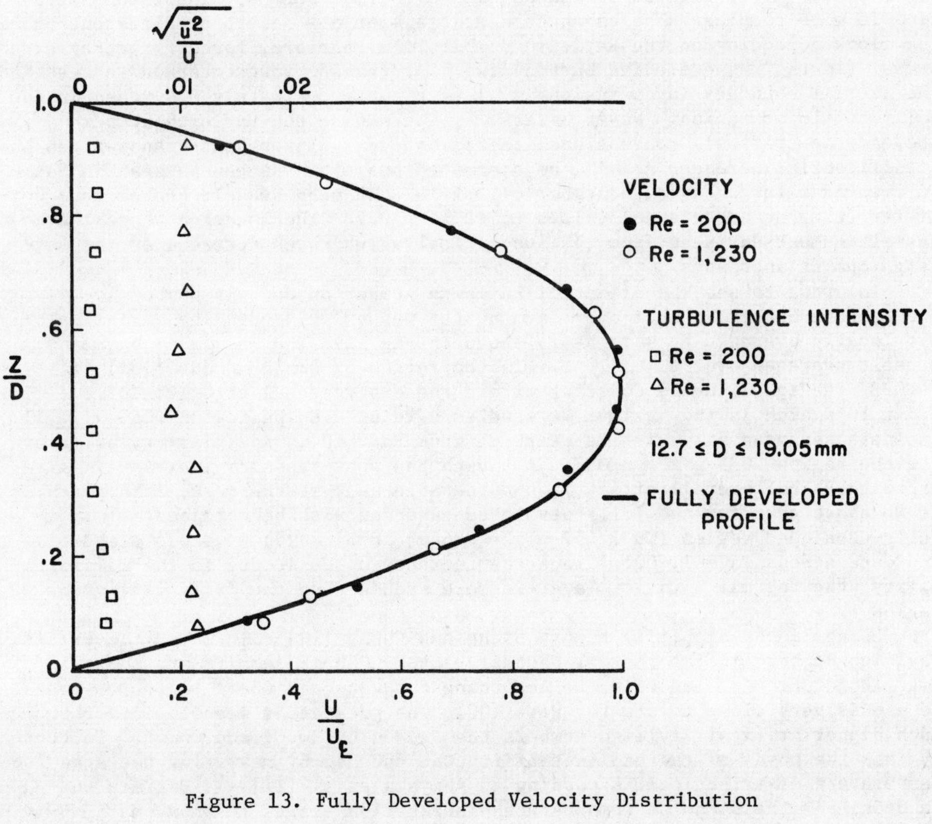

Figure 13. Fully Developed Velocity Distribution

velocity distributions were measured at the mid-height. One example is shown in Fig. 14 for Re = 200. The profile was very flat over most of the test section and the side wall boundary layers did not extend to the region covered by the main heaters so that their effect on the heat transfer measurements was considered to be small. The low turbulence level is also indicated on the figure.

Temperature distributions were measured at various axial locations in the boundary layer for subcritical temperature differences. The results are shown in Fig. 15. The axial distance from the start of the heated section was non-dimensionalized as $x' = x/DPe$ where $Pe = Pr\,Re$. The present data agree very well with the theoretical results given by Hwang and Cheng [12]. It is noted that the temperature field becomes fully-developed at around $x' = 0.4$. The same degree of two-dimensionality was observed for the temperature distribution as for the velocity distribution.

Heat Transfer Rate and Temperature Distribution

Fig. 16 shows the variation of mean (over a given length) Nusselt number with $x'$ for $10^3 < Ra < 3.0 \times 10^4$ and $30 < Re < 1,100$. The solid line is the theoretical curve given by Hutton and Turton [17] for the sub-critical temperature difference. When $Ra < 1,700$ (the critical Rayleigh number for the fully-developed region) the mean Nusselt number variation with $x'$ agrees very well with the theoretical result. For fully-developed flow and temperature fields it was shown in Fig. 5 that there is a sharp increase in the Nusselt number when the Rayleigh number exceeds 1700. Although the same qualitative trend can be observed in Fig. 16 for cases with a thermal boundary layer the situation is more complex. The enhanced heat transfer now occurs at different axial locations depending on the Rayleigh number. Furthermore, for a fixed Rayleigh number (in the range studied herein) the heat transfer augmentation is less in the thermal boundary layer region ($x'$ less than approximately 0.4) than in the fully-developed region. Nevertheless, by increasing the Rayleigh number (beyond the critical) conventional thermal boundary layer heat transfer can be significantly increased as will be discussed shortly. As can be seen in Fig. 16 even at a large Rayleigh number of $3 \times 10^4$ the mean Nusselt number does not differ from the subcritical values up to $x' = 0.7$. The location where the mean Nusselt number deviates from the subcritical value moves upstream as the Rayleigh number increases.

In order to see the effect of Rayleigh number on Nusselt number locally the average Nusselt number over each heater was calculated. The variation of the locally averaged Nu with Ra is shown in Fig. 17. Nu [$a \leq x' \leq b$] means Nusselt averaged over $a \leq x' \leq b$. In the region of small $x'$ (Nu [$.011 \leq x' \leq .023$] in Fig. 7) Nu is constant within the experimental error up to $Ra = 3.1 \times 10^4$ which is the maximum Ra studied herein. In the region $.080 \leq x' \leq .113$, Nu increases gradually with Ra starting from $Ra \doteq 10^4$. A similar trend occurs for the region $.035 \leq x' \leq .057$. Although the heat transfer increase is gradual with Ra the overall effect is considerable and, furthermore, it appears as if Nu above that for the fully-developed case can also be obtained. In the fully-developed region (Nu [$.457 \leq x' \leq .649$] in Fig. 17) Nu shows a sharp increase across $Ra = 1,700$. Since the increase of Nu is due to thermal instability, the thermal boundary layer is more stable than the fully-developed region.

Using linear stability theory Hwang and Cheng [12] calculated the critical Rayleigh numbers in the thermal boundary layer. Their results are shown in Fig. 18 for $Pr = .7$ and $Pe = \infty$. According to Hwang and Cheng the curve for $Pe = \infty$ is very close to one for $Pe > 100$. The present Nu measurements show much higher critical Rayleigh numbers than given by Hwang and Cheng. To verify this the onset of thermal instability was checked by measuring the spanwise temperature distributions. According to Akiyama et al [11] and Ostrach and Kamotani[13] the spanwise temperature distribution starts to show small

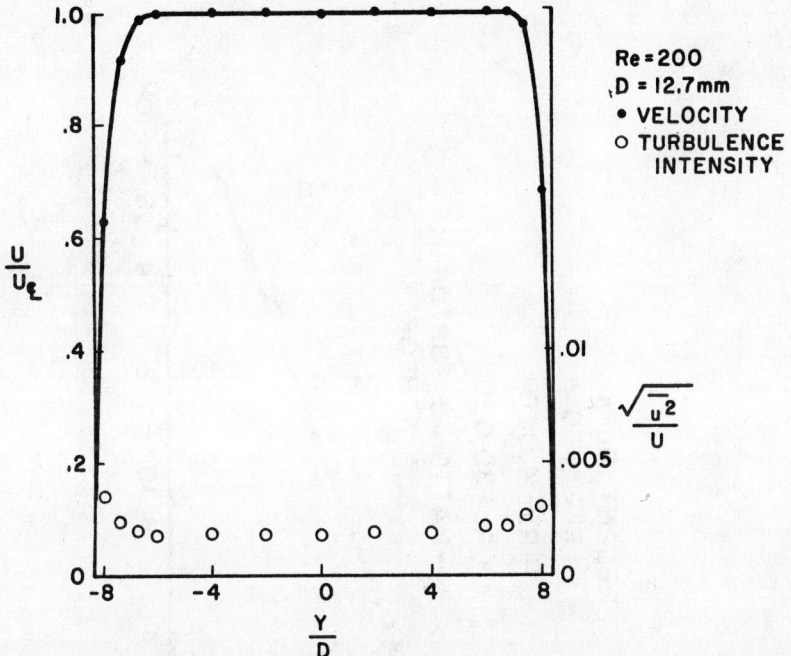

Figure 14. Spanwise Velocity Distribution

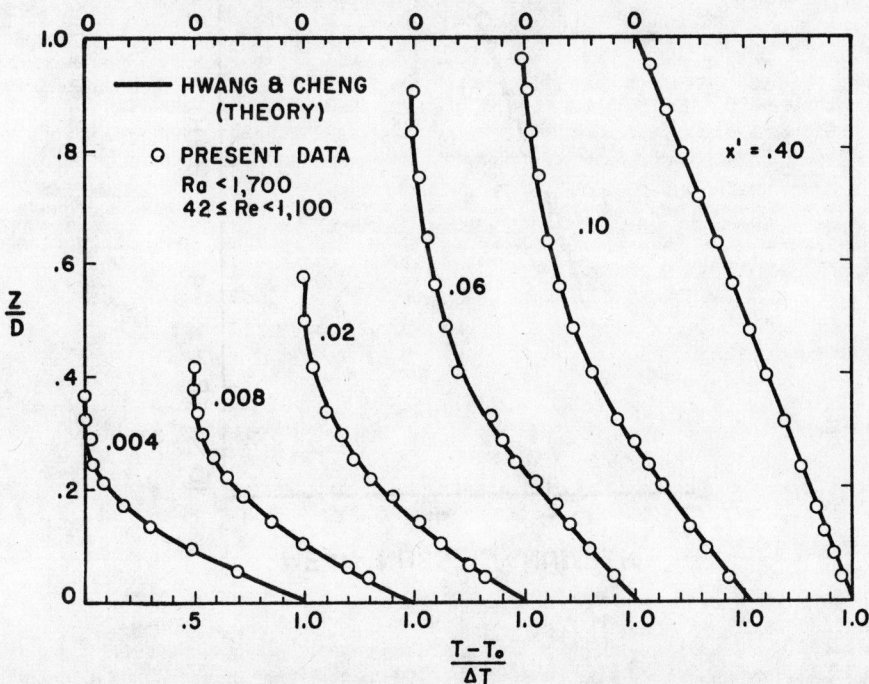

Figure 15. Sub-critical Temperature Distribution

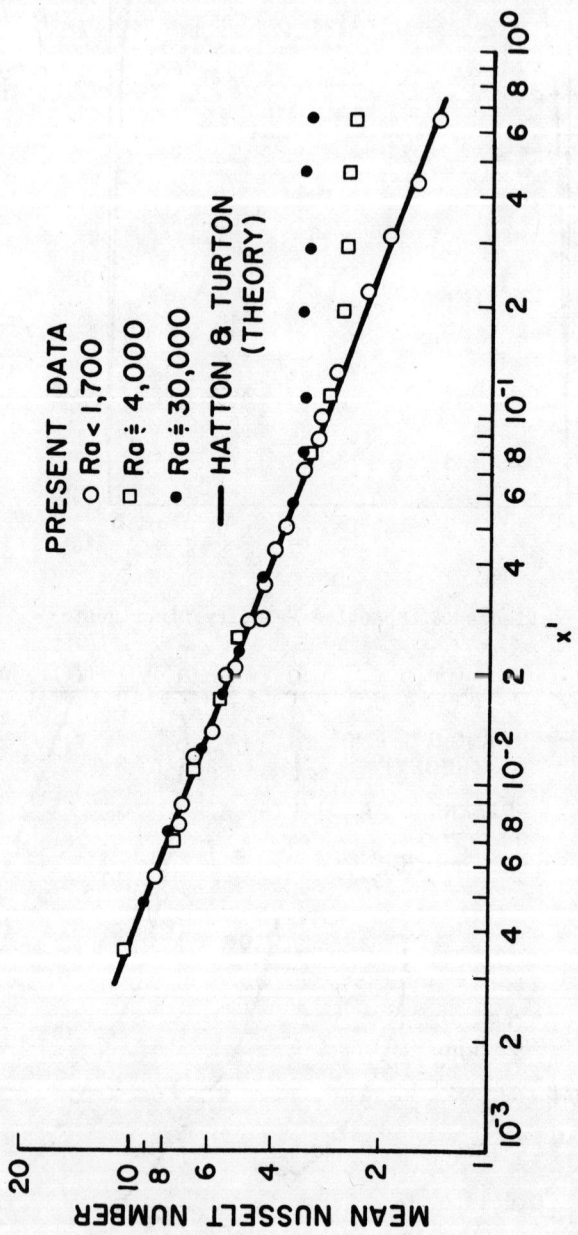

Figure 16. Mean Nusselt Number

# THE EFFECT OF THERMAL INSTABILITY ON LAMINAR CHANNEL FLOW

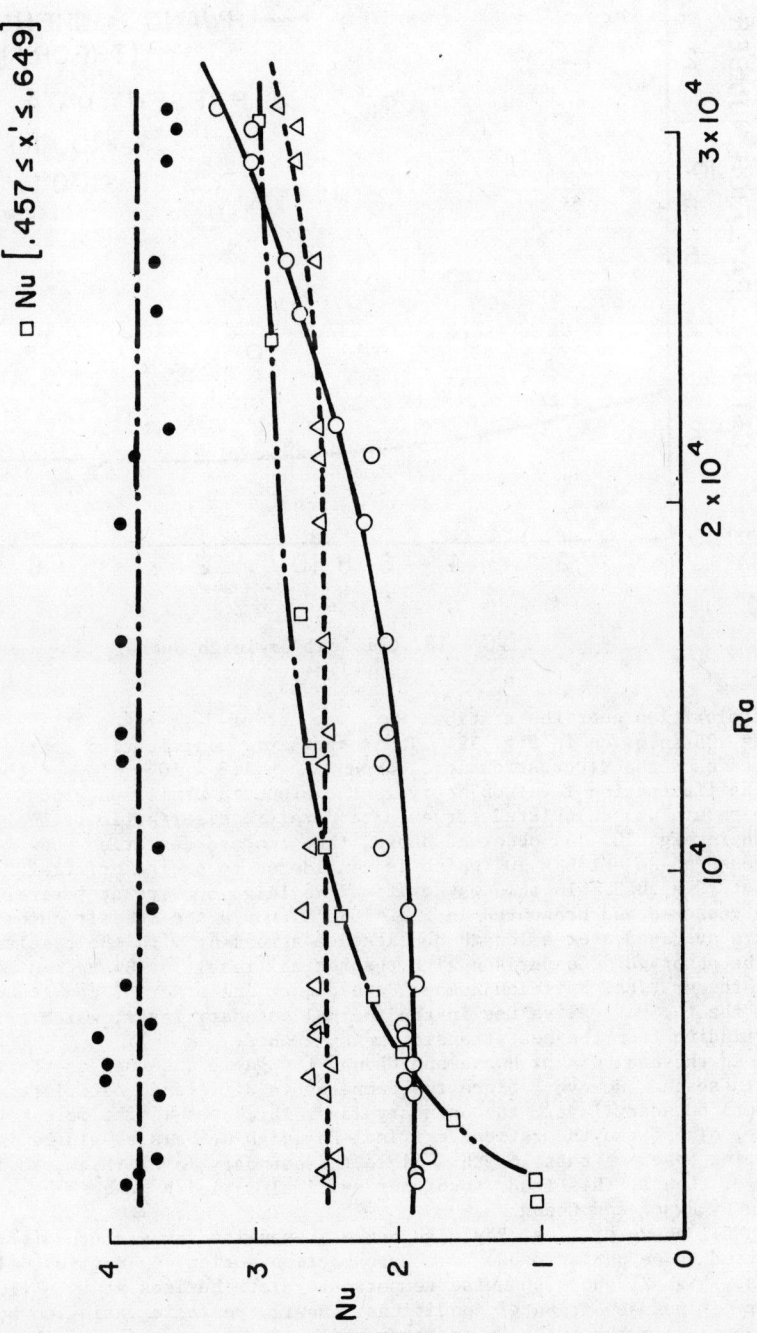

Figure 17. Locally Averaged Nusselt Number

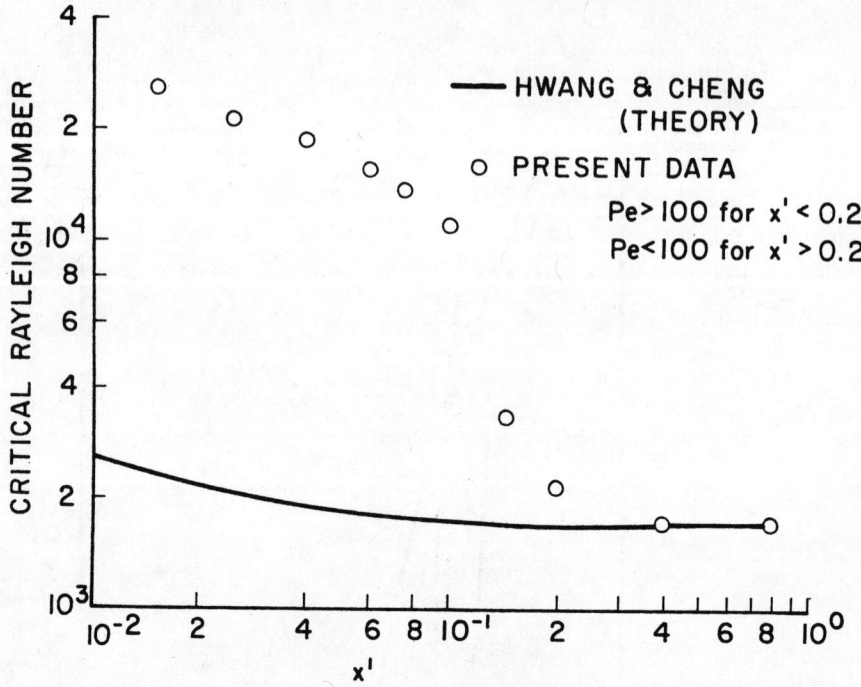

Figure 18. Critical Rayleigh Number

irregularities near the critical Ra. One set of data taken at $x' = .040$ and $z/D = .25$ is given in Fig. 19. The temperature distribution starts to show an increase of the fluctuation level around $Ra = 1.8 \times 10^4$. To see the increase of the fluctuation level objectively the standard deviation around the mean temperature was calculated for each temperature distribution. The results are shown in Fig. 20. As observed above, the standard deviation show a sharp increase at $Ra = 1.8 \times 10^4$ which is considered to be the critical Rayleigh number at $x' = .040$. In this way critical Rayleigh numbers at several locations were measured and presented in Fig. 18. Although the Nusselt numbers in Fig. 17 are averaged over a length qualitative agreement with the results of Fig. 18 can be observed. Comparison with theoretical result of Hwang and Cheng shows that the critical Rayleigh numbers are almost one order of magnitude higher than the theoretical values in the thermal boundary layer, which agrees with the finding from the heat transfer measurements.

In the analysis of Hwang and Cheng, the gap D was used as the vertical length scale. However, since the temperature difference $\Delta T$ exists only in the thermal boundary layer, the boundary-layer thickness is the more appropriate scale. Thus, the theoretical critical Rayleigh numbers should be interpreted as being based on that length. Since the boundary layer thickness is always smaller than D, this leads to higher critical Rayleigh numbers based on D than given by Hwang and Cheng.

Well above critical Rayleigh numbers spanwise temperature distributions are expected to be periodic due to the convection motion of longitudinal vortex rolls. Fig. 21 shows spanwise temperature distributions at $x' = .10$ measured under various experimental conditions. Nearly periodic variation becomes apparent above $Ra = 2 \times 10^4$. It is interesting to note that the pitch of vortex rolls is close to D instead of 2D which was observed in the fully-developed region in the range $1,700 < Ra < 8,000$ (Fig. 8). As was already discussed in the

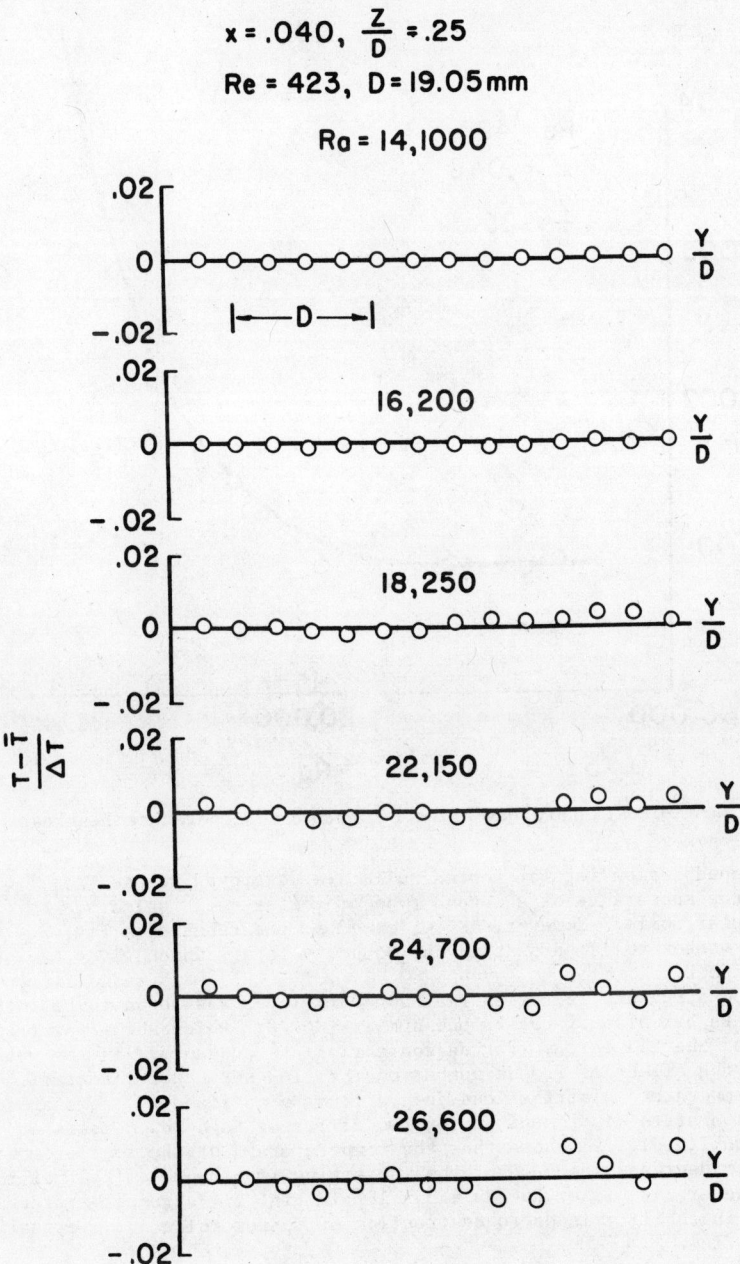

Figure 19. Spanwise Temperature Distribution Near Critical Rayleigh Number

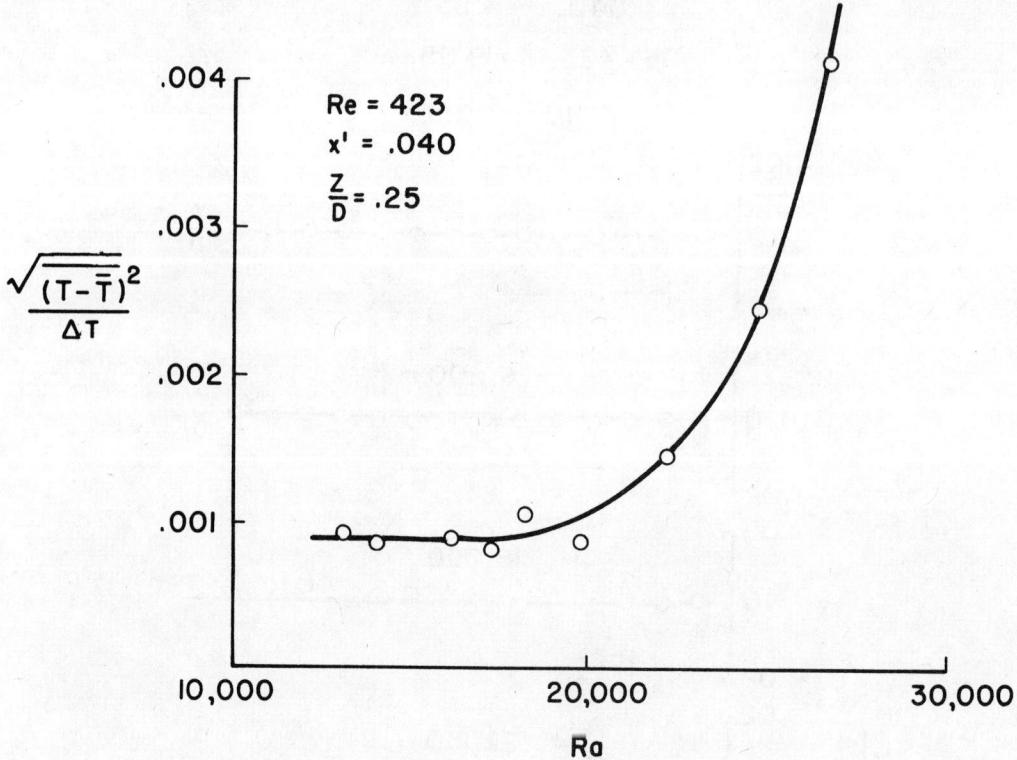

Figure 20. Standard Deviation of Spanwise Temperature Fluctuation

fully-developed region regular vortex rolls are destroyed beyond Ra = 8,000 because of the appearance of a second type of vortex rolls which has half the size if regular rolls. However, as evidenced by the results in Fig. 21, the second type vortex rolls persist in the boundary layer, which gives the periodicity equal to D.

The flow structure in the thermal boundary layer was found to be influenced not only by Ra but also by the Froude number ($Re^2/Gr$) which shows the relative importance of inertia forces (forced convection) to buoyancy. In the present experiments the effect of Froude number on the flow structure was studied by measuring temperature distributions for various Re at fixed Ra. Comparison of spanwise temperature distributions for two different Reynolds numbers at Ra $\doteq$ 2.1 x $10^4$ in Fig. 21 shows that the temperature distribution becomes more orderly as Fr decreases because of the increasing importance of thermal instability. However, at higher Ra (Ra $\doteq$ 3 x $10^4$ in Fig. 21) stronger effects of the thermal instability leads to destruction of vortex rolls and eventually to a turbulent flow.

In addition to spanwise temperature distributions, vertical temperature distributions were also measured. Fig. 22 shows the distributions of the mean temperature $\bar{T}(x,z)$ which was previously defined.

At Ra = 2 x $10^4$ even though one observes well defined vortex rolls as explained above, the mean vertical temperature distribution is not much different from the sub-critical temperature distribution at the same axial location. The vertical temperature distribution is modified gradually with increasing Ra. As a result, the local Nu increased gradually with Ra as can be seen in Fig. 17,

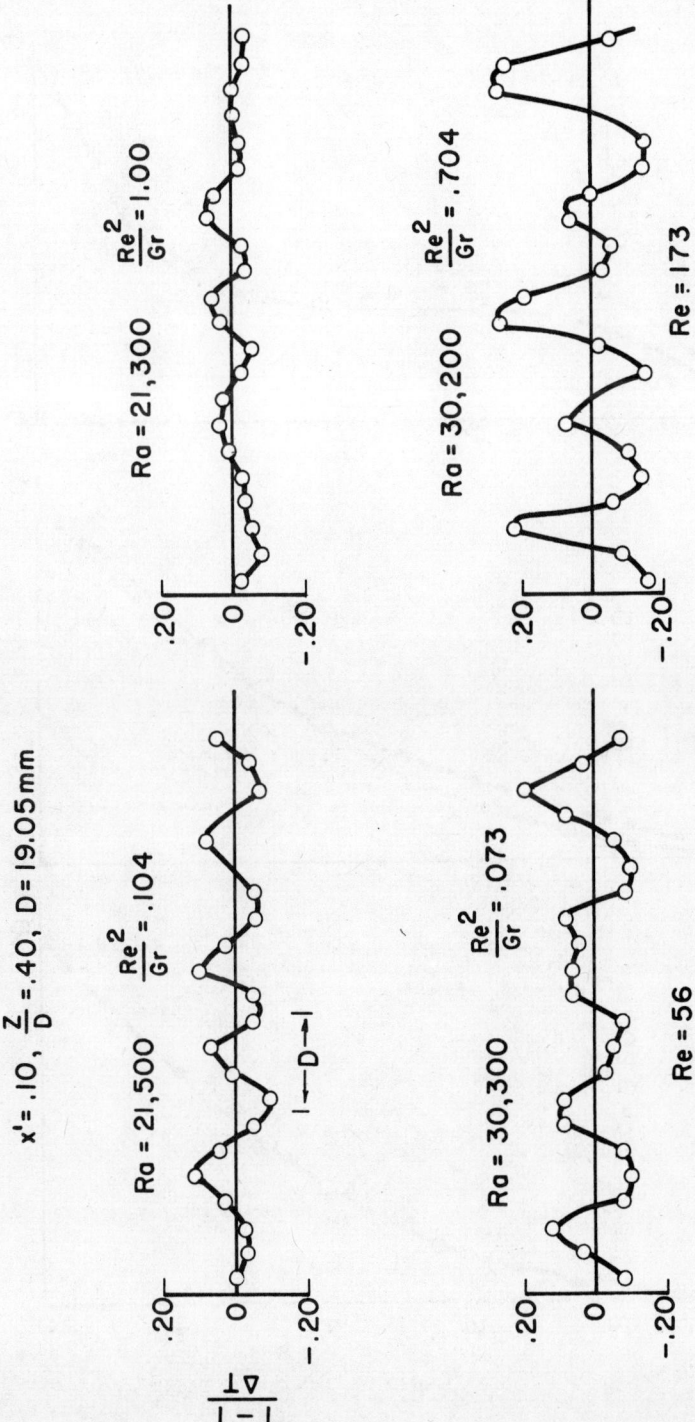

Figure 21. Spanwise Temperature Distribution

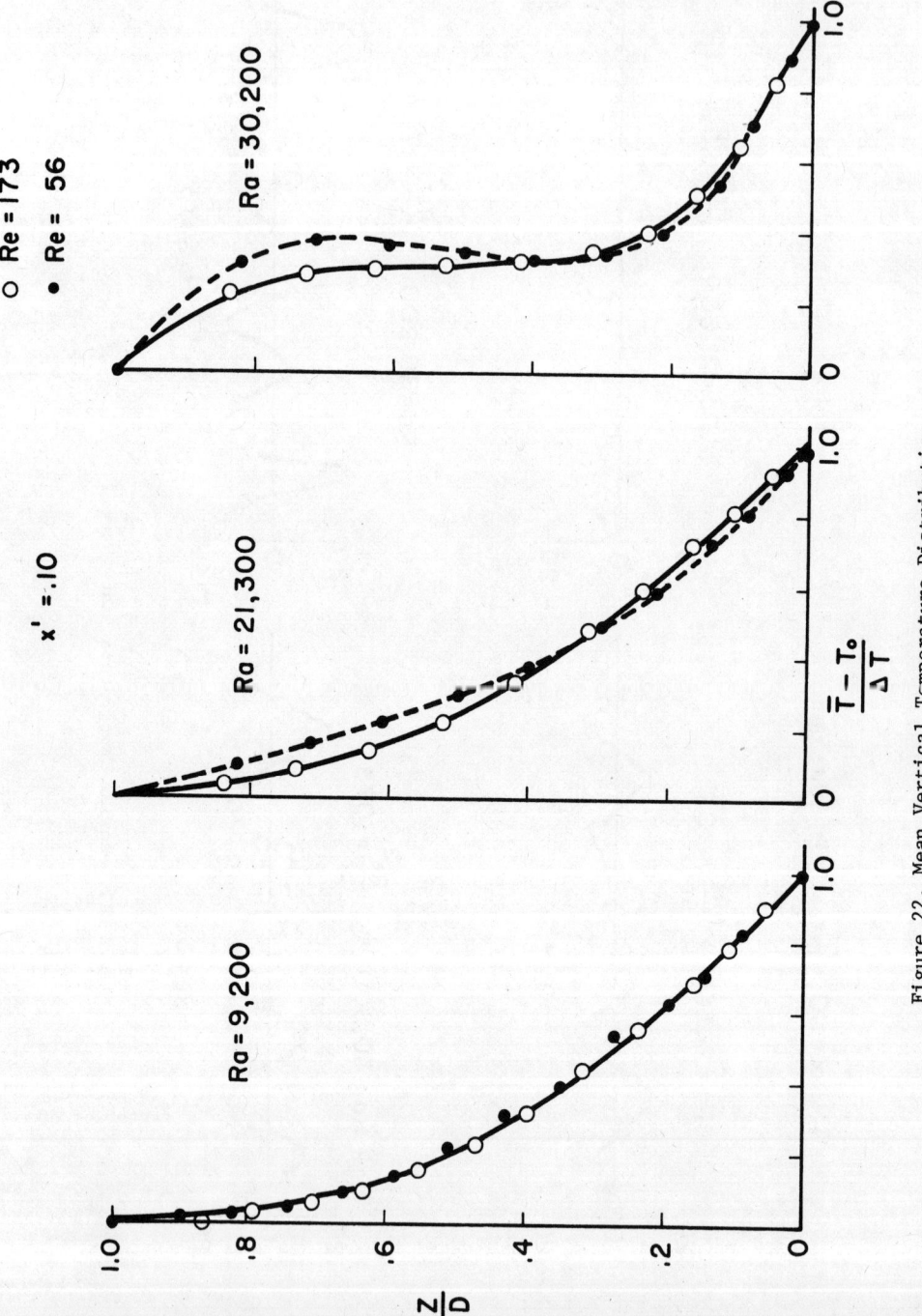

Figure 22. Mean Vertical Temperature Distribution

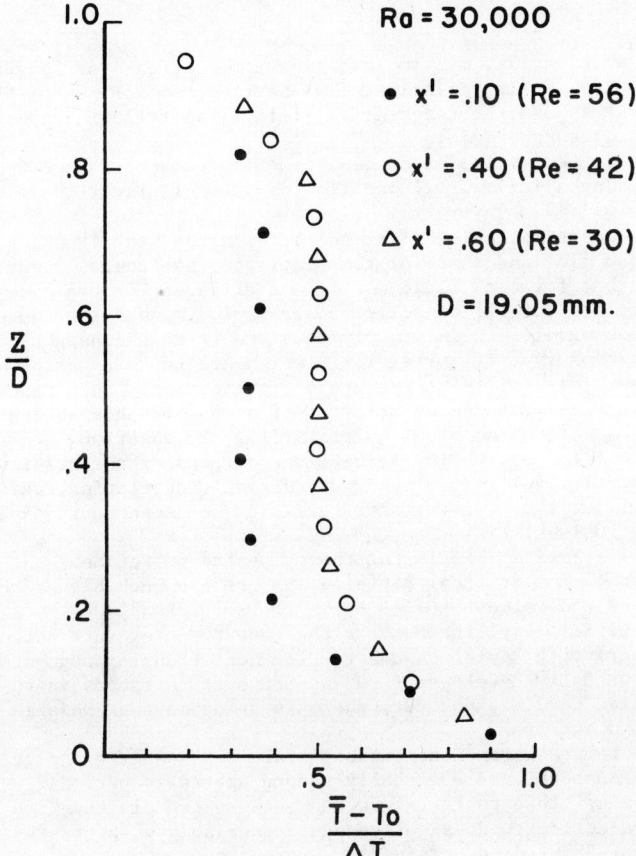

Figure 23. Mean Vertical Temperature Distribution

which is quite a contrast to a sharp increase of Nu across the critical Ra in the fully-developed region. The vortex motion does not reach the upper wall at Ra ≑ 2 x 10⁴, but its influence reaches further from the lower wall as $Re^2/Gr$ decreases (Fig. 22). At Ra = 3 x 10⁴ the vortex motion reaches the upper wall and substantial heat transfer takes place at the upper wall surface. The heat transfer rate at the upper wall increases as $Re^2/Gr$ decreases.

The effect of Froude number on the heat transfer rate at the lower plate surface was not clear because of the limited range of Fr number covered by the present experiments. However, it appears that the Nusselt number is not very sensitive to Fr, because even though Fr was changed as much as a factor of 50 (for a fixed Ra) in the present experiments, no appreciable change of the mean temperature gradient at the lower plate surface was observed (see Fig. 22).

It was found that the vertical temperature distribution profile for Ra = 3 x 10⁴ becomes independent of the axial distance at about x´ = .40 (Fig. 23). Since this is almost the same as the entrance length of the sub-critical flow, Rayleigh number does not seem to affect the entrance length at least in the range of Ra studied herein.

CONCLUSIONS

Experiments were carried out to investigate the effects of longitudinal vortex rolls in fully developed laminar flow between two horizontal plates on the heat transfer rate and the temperature field. The following conclusions were obtained from the experiments.
1. Appreciable heat transfer augmentation (Nu greater than 2.5) is obtained by superposing a fully-developed flow on the cellular flow generated by heating a horizontal fluid layer from below.
2. There is no appreciable difference between the heat transfer rate for the fully developed flow and that for the stationary horizontal layer of fluid despite the fact that forms of fluid motion are different in two cases.
3. The first type of vortex rolls create periodic spanwise temperature distributions whose wavelength is exactly equal to twice the height of the passage. However, second type of vortex rolls which are found to exist even at low Rayleigh numbers distort the temperature distribution. Such distortion increases as Ra increases, and eventually the flow becomes unstable at around Ra = 8,000. For Ra > 8,000 no stable vortex rolls are observed.

Experiments carried out to investigate the effects of thermal instability in the hydrodynamically fully-developed but thermally developing region of a channel flow of air on the heat transfer rate and the temperature field yielded the following conclusions:
1. The flow is more stable in the thermal entrance region than the fully developed region. The critical Rayleigh numbers are much higher than the theoretically predicted values.
2. Beyond critical Rayleigh numbers the heat transfer rate increases gradually with increasing Rayleigh number. The heat transfer augmentation was lower than that for fully-developed flow and temperature fields in the Rayleigh number range studied herein. However, there is evidence that this is different for larger Rayleigh numbers.
3. Spanwise temperature distributions start to show irregularities across critical Rayleigh numbers, and eventually become spatially periodic, the dominant wavelength being close to the gap width between two plates.
4. The entrance length does not change appreciably with the Rayleigh number in the range of Ra studied herein.

ACKNOWLEDGEMENT

This research was supported by the U. S. Air Force Office of Scientific Research under Grant AFOSR - 72-2342C.

REFERENCES

1. Chandra, K., "Instability of Fluid Heated from Below", Proceedings of the Royal Society (London), Series A, Vol. 164, 1938, pp. 231-242.

2. Mori, Y. and Uchida, Y., "Forced Convective Heat Transfer between Horizontal Plates", International Journal of Heat and Mass Transfer, Vol. 9, 1966, pp. 803-817.

3. Jakob, M., "Heat Transfer", John Wiley & Sons, Inc., Vol. 1, 1949.

4. Malkus, W.V.R., "Discrete Transitions in Turbulent Convection", Proceedings of the Royal Society (London), Series A, Vol. 225, 1954, pp. 185-195.

5. Silveston, P. L., "Warmedurchgang in waagerechten Flussigkeitsschichten", Forschung auf dem Gebiete des Ingenieurwesens, Vol. 24, No. 2, 1958, pp. 59-69.

6. Globe, S. and Dropkin, D., "Natural-Convection Heat Transfer in Liquids Confined by Two Horizontal Plates and Heated from Below", Journal of Heat Transfer, Transactions of the ASME, Series C, Vol. 8, 1959, pp. 24-28.

7. Plows, W.H., "Some Numerical Results for Two-Dimensional Steady Laminar Bénard Convection", The Physics of Fluids, Vol. 11, 1968, pp. 1593-1599.

8. Ogura, Y. and Yagihashi, A., "A Numerical Study of Convection Rolls in a Flow between Horizontal Parallel Plates", Journal of the Meteorological Society of Japan, Vol. 47, 1969, pp. 205-217.

9. Hwang, G. J. and Cheng, K. C., "A Boundary Vorticity Method for Finite Amplitude Convection in Plane Poiseuille Flow", Developments in Mechanics, Proceedings of the 12th Midwestern Mechanics Conference, Vol. 6, 1971, pp. 207-220.

10. Nakayama, W., Hwang, H. J. and Cheng, K. C., "Thermal Instability in Plane Poiseuille Flow", Journal of Heat Transfer, Transactions of the ASME, Series C, Vol. 92, No. 1, 1970, pp. 61-68.

11. Akiyama, M., Hwang, G. J. and Cheng, K. C., "Experiments on the Onset of Longitudinal Vortices in Laminar Forced Convection between Horizontal Plates", Journal of Heat Transfer, Transactions of the ASME, Series C, Vol. 93, No. 4, 1971, pp. 33-341.

12. Hwang, G. J. and Cheng, K. C., "Convective Instability in the Thermal Entrance Region of a Horizontal Parallel-Plate Channel Heated from Below", Journal of Heat Transfer, Transactions of ASME, Series C, Vol. 95, No. 1, 1973, pp. 72-77.

13. Ostrach, S. and Kamotani, Y., "Heat Transfer Augmentation in a Laminar Fully-Developed Channel Flow by Means of Heating from Below", Journal of Heat Transfer, Vol. 97, Series C, No. 2, May 1975.

14. Schlüter, A., Lortz, D. and Busse, F., "On the Stability of Steady Finite Amplitude Convection", Journal of Fluid Mechanics, Vol. 23, 1965, pp.129-144.

15. Segel, L. A. and Stuart, J. T., "On the Question of the Preferred Mode in Cellular Thermal Convection", Journal of Fluid Mechanics, Vol. 13, 1962, pp. 289-306.

16. Churchill, S. W. and Usagi, R., "A General Expression for the Correlation of Rates of Transfer and Other Phenomena", AIChE Journal, Vol. 18, No. 6, 1972, pp. 112-1128.

17. Hatton, A. P. and Turton, J. S., "Heat Transfer in the Thermal Entry Length with Laminar Flow between Parallel Walls at Unequal Temperatures", International Journal of Heat and Mass Transfer, Vol. 5, 1962, pp. 673-679.

# ENHANCEMENT OF TURBULENT HEAT TRANSFER DUE TO BUOYANCY FOR DOWNWARD FLOW OF WATER IN VERTICAL TUBES

J. D. JACKSON and J. FEWSTER

Engineering Laboratories, University of Manchester,
United Kingdon

## ABSTRACT

Previous investigations of turbulent mixed convection to water in vertical tubes with buoyancy opposing the applied flow are reviewed. Some recent experimental work by the authors, which considerably extends the range of variables covered so far, is presented. These results confirm that the influence of buoyancy is to enhance heat transfer, the effect building up steadily as the free convective component increases in relation to the forced one. An explanation of this enhancement effect is given and theoretical considerations are presented which lead to a correlation of our data and to a correlation equation. The paper concludes with comparisons between the present method of correlation and those of earlier investigations.

## NOMENCLATURE

$c_f$    Friction coefficient, $c_f = \tau_w / \tfrac{1}{2} \rho \, u_{mean}^2$
$C_p$    Specific heat at constant pressure (J/kgK)
$d$    Tube internal diameter ( m )
$G_r$    Grashof number, $G_r = (\rho_b - \rho_w) g d^3 \mu^2 / \rho_b^2$
$\overline{G_r}$    Grashof number, $\overline{G_r} = (\rho_b - \bar{\rho}) g d^3 \bar{\mu}^2 / \rho_b^2$
$g$    Acceleration due to gravity (m/s$^2$)
$k$    Thermal conductivity (W/mK )
$K_1$    Constant in equation 8
$K_2$    Constant in equation 10
$L$    Length of heated tube ( m )
$\dot{m}$    Mass flow rate (kg/s)
$Nu$    Nusselt number, $N_u = \alpha d/k$
$Nu_o$    Nusselt number for forced convection
$\overline{Nu}_o$    Forced convection Nusselt number based on $\overline{P}_r$
$P_r$    Prandtl number, $P_r = \mu C_p / k$
$\overline{P}_r$    Mean of Prandtl numbers evaluated at $T_b$ and $T_w$
$\dot{q}_w$    Heat flux at wall (W/m$^2$)
$R_e$    Reynolds number, $R_e = 4 \dot{m} / \pi \mu d$
$R_e'$    Reynolds number corresponding to shear stress $\tau_w'$ for forced flow
$T$    Temperature (°C)
$T_b$    Bulk temperature (°C)
$T_w$    Temperature at wall (°C)
$\Delta T$    Temperature difference, $\Delta T = (T_w - T_b)$ (K)
$u$    Velocity (mean value, $u_{mean} = 4\dot{m}/\rho \pi d^2$) (m/s)
$x$    Downstream coordinate (m)
$y$    Transverse coordinate (m)
$y^+, u^+$    Universal parameters, $u^+ = u/(\tau_w/\rho)^{1/2}$; $y^+ = y(\tau_w \rho)^{1/2}/\mu$

Greek

$\alpha$     Heat transfer coefficient, $\alpha = \dot{q}_w/\Delta T$ ( $W/m^2$ K )
$\alpha_o$     Heat transfer coefficient for forced convection ( $W/m^2$ K )
$\alpha_o'$     Heat transfer coefficient for Reynolds number $R_e'$ ( $W/m^2$ K )
$\alpha_f$     Heat transfer coefficient for free convection ( $W/m^2$ K )
$\delta_B$     Buoyant layer thickness ( m )
$\delta_M$     Thickness of combined sub-layer and buffer layer ( m )
$\delta_M^+$     Dimensionless version of $\delta_M$, $\delta_M^+ = \delta_M (\tau_w \rho)^{1/2}/\mu$
$\delta_T$     Thickness of thermal layer (m)
$\mu$     Viscosity (kg/ms)
$\rho_w$     Density evaluated at $T_w$ (kg/m$^3$)
$\rho_b$     Density evaluated at $T_b$ (kg/m$^3$)
$\bar{\rho}$     Integrated mean density, $\bar{\rho} = \frac{1}{\Delta T}\int_{T_b}^{T_w} \rho\, dT$   (kg/m$^3$)
$\tau_w$     Shear stress at the wall (N/m$^2$)
$\tau_w'$     Wall shear stress for a forced flow rate adjusted to increase the stress level in the buffer layer to that of the mixed convection case (N/m$^2$)
$\Delta \tau_{sB}$     Increase in shear stress across buoyant layer (N/m$^2$)

## 1. INTRODUCTION

In this paper we consider the problem of mixed convection in a vertical tube for the case where buoyancy forces oppose the applied flow, ie. upward flow in a cooled tube or downward flow in a heated tube. For this configuration the distortion of the velocity profile due to buoyancy is such as to cause turbulence generation at Reynolds numbers for which laminar flow would prevail under buoyancy-free conditions. Thus, we are concerned here exclusively with turbulent flow even though in some of our experiments the Reynolds numbers are near to the usual transition value.

There have been surprisingly few studies of the present problem, in spite of the fact that it is now nearly forty years since it was first investigated by Watzinger and Johnson (1939)[1]. In their experiments warm water flowed upwards in a tube which was externally cooled by a counterflow of cold water. Subsequent experiments have all involved downward flow in heated tubes. For water at atmospheric pressure there have been five investigations; Petukhov and Nolde (1959)[2], Petukhov and Strigin (1968)[3] and Herbert and Sterns (1968 and 1972)[4,5]. Table I below shows the test section dimensions and the range of conditions covered.

Table I. Previous investigations with water at atmospheric pressure

| Reference | d(mm) | L/d | $R_e$ | $G_r$ (maximum) |
|---|---|---|---|---|
| Watzinger and Johnson (1939)[1] | 50 | 20 | 1600–10000 | $4.10^7$ |
| Petukhov and Nolde (1959)[2] | 16 | 130 | 2000–18000 | $3.10^6$ |
| Petukhov and Strigin (1968)[3] | 19 | 20 | 300–14000 | $5.10^6$ |
| " " " " | 50 | 80 | 300–30000 | $7.10^7$ |
| Herbert and Sterns (1968)[4] | 24 | 95 | 6000–65000 | $2.10^7$ |
| " " " (1972)[5] | 23 | 80 | 6000–65000 | $2.10^7$ |

In the above investigations the Prandtl number was mainly in the range 2 to 5.

Mixed convection heat transfer can be thought of in terms of two components, one due to the forced flow and one due to the buoyancy induced flow. The former depends on Reynolds number and the latter on Grashof number. The lower the Reynolds number and the bigger the Grashof number the more important is the free convection component in relation to the forced one. Very significant departures from the forced convection condition are evident in the data of some of the investigations listed in Table I, the effect being an enhancement of heat transfer relative to that for forced flow alone. The general behaviour is illustrated in Table II. Note that the figures quoted are by no means precise but serve to illustrate the order of magnitude of the enhancement effect and the trends with $R_e$ and $G_r$.

Table II. Approximate figures for enhancement of heat transfer due to buoyancy based on data from previous investigations

| Source | Grashof Number | Reynolds Number | | | |
|---|---|---|---|---|---|
| | | 6000 | 10000 | 18000 | 30000 |
| Petukhov and Strigin[3] | $5.10^6$ | 25% | 10% | - | - |
| Herbert and Sterns[5] | $2.10^7$ | 40% | 20% | 0 | 0 |
| Watzinger and Johnson[1] | $3.10^7$ | 70% | - | - | - |
| Petukhov and Strigin[3] | $7.10^7$ | 190% | 100% | 50% | 20% |

In the investigation of Petukhov and Nolde[2] ($G_r = 3.10^6$, $2000 < R_e < 18000$) no significant enhancement was observed.

So far we have concentrated our attention on data for water at atmospheric pressure. Two recent papers, Alferov et.al. (1973)[6] and Ikryannikov et.al.(1972)[7], have reported strongly enhanced heat transfer for downward flow of fluid at supercritical pressure in heated tubes (water at about 240 bar in one case and carbon dioxide at 76 bar in the other). The form in which the results are published prevents direct comparison with the atmospheric pressure water data but the trends are clearly similar.

For air the problem of mixed convection with opposed forced and free components has been studied by Eckert and Diaguilla (1954)[8], Brown and Gauvin (1966)[9], Hall and Price (1971)[10], Khosla et.al.(1974)[11] and Axcell (1975)[12]. Data from three of these sources (references 8,10 and 12) provide clear evidence of enhancement of heat transfer but only the last-mentioned investigation has produced data suitable for comparison with that reviewed earlier here. Again the trends are clearly similar to those evident in Table II.

Finally we should mention an interesting aspect of the problem which has been examined recently by Khosla et.al.[11] using the electrochemical indicator technique of Baker [13]. This concerns the fact that as $R_e$ and $G_r$ are varied so as to cause a build up of enhancement a stage is reached beyond which the flow direction in the wall layer region reverses. Khosla et.al. performed tests with upward flow of water in an externally cooled glass tube and made some measurements of the conditions under which the flow reversal takes place. The fact that a stalled flow exists under these conditions is not evident from heat transfer observations; the enhancement effect varies steadily as the changeover condition is passed through.

In the next section some recent experiments are described on downward flow of atmospheric pressure water in a heated tube of 98.4 mm internal diameter. These tests represent a significant extension of the range of parameters covered so far.

## 2. EXPERIMENTAL INVESTIGATION

### 2.1 Apparatus and procedure

The experimental circuit and test section are shown diagrammatically in figure 1. The test section is fed from a header tank situated 40 m above floor level, via valves which direct the flow in the test section to be either upwards or downwards. The water is discharged into a reservoir tank via a flow control valve and metering section and then pumped back independently to the header tank. The flow rate is measured by an orifice plate and two manometers; a mercury/water manometer for the higher flow rates and an air/water manometer for lower flow rates. The orifice plate has been accurately calibrated by collecting water at the reservoir and measuring the time taken for a vessel of known volume to be filled. The main uncertainty in flow rate is associated with reading the manometer and is estimated to be less than $\pm 1\%$.

The test section consists of a stainless steel tube of inside diameter 98.4 mm and length 9 m. Mixing boxes of diameter 200 mm having radial entry/exit connections are situated at each end of the test section.

The test section is resistance heated, the power being transmitted to the tube by copper bars and clamps from a step-down transformer which is supplied with A.C from the mains via a variable auto-transformer. The power supply is calculated from measurements of test section current and potential difference, the instruments used being accurate to within $\pm 1\%$ in the working range. Rigid fibreglass insulation is used to ensure negligible heat loss from the test section.

Chromel alumel thermocouples resistance welded to the outside of the test section are situated every 2 diameters along the heated length of 50 diameters and also along the inlet/outlet lengths of 10 diameters. In addition, at five stations on the heated length four thermocouples are situated at positions around the circumference 90° apart. The thermocouple e.m.f. is measured using a digital voltmeter with a high degree of A.C. rejection and a resolution equivalent to 0.05°C. Samples of the thermocouple wire showed no significant discrepancy when compared against accurately calibrated standard thermocouples.

Inside wall temperature is calculated from outside wall temperature knowing the heat generation rate and wall thickness. The uncertainty in this correction is estimated to be less than $\pm 0.35$°C for the worst case. Thus, an upper limit on uncertainty for inside wall temperature is $\pm 0.4$°C.

Inlet bulk temperature is known to $\pm 0.05$°C from measurements of wall temperature on the unheated tube up-stream of the test section. The fluid is supplied to the test section at near ambient temperature.

The experimental procedure is first to set the flow rate to the required value by means of the flow control valve, and then, using the autotransformer control to supply power to the test section at a rate which causes the wall temperature distribution to achieve the required level. The apparatus responds quickly to the controls and steady distributions of wall temperature are arrived at after a few minutes. Very repeatable readings can be obtained and the results are completely insensitive to the manner in which experimental conditions are approached.

Ten flow rates in the range 0.0747 kg/s to 3.25 kg/s were chosen for the tests. Wall temperature distributions were recorded for each of six heat fluxes, up to a maximum of 26 kW/m$^2$, for each flow rate. The fluid inlet temperature was between 21°C and 23°C and the bulk temperature at outlet at the condition of lowest flow rate and highest heat flux rose to a maximum value of about 65°C. The maximum difference between wall and

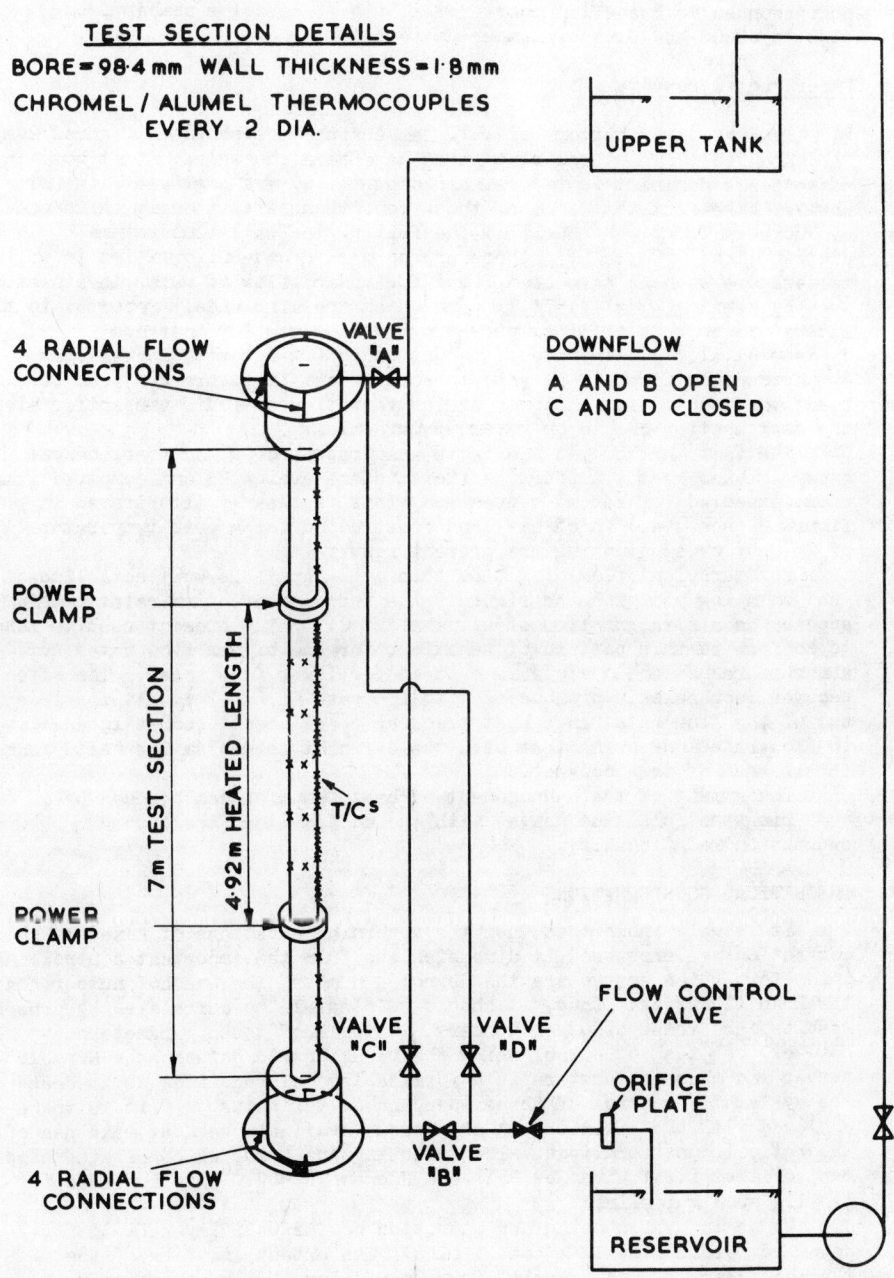

FIGURE 1. EXPERIMENTAL CIRCUIT AND TEST SECTION.

bulk temperature was about 30°C.

In terms of dimensionless parameters the conditions of the tests corresponded to Prandtl number from 2.5 to 7, Reynolds number from 1000 to 40000 and Grashof number up to $3.10^8$.

## 2.2 Experimental results

Typical distributions of wall temperature are shown in figures 2(a), 2(b) and 2(c). The three flow rates have been chosen such that buoyancy effects are dominant in (a), significant in (b) and negligible in (c). However, the fact that some of these conditions are strongly influenced by buoyancy is not obvious from the form of the wall temperature distributions. Certainly, there are no marked non-uniformities in wall temperature such as have been found for upward flow of water in a heated tube by Kenning et al (1974)[14] (and which are also widely reported in the literature on supercritical pressure fluids - see for instance Jackson et.al.[15]). The modest reduction in $(T_w-T_b)$ with increase in x/d evident in figure 2(a) (and to some extent in figure 2(b)) can be readily accounted for in terms of the variation of fluid properties along the test section due to bulk temperature rise.

The fact that buoyancy effects are present does, however, become apparent when heat transfer coefficients are evaluated and compared with those expected for forced convection. This is clearly illustrated in figure 3 (see over) which has been constructed for a bulk temperature of 25°C by cross-plotting the present results.

For turbulent flow of fluids such as water it is well established that on a log plot such as figure 3 the forced convection relationship appears as a straight line of slope about 0.8. The present results tend to conform to this pattern of behaviour for the higher flow rates but a marked reduction in slope is evident for lower flow rates. The effect becomes increasingly pronounced with increase in $(T_w-T_b)$. At the lower end of the flow rate range heat transfer coefficient becomes insensitive to flow rate indicating that here the dominant mechanism for heat transfer is that of free convection.

The extent of the enhancement of heat transfer can be seen by comparing the full line curves with the dotted line (the forced convection relationship).

## 3. THEORETICAL CONSIDERATIONS

It is well known that when the governing equations of mixed convection are expressed in dimensionless form the important controlling parameters which emerge are the Reynolds number, the Grashof number and the Prandtl number. Thus, it should be possible to correlate the present results, and those of other workers, in terms of these parameters. However, the task of establishing the relationship between the Nusselt number and these parameters is a formidable one, requiring as it does the systematic control of three independent variables. If it is to be achieved with the limited data at present available we must make use of theoretical considerations. The following analysis, which is a development of that first given by Hall and Jackson (1969)[16], is intended to provide such guidelines.

We begin by focussing our attention on the wall layer fluid for downward forced flow in a heated pipe. The extent and form of the thermal layer in such a system is mainly determined by the manner in which the turbulent diffusion properties build up across the boundary layer. For the range of Prandtl under consideration here the thermal layer is

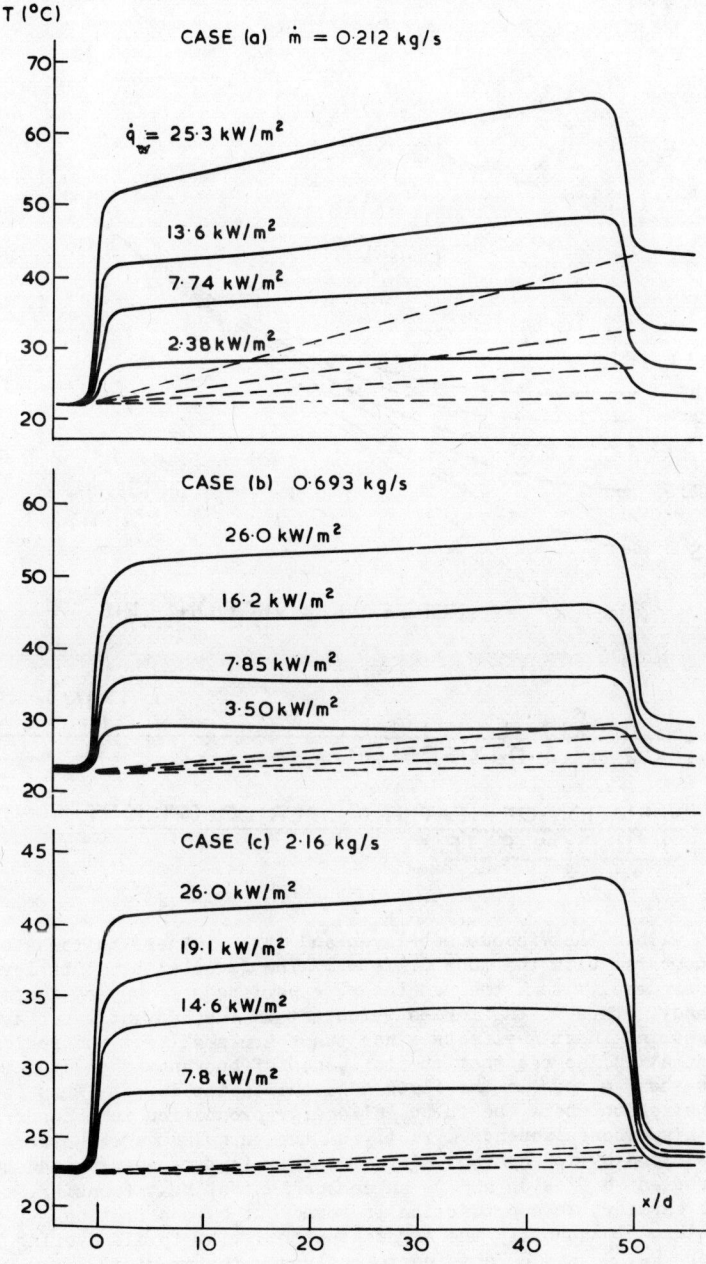

FIGURE 2. TYPICAL DISTRIBUTIONS OF WALL TEMPERATURE (DOTTED LINES SHOW THE CALCULATED DISTRIBUTIONS OF BULK TEMPERATURE)

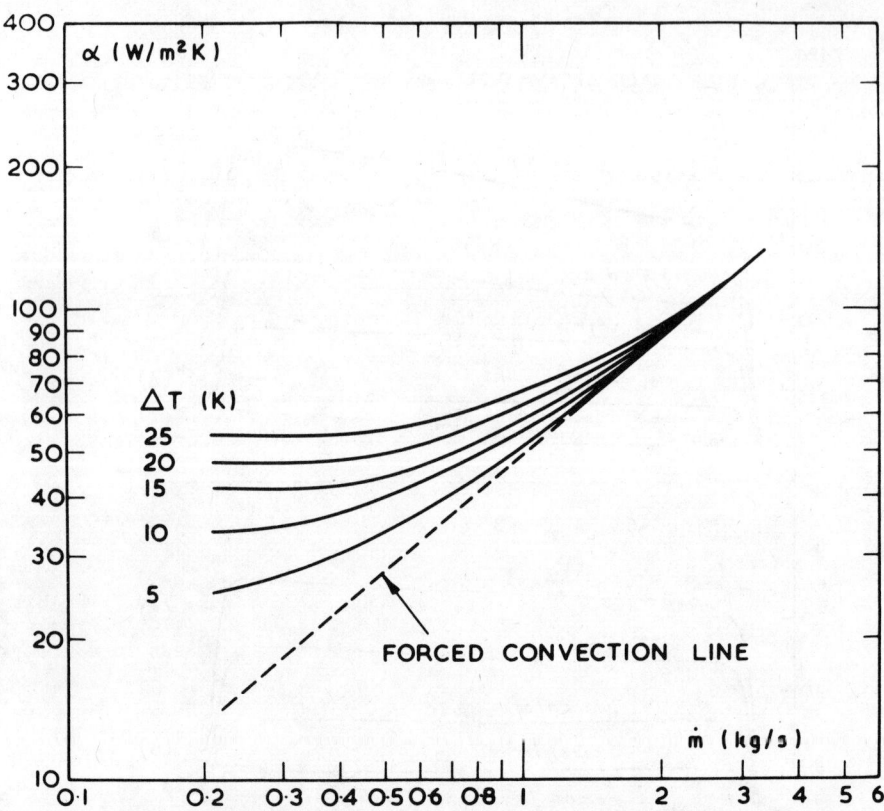

FIGURE 3. VARIATION OF HEAT TRANSFER COEFFICIENT WITH FLOW RATE.

concentrated within the viscous sub-layer and its thickness is therefore very small compared with the tube diameter. The fluid within this layer, being of lower density than the remainder, experiences an upward force due to buoyancy. Thus if the forces which act on the buoyant layer are balanced (ignoring inertia effects since these are small for the region under consideration) we see that the influence of buoyancy should be to increase the shear stress in the fluid adjacent to the layer. This is precisely the region where the turbulent energy production is concentrated and, since this process depends directly on the turbulent shear stress, it should be aided by the buoyancy effect. This in turn should lead to improved turbulent diffusion and to an enhancement of heat transfer. As we have seen earlier, this prediction is borne out by observation. Below, we attempt to quantify the arguments with a view to discovering how the parameters $R_e$, $G_r$ and $P_r$ combine to influence the problem.

Equating the total buoyancy force acting a layer of thickness $\delta_B$ to the net shear force, the increase in shear stress across the buoyant layer is obtained as

$$\Delta T_{\delta_B} = \int_0^{\delta_B} (\rho_b - \rho) g \, dy \qquad (1)$$

Noting that the thermal layer thickness $\delta_T$ is effectively equal to $\delta_B$ and approximating the temperature gradient in the thermal layer by $(T_w - T_b)/\delta_T$ we obtain

$$\Delta T_{\delta_B} \simeq \delta_B g \int_{T_b}^{T_w} (\rho_b - \rho) \, dT / (T_w - T_b) \qquad (2)$$

Introducing an integrated density $\bar{\rho}$, defined by

$$\bar{\rho} = \int_{T_b}^{T_w} \rho \, dT / (T_w - T_b) \qquad (3)$$

we obtain

$$\Delta T_{\delta_B} \simeq (\rho_b - \bar{\rho}) g \, \delta_T. \qquad (4)$$

Now $\delta_T$ is a measure of the thermal resistance of the boundary layer and, is, therefore inversely proprtional to the heat transfer coefficient. Noting that for forced convection the latter is proportional to Prandtl number raised to a power of about one half, and remembering that for a Prandtl number of unity $\delta_T$ becomes identical with the combined sub-layer and buffer layer thickness, $\delta_M$, we conclude that

$$\delta_T = \delta_M / P_r^{1/2} \qquad (5)$$

Hence, combining equations 4 and 5 and introducing the dimensionless form of $\delta_M$, we obtain

$$\Delta T_{\delta_B} \simeq (\rho_b - \bar{\rho}) g \mu \, \delta_M^+ / (\tau_w \rho P_r)^{1/2} \qquad (6)$$

When expressed completely in terms of dimensionless parameters this equation becomes

$$\Delta T_{\delta_B} / T_w \simeq 2\sqrt{2} \, \delta_M^+ \, \bar{G}_r / R_e^3 c_f^{3/2} P_r^{1/2} \qquad (7)$$

Data for near wall velocity distributions in forced flow are known to correlate well in terms of the universal parameters $u^+$ and $y^+$ and so $\delta_M^+$ is insensitive to Reynolds number. Thus for the range of Reynolds number under consideration here we can reasonably take $\delta_M^+$ to be a constant. Furthermore, friction factor for forced flow is known to be inversely related to Reynolds number raised to a power of about one quarter. Thus with the above empirical ideas injected into the analysis the parameter $\Delta T_{\delta_B} / T_w$, representing buoyancy induced distortion of the shear stress distribution, is given by

$$\Delta T_{\delta_B} / T_w \simeq K_1 \bar{G}_r / R_e^{21/8} P_r^{1/2} \qquad (8)$$

where $K_1$ is a constant. This result provides a useful guideline as to the manner in which $\bar{G}_r$, $R_e$ and $P_r$ combine to influence the problem at the forced convection end of the range. We can also show that this particular combination of $\bar{G}_r$, $R_e$ and $P_r$ emerges when the other end of the mixed convection range is considered.

For free convection inside a long heated tube it can be argued that, provided the diameter is large compared with $\delta_M$, as it certainly is for turbulent flow, tube diameter should have negligible influence on the heat transfer process. As a result, we can infer that for turbulent natural convection inside a long tube Nusselt number should vary directly as Grashof number raised to the power one third. It is widely assumed that it also varies with Prandtl number to the same power. Accepting this and utilising the fact that $\delta_T$ is inversely related to heat transfer coefficient

$$\delta_T = d/\text{const}\ \bar{G}_r^{-1/3} P_r^{1/3} \tag{9}$$

If this expression for $\delta_T$ is substituted into equation 4 and $\Delta T_{\delta_B}/T_w$ is expressed in terms of $R_e$, $\bar{G}_r$ and $P_r$ we find that

$$\Delta T_{\delta_B}/T_w \simeq K_2 (\bar{G}_r / R_e^{21/8} P_r^{1/2})^{2/3} \tag{10}$$

where $K_2$ is a constant. This result, viewed in conjunction with equation 8, suggests that the group $\bar{G}_r/R_e^{21/8}P_r^{1/2}$ might have some significance throughout the mixed convection range. Thus, if, as was argued at the beginning of the present section, buoyancy induced enhancement of heat transfer can be associated directly with the form of the shear stress distribution the above analysis suggests a correlation of heat transfer data of the form *

$$Nu/Nu_0 = \phi (\bar{G}_r / R_e^{21/8} P_r^{1/2}) \tag{11}$$

$Nu_0$ is the Nusselt number for forced convection evaluated at the Reynolds number and Prandtl number of the mixed convection data.

## 4. CORRELATION OF DATA

### 4.1 The present results

In order to express our results in terms of $Nu_0/Nu_0$ the forced convection Nusselt number $Nu_0$ must be evaluated for each experimental condition. The present results differ markedly from those for forced convection at all except our highest value of Reynolds number, so much so that the required values of $Nu_0$ cannot be obtained reliably by extrapolation to zero temperature difference. An alternative approach involving the use of a correlation equation for forced convection must therefore be employed. For this purpose we have chosen the equation of Petukhov and Kirrilov (1958)[17].

$$Nu_0 = R_e P_r (C_f/2)/(12.7(C_f/2)^{1/2}(P_r^{2/3} - 1) + 1.07) \tag{12}$$

in which $C_f = 1/(3.64 \log_{10} R_e - 3.28)^2$

---

* The arguments leading to the particular form $\bar{G}_r/R_e^{21/8} P_r^{1/2}$ have drawn on empirical relationships for heat transfer and friction and have involved the assumption that $\delta_M^+$ can be regarded as constant. The use of alternative empirical equations and refinements which relate $\delta_M^+$ to $R_e$ can lead to indices for $R_e$ and $P_r$ which differ slightly from the values 21/8 and $\frac{1}{2}$ quoted above. In the case of the $R_e$ index, alternative approaches lead to values between 2.57 and 2.7.

A recent critical evaluation of equation 12 by Webb[18] has shown it to be far superior to the simpler equations of Dittus Boelter type which are commonly used, and the work of Notter and Sleicher[19] (since extended to the case of variable properties by Sleicher and Rouse [20]) substantiates it further.

There is a systematic discrepancy of about 16% between our high Reynolds number results (for which buoyancy effects are small) and equation 12. It is probable that this is partly attributable to incompletely developed fluid flow and disturbances associated with entry. In order to take account of this discrepancy we have applied a factor of 1.16 to the values of $Nu_o$ given by equation 12. We have used a similar approach of matching at the high Reynolds number end of the range when evaluating $Nu_o$ for data from earlier investigations [1,3,5] (see section 4.3).

In figure 4(a) the present results are shown plotted in the form $Nu/Nu_o$ against $G_r/R_e^{21/8} P_r^{1/2}$. It can be seen that the data are well correlated by these parameters. Note that all fluid properties (apart from $\bar{\rho}$) were evaluated at the local bulk temperature. The theoretical considerations given earlier are not sufficiently precise to indicate whether the properties should be evaluated at this temperature or some other temperature.

Figure 4(b) shows an alternative scheme in which a mean Prandtl number formed by averaging the wall and bulk values was used. The forced convection Nusselt number $\overline{Nu}_o$ was found by using this mean Prandtl number in equation 12. Comparison between figures 4(a) and 4(b) suggest that, if anything, this second approach is less effective than the first.

A number of other approaches were tried; for instance, a Reynolds number index of 27/10 rather than 21/8 was used (this value is arrived at if a friction factor relationship involving $R_e^{-1/5}$ rather than $R_e^{-1/4}$ used in the analysis. None of the alternative schemes proved to be any better than the first one (figure 4(a)).

## 4.2 Correlation equation

The analysis of Section 3, in which we viewed the buoyancy effect as a perturbation on the forced flow, can be taken further to provide a framework for correlation equation. We liken the buoyancy influenced condition, with its augmented shear stress, to one of forced flow in the same tube but a some Reynolds number $R_e'$ which is higher than the one which actually prevails. The corresponding values of wall shear stress and heat transfer coefficient are given the symbols $\tau_w'$ and $\alpha_o'$. A simple relationship between $\tau_w$ and $\alpha_o$ for forced flow in a given tube can be deduced from the known variations of $Nu_o$ with $R_e$ and $c_f$ with $R_e$.

With $Nu_o$ assumed proportional to $R_e^{0.82}$ and $c_f$ inversely proportional to $R_e^{1/4}$ we find that

$$\alpha_o' / \alpha_o = (\tau_w' / \tau_w)^{0.47} \qquad (13)$$

On the assumption that the influence of buoyancy on wall shear stress is small compared with $\Delta \tau_{SB}$ we can approximate $(\tau_w' / \tau_w)$ by $(1 + \Delta \tau_{SB}/\tau_w)$. Hence we arrive at

$$Nu/Nu_o \simeq \left[1 + \Delta \tau_{SB}/\tau_w\right]^{0.47} \qquad (14)$$

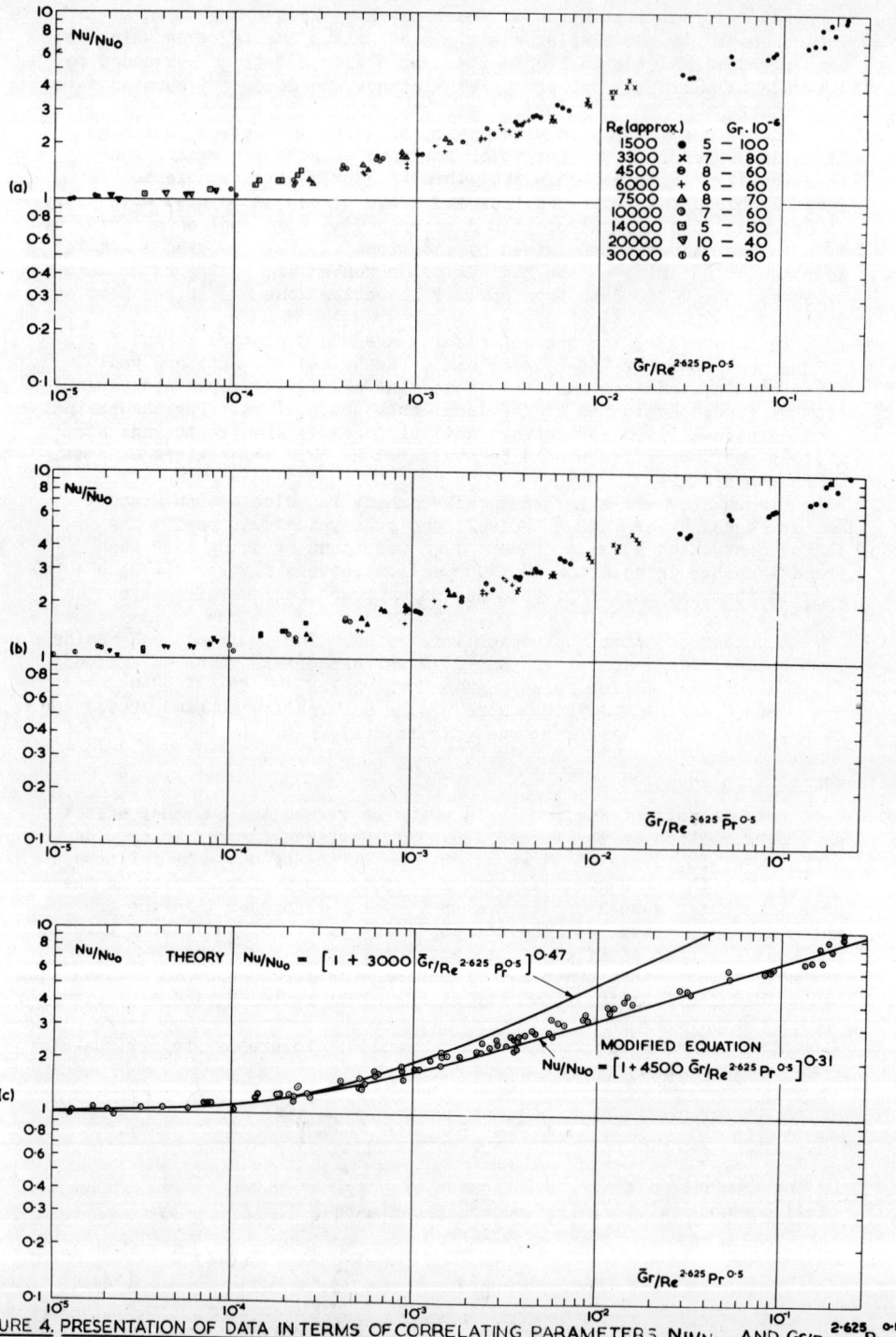

FIGURE 4. PRESENTATION OF DATA IN TERMS OF CORRELATING PARAMETERS $Nu/Nu_0$ AND $\overline{Gr}/Re^{2.625} Pr^{0.5}$

Finally, using equation 8 for $\Delta T_{\delta_B}/T_w$ we obtain

$$Nu/Nu_0 \simeq [1 + K_1 \bar{G}_r/R_e^{21/8} P_r^{1/2}]^{0.47} \quad (15)$$

The constant of $K_1$ depends on the value of $\delta_M^+$ and the latter is between about 15 and 25. Corresponding rounded values for $K_1$ are 2000 and 3000.

Equation 15 is compared with the present data in figure 4(c) for the case $K_1 = 3000$, this being the one which leads to the largest discrepancy between theory and experiment. The equation provides a very satisfactory description of the data for $\bar{G}_r/R_e^{21/8} P_r^{1/2}$ less than $3.10^{-4}$ (at which condition there is about 30% enhancement of heat transfer). The rapid divergence between theory and experiment beyond this point is only to be expected in view of the assumptions made in the analysis.

The discrepancy cannot be removed by simply adjusting $K_1$. The predicted behaviour is such that Nu does not become independent of $R_e$ for large values of $\bar{G}_r/R_e^{21/8} P_r^{1/2}$ whereas experiment shows this to be the case and common sense requires that it should be so. If the index 0.47 in equation 15 is changed to 0.31 then Nu does become asymtotically independent of $R_e$. With this in mind we have also shown on figure 4(c) a curve for the equation

$$Nu/Nu_0 = [1 + 4500 \bar{G}_r/R_e^{21/8} P_r^{1/2}]^{0.31} \quad (16)$$

which incorporates the required asymptotic behaviour and has, by virtue of the choice of constant 4500, been made to fit the experimental data. This equation, provides a simple and accurate description of the present data. It has been used to obtain the values of percentage enhancement of heat transfer shown below in Table III.

Table III. Figures for enhancement of heat transfer based on the present data (from equation 16 with $P_r = 2.25$)

| Grashof Number | Reynolds Number | | | | |
|---|---|---|---|---|---|
| | 6000 | 10000 | 18000 | 30000 | 50000 |
| $1.10^6$ | 5.3% | 1.4% | 0.3% | 0.1% | 0 |
| $3.10^6$ | 14.5% | 4.2% | 0.9% | 0.2% | 0.1% |
| $5.10^6$ | 22.3% | 6.9% | 1.7% | 0.4% | 0.2% |
| $1.10^7$ | 61.0% | 23.1% | 5.7% | 1.8% | 0.5% |
| $3.10^7$ | 78.5% | 31.7% | 8.7% | 2.4% | 0.7% |
| $7.10^7$ | 125% | 57.6% | 18.2% | 5.5% | 1.6% |
| $1.10^8$ | 150% | 72.0% | 24.5% | 8.9% | 2.3% |
| $5.10^8$ | 307% | 170% | 75.5% | 30.0% | 10.3% |

These figures confirm the trends evident in those based upon earlier data presented in Table II.

4.3 Comparison with earlier data and correlations

Only a very limited amount of data is available in a form in which it can be compared directly with that of the present investigation. Figure 5 shows data of Watzinger and Johnson[1], and Herbert and Sterns[5] along with the curve which represents our data (equation 16). The points lie consistently below the curve but show a similar pattern of behaviour. The discrepancy could be due in part to differences in thermal boundary conditions but is more likely attributable to uncertainties in the calculation of $Nu_o$ (particularly in the case of the data of Watzinger and Johnson).

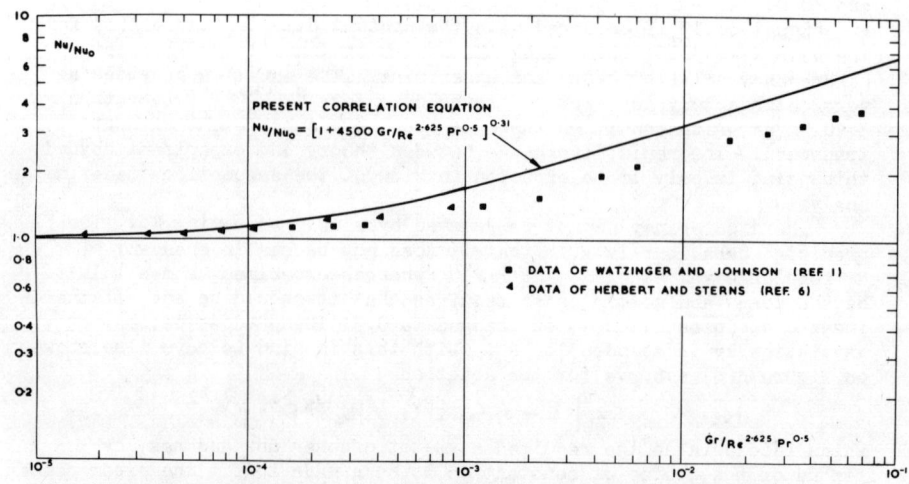

FIGURE 5. COMPARISON BETWEEN PRESENT CORRELATION AND OTHER DATA.

The investigation which yielded the largest body of data prior to the present one was that of Petukhov and Strigin [3]. They presented their data in terms of parameters $Nu/Nu_o$ and $Gr Nu/4Re^2$. The points fit the equation

$$Nu/Nu_o = [1 + 0.5\, Gr Nu/4Re^2]^{1/3} \quad (17)$$

to within $\pm$ 10% for $1 < Gr Nu/4Re^2 < 1000$. This range of conditions extends well into the region where buoyancy effects completely dominate. If the asymptotic behaviour of equation 17 is examined using a turbulent forced convection relationship for $Nu_o$ it is found that $Nu$ does not become independent of $R_e$. The situation made is complicated by the fact that the experiments extend to values of Reynolds number as low as 200 and for such cases it is quite inappropriate to use a turbulent forced convection relationship for $Nu_o$.

Unfortunately, there is insufficient information in reference 3 to enable the data of Petukhov and Strigin to be expressed in terms of the group $G_r/R_e^{2\cdot1/8}Pr^{1/2}$. Thus, in order to make a comparison between their data and ours we have plotted, our results (and those of Watzinger and Johnson[1] and Herbert and Sterns[5]) in terms of $Nu/Nu_o$ and $Gr Nu/4Re^2$ in figure 6. It is clear that our results do not correlate satisfactorily on this basis and they lie well below the Petukhov and Strigin correlation curve.

Alferov et al.[6,21] present their supercritical pressure water data (and some of Shitsman[27]) in terms of parameters $\alpha/\alpha_o$ and $\alpha_o/\alpha_f$ where $\alpha_o$ and $\alpha_f$ are coefficients for forced and free convection calculated from established correlations for these modes of heat transfer. Although

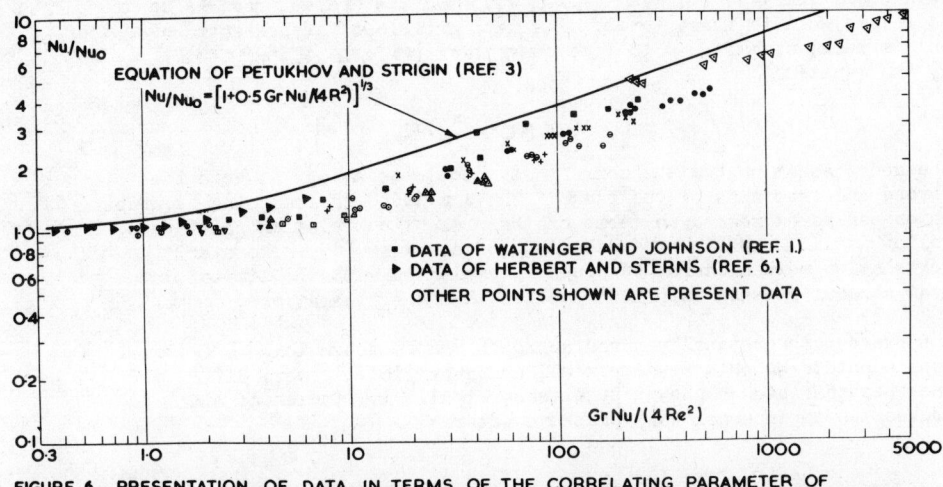

FIGURE 6. PRESENTATION OF DATA IN TERMS OF THE CORRELATING PARAMETER OF PETUKHOV AND STRIGIN (REF. 3) AND COMPARISON WITH THEIR CORRELATION EQUATION.

their method appears to have been arrived at without the aid of theoretical guidelines it can be shown to be effectively identical to that which has been developed here. This becomes clear if $Nu$ is taken to be proportional to $R_e^{0.82} P_r^{0.5}$ for forced convection and $\bar{G}_r^{1/3} P_r^{1/3}$ for free convection in forming the ratio $\alpha_f / \alpha_o$. The latter turns out to be proportional to $(\bar{G}_r / R_e^{2.46} P_r^{0.5})^{1/3}$. The correlation equation proposed by Alferov et al, namely

$$\alpha/\alpha_o = 1.65 (\alpha_f/\alpha_o)^{0.68} \qquad \text{for } \alpha_f/\alpha_o > 0.5 \qquad (18)$$

provides only a very crude description of the data and there are insufficient points to provide a stringent test of the correlation method.

Ikryannikov et. al.[7] present their data for supercritical pressure carbon dioxide in terms of parameters $Nu/Nu_o$ and $G_r/R_e$. There is a very considerable spread of points on their plot and it can in no sense be regarded as a justification of $G_r/R_e$ as being an appropriate correlating parameter.

CONCLUSIONS

The present measurements of heat transfer to water for downward flow in a tube of 98.4 mm internal diameter show strong enhancement due to buoyancy and as such are consistent with earlier measurements on tubes of smaller diameter [1,3,4,5]

Our results are not correlated by the parameters $Nu/Nu_o$ and $Gr Nu/4R_e^2$ which Petukhov and Strigin have employed in an earlier investigation [3]

Theoretical considerations presented here suggest that the appropriate parameters are $Nu/Nu_o$ and $\bar{G}_r/R_e^{21/8}P_r^{1/2}$ and the present results do correlate well in these terms. The data are closely represented over the entire range from forced to free convection ($10^{-5} < \bar{G}_r/R_e^{21/8}P_r^{1/2} < 0.2$) by the equation

$$Nu/Nu_o = \left[1 + 4500\,\bar{G}_r/R_e^{21/8}P_r^{1/2}\right]^{0.31} \tag{19}$$

The above equation has the correct asymptotic behaviour at both the forced and free ends of the range. The limited data from other sources[1,5], which can be expressed in terms of the present correlating parameters, follow a similar pattern of behaviour to equation (19). However, the correlation method cannot be properly evaluated with the limited data available at present for tubes smaller than the present one.

The present theoretically based approach can be shown to be effectively equivalent to an entirely empirical and superficially very different one which has been employed by Alferov et.al.[6] in studies of mixed convection to supercritical pressure water.

REFERENCES

1. WATZINGER, A and JOHNSON D.G. (1939) 'Warmeübertragung von Wasser an Rohrwand bei senkrechter Stromung im Übergangsgebiet zwischen laminarer und turbulenter Strömung' Forschung auf dem Gebiete des Ingenierwesens, Vol. 10, pp 182-196.
2. PETUKHOV, B.S. and NOLDE, L.D. (1959) 'Heat transfer to water in a vertical heated tube for upward and downward flow'. Teploenergetika, Vol. 6 pp 72-80.
3. PETUKHOV, B.S. and STRIGIN, B.K. (1968) 'Experimental investigation of heat transfer with viscous inertial-gravitational flow of a liquid in vertical tubes' Teplofizika Vysokikh Temperatur, Vol. 6, No. 5, pp 933-937.
4. HERBERT, L.S. and STERNS, U.J. (1968) 'An experimental investigation of heat transfer to water in film flow' Part I - Non boiling runs with and without induced swirl. Part II - Boiling runs with and without induced swirl. The Canadian Journal of Chemical Engineering, Vol. 46, pp 401-412.
5. HERBERT, L.S. and STERNS, U.J. (1972) 'Heat transfer in vertical tubes - interactions of forced and free convection' Chemical Engineering Journal, Vol. 4.
6. ALFEROV, N.S., BALUNOV, B.F. and RYBIN, R.A. (1973) 'Reduction in heat transfer in the region of supercritical state variables of a liquid' Heat Transfer Soviet Research, Vol. 5, No. 5, September-October.
7. IKRYANNIKOV, N.P., PETUKHOV, B.S. and PROTOPOPOV, V.S. (1972). 'An experimental investigation of heat transfer in the single-phase near-critical region with combined forced and free convection', Teplofizika Vysokikh Temperatur, Vol. 10, No. 1, pp 96-100.
8. ECKERT, E.R.G. and DIAGUILLA, A.J. (1954) 'Convective heat transfer for mixed free and forced flow in tubes' Trans. A.S.M.E., Vol. 76, pp 497-504.

9. BROWN, C.K. and GAUVIN, W.H. (1966) 'Temperature profiles and fluctuations in combined free and forced convection flows', Chem. Eng. Sci. Vol. 21, pp 961-970.
10. HALL, W.B. and PRICE, P.H. (1971) 'Interaction between a turbulent free convection layer and a downward forced flow', Paper C113/71, I.Mech.E. Symposium on Heat and Mass Transfer by Combined Forced and Natural Convection, Manchester.
11. KHOSLA, J., HOFFMAN, T.W., and POLLOCK, K.C. (1974) 'Combined forced and natural convective heat transfer to air in a vertical tube' Paper N C 4.4, Proc. 5th Int. Heat Transfer Conf. Tokyo.
12. AXCELL, B.P. (1975) 'The effect of buoyancy on turbulent forced convection', Ph.D. thesis, Manchester University.
13. BAKER, D. (1966) 'A technique for the precise measurement of small fluid velocities' Journal of Fluid Mechanics, Vol. 26, Part 3, p 573.
14. KENNING, D.B.R. SCHOCK, R.A.W. and POON J.Y.M. (1974) 'Local reduction in heat transfer due to buoyancy effects in upward turbulent flow' Paper N.C. 4.3, Proc. 5th Int. Heat Transfer Conf., Tokyo.
15. JACKSON, J.D. (1975) 'Heat transfer to supercritical Pressure Fluids' HTFS Design Report No 34 U.K.A.E.A. Harwell AERE - R 8158.
16. HALL, W.B. and JACKSON, J.D. (1969) 'Laminarisation of a turbulent pipe flow by buoyancy forces' Am.Soc.Mech.Engrs. Paper No 69-HT-55.
17. PETUKHOV, B.S. and KIRILLOV, V.V. (1958) 'The problem of heat exchange in the turbulent flow of liquids in tubes'. Teploenergetika No. 4, pp 63-68.
18. WEBB, R.L. (1971) 'A critical evaluation of analytical solutions and Reynolds analogy equations for turbulent heat and mass transfer in smooth tubes' Warme-u. Stoffübertragung, Vol. 4, pp 197-204.
19. NOTTER, R.H. and SLEICHER, C.A. (1972) 'A solution to the turbulent Graetz problem' Part III 'Fully developed and entry region heat transfer rates' Chem. Engng. Sci. Vol. 27, pp 2073-2093.
20. SLEICHER, C.A. and ROUSE, M.W. (1974). 'A convenient correlation for heat transfer to constant and variable property fluids. in turbulent pipe flow' Int. J. Heat Mass Transfer, Vol. 18, pp 677-683.
21. SHITSMAN, M.E. (1968) 'Temperature conditions in tubes at supercritical pressures' Teploenergetika, Vol. 15 No. 5, pp 57-61.

# AIR AND SMOKE MOVEMENTS IN BUILDING FIRES

PAUL GERHARD SEEGER

*Forschungsstelle für Brandschutztechnik*
*an der Universität Karlsruhe, Karlsruhe, Germany*

ABSTRACT

In the event of a fire in a building the smoke often creates greater hazards to occupants attempting to escape and more difficulties to fire fighters, than does the heat, because the smoke may enter the escape routes in such a quantity that after a short time it is impossible to pass them. The mechanisms being responsible for the air and smoke movements are the expansion of the gases due to temperature rise in the environment of the fire, the stack effect and wind effects on the building. These effects can be calculated by aerodynamic and thermodynamic relationships. Fore some time past there are efforts to develop methods by which the smoke can be prevented from penetrating the escape routes.

NOMENCLATURE

- $a$    leakage area per unit height ($m^2/m$)
- $A$    opening area, surface ($m^2$)
- $b$    breadth (m)
- $C$    constant (-)
- $g$    gravitational acceleration ($m/s^2$)
- $h$    height (m)
- $H$    height of building and/or window (m)
- $n$    number of moles (-)
- $\dot{m}$    mass flow (kg/s)
- $p$    pressure ($N/m^2$)
- $\Delta p$    pressure difference ($N/m^2$)
- $q$    velocity pressure ($N/m^2$)
- $R$    Gas constant (J/mol K)
- $T$    absolute temperature (K)
- $v$    velocity of flow (m/s)
- $V$    volume ($m^3$)
- $\alpha$    discharge coefficient (-)
- $\beta$    wind pressure coefficient (-)
- $\gamma$    angle between wind direction and the normal on a wall (rad)
- $\lambda$    air ratio (-)
- $\rho$    density ($kg/m^3$)
- $\phi$    ventilation parameter (kg/s)

Copyright © 1977 by Hemisphere Publishing Corporation

Subscripts

| | |
|---|---|
| o | standard state, $0\ °C$, 1013 mbar |
| 1 | beneath the neutral pressure plane |
| 2 | above the neutral pressure plane |
| a | air |
| e | exterior |
| f | fuel |
| i | interior |
| min | stoichiometric |
| w | window, wind |

INTRODUCTION

In the event of a fire in a building, smoke produced often creates greater hazards to occupants attempting to escape and more difficulties to fire fighters and rescue squads entering the building on fire than does the heat. This results from the fact that already during the beginning of a fire smoke will spread within the building and enter escape routes very quickly, due to the different temperatures inside and outside the building and corresponding pressure differences, thus preventing already after a short time the use of escape routes.

In most cases it is less the heat carried by the smoke that endangers people but the solid and gaseous components contained in the smoke. It may, for instance, happen that soot particles and other vaporous components of the smoke completely obscure the sight on escape routes or that people trying to pass through the smoke will be adversely affected by toxic components of smoke — in particular carbon monoxide — combined with reduced oxygen content of the atmosphere to an extent which may be fatal after a short period.

Since on the one side together with the growing height of buildings these dangers rapidly increase already for the only reason that increased building heights automatically include increased lengths of escape routes, and on the other side buildings of growing height are being erected all over the world, efforts have been made everywhere to develop technical methods allowing to keep escape routes free from smoke over a sufficient length of time to render them usable.

MECHANISM OF AIR AND SMOKE FLOW IN A BUILDING

For the air and smoke flow in a building pressure differences are responsible which are the result of the expansion of gases produced by the rise of temperature in the environment of a fire, of the buoyancy of the air column which in the building is warmer than outside and of wind effects on the building. These pressure differences based on natural phenomena may be superimposed by pressure differences caused for instance by air conditioning installations.

FLOW CAUSED BY GAS EXPANSION

If a fire develops inside a room, pressure and volume of gases in this room will increase due to the rising temperature in accordance with the universal gas law for ideal gases:

$$pV = nRT. \tag{1}$$

With normal fires, i.e. such fires where no explosive burning takes place, pressure changes in the fire room due to the expansion of gases will only be limited ones since gases will discharge through normally existing leakage openings into rooms adjoining the fire room as result of the arising pressure differences. Related to the absolute atmospheric pressure, however, these unimportant pressure changes may be neglected so that - as the absolute pressure may be considered a constant value - the volume of gases will increase directly in proportion to the temperature. Consequently, the volume of gases in a room where, for instance, the temperature rises from 300 K to 900 K because of a fire, will increase by the threefold which will lead to the spread of gases into adjoining rooms.

Since due to pressure differences caused by expansion a gas flow will only last as long as the temperature rises, it may, generally, be neglected in comparison to the air and smoke flow originating from buoyancy which permanently continues as long as pressure differences caused by differences of density and temperature respectively between fire room and other rooms and/or inside and outside the building exist.

## FLOW CAUSED BY BUOYANCY

The atmospheric pressure within a gas column will decrease according to height. The extent of pressure decrease $\Delta p$ per unit of height depends on the density of the gas column $\rho$ and its temperature T respectively, i.e.:

$$\Delta p = hg\rho \tag{2}$$

and/or

$$\Delta p = hg\rho_o \frac{T_o}{T} \tag{3}$$

since according to the universal gas law for ideal gases, for the density the following relation applies:

$$\rho = \rho_o \frac{T_o}{T} \frac{p}{p_o}. \tag{4}$$

Here again, small pressure differences related to the absolute pressure may be neglected, i.e. the pressure may be considered a constant one.

Therefore, if, for example, a gas is present in a stairwell shaft of a building with an opening to outside at the bottom of the shaft, and the values for the density and/or temperature of this gas are lower and/or higher than those for the ambient atmosphere, a pressure difference will show due to the different pressure decrease in both gas columns as illustrated in Fig. 1 which increases according to height and can be calculated as follows:

$$p_i - p_e = \Delta p = hg(\rho_e - \rho_i) \tag{5}$$

and/or

$$p_i - p_e = \Delta p = hgT_o\left(\frac{\rho_{oe}}{T_e} - \frac{\rho_{oi}}{T_i}\right). \tag{6}$$

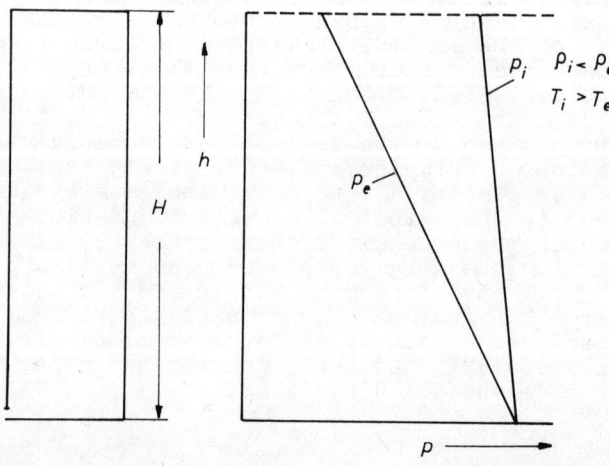

Fig. 1. Pressure pattern inside and outside a vertical shaft with an opening to outside at the bottom.

If inside and outside the building the type of the gas is the same, Equation (6) will be simplified, as shown under:

$$p_i - p_e = \Delta p = hgT_o\rho_o\left(\frac{1}{T_e} - \frac{1}{T_i}\right). \tag{7}$$

In case that at the height H a second opening between stairwell shaft and outside will be provided, an upward flow will set in as result of the buoyancy of the lighter gas column in the shaft. From the opening at the top of the shaft, gas will discharge which will be replaced by cooler follow-on gases entering the shaft through the opening at the bottom. As these phenomena are just the same as those in a chimney, one often speaks in this connection of the "chimney effect" or "stack effect". Because of this flow, the plane where the absolute pressure in both gas columns inside and outside the shaft is the same, will shift from the lower opening to a point between the top and the bottom opening. Underneath this so-called neutral pressure plane there will be an underpressure, above it an overpressure in the shaft compared to the outside pressure as can be seen on Fig. 2. If $A_1$ and $A_2$ represent the areas of both openings in the shaft and $v_1$ and $v_2$ the relevant flow velocities at those openings, it will be possible to calculate the position of the neutral pressure plane by means of the equation of continuity for steady incompressible flows:

$$A_1 v_1 \rho_e = A_2 v_2 \rho_i . \tag{8}$$

In this case, according to the Bernoulli's relationship and

# AIR AND SMOKE MOVEMENTS IN BUILDING FIRES

Equations (4) and (7)

$$v_1 = \alpha \left[\frac{2\Delta p_1}{\rho_e}\right]^{1/2} = \alpha \left[2h_1 g\left(1 - \frac{T_e}{T_i}\right)\right]^{1/2} \qquad (9a)$$

and/or

$$v_2 = \alpha \left[\frac{2\Delta p_2}{\rho_i}\right]^{1/2} = \alpha \left[2h_2 g\left(\frac{T_i}{T_e} - 1\right)\right]^{1/2} . \qquad (9b)$$

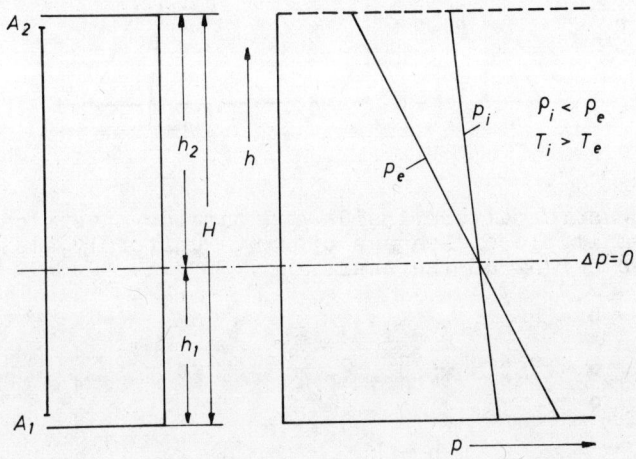

Fig. 2. Pressure pattern inside and outside a vertical shaft with one opening to outside at the bottom and at the top.

If Equations (9a) and/or (9b) will be inserted in Equation (8), it follows that:

$$\frac{h_2}{h_1} = \frac{A_1^2 T_i}{A_2^2 T_e} . \qquad (10)$$

The ratio of the heights above and beneath the neutral pressure plane, thus, is inversely proportional to the ratio of the squares of the opening areas and directly proportional to the ratio of the temperatures inside and outside a building shaft [1, 2].

If leakages in the shaft are not limited to one opening above and beneath the neutral pressure plane of the building shaft, but a multiple of leakage openings exists which are uniformly distributed over the height of the shaft, as shown in Fig. 3, the relationship for the position of the neutral pressure plane may be expressed according to Equation (8) and to [3] as follows:

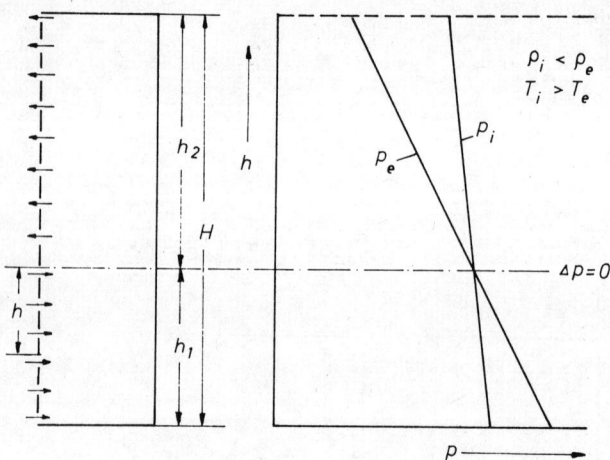

Fig. 3. Pressure pattern inside and outside a vertical shaft with a multiple of leakage openings uniformly distributed over the height of the shaft.

$$\int_{h=0}^{h=h_1} \alpha a[2g(1-\frac{T_e}{T_i})]^{1/2} \rho_o \frac{T_o}{T_e} h^{1/2} dh =$$

$$\int_{h=0}^{h=h_2} \alpha a[2g(\frac{T_i}{T_e}-1)]^{1/2} \rho_o \frac{T_o}{T_i} h^{1/2} dh. \quad (11)$$

In this relationship, factor $a$ is the leakage area per unit height. The integration over the height $h_1$ and/or $h_2$ as follows:

$$\frac{2}{3}\alpha a[2g(1-\frac{T_e}{T_i})]^{1/2} \rho_o \frac{T_o}{T_e} h_1^{3/2} =$$

$$\frac{2}{3}\alpha a[2g(\frac{T_i}{T_e}-1)]^{1/2} \rho_o \frac{T_o}{T_i} h_2^{3/2} \quad (12)$$

results in the following ratio of the heights above and beneath the neutral pressure plane:

$$\frac{h_2}{h_1} = (\frac{T_i}{T_e})^{1/3} \quad (13)$$

Provided $T_i > T_e$, also in this case a gas flow exists through leakage openings which above the neutral pressure plane is flowing out of the shaft and beneath the neutral pressure plane is flowing into the shaft.

If there is a compartment, which has an opening of considerable size, e.g. window or door opening, as shown in Fig. 4, and the gas temperature is the same all over this compartment but higher than outside the compartment, the position of the neutral pressure plane can also be calculated according to Equation (12), provided, however, that the pressure differences are only caused by buoyancy and no other effects exist, thus precluding any mixing of both gas flows near the opening. In this case, the value

$$H = h_1 + h_2 \tag{14}$$

represents the height of the opening.

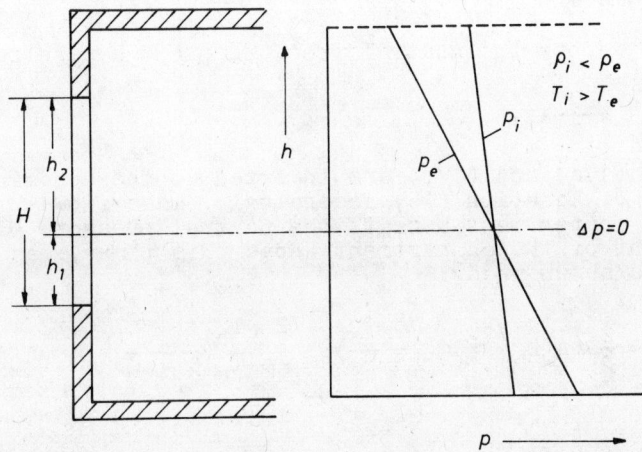

Fig. 4. Pressure pattern inside and outside a room with a large opening.

If in this compartment, in addition, a burning process takes place, part of the air flow entering the compartment beneath the neutral pressure plane will, together with the fuels available in this compartment, transform to smoke gas on the surface of these fuels so that above the neutral pressure plane not only the heated air but also smoke gas equivalent to the mass loss of fuels per unit time will flow out of the compartment. To determine the position of the neutral pressure plane, the following expression may be derived according to [4] provided that the same conditions apply as for the case without burning:

$$\frac{\dot{m}_f + \dot{m}_a}{\dot{m}_a} = \frac{\frac{2}{3} \alpha b [2g(\frac{T_i}{T_e} - 1)]^{1/2} \rho_o \frac{T_o}{T_i} h_2^{3/2}}{\frac{2}{3} \alpha b [2g(1 - \frac{T_e}{T_i})]^{1/2} \rho_o \frac{T_o}{T_e} h_1^{3/2}} \tag{15}$$

where b is the breadth of the opening. It is presumed in this expression that the density of the outflowing smoke $\rho_{oi}$ equals the density of the entering air flow $\rho_{oe}$ which is allowable considering the high proportion of nitrogen content. The ratio of the heights above and beneath the neutral pressure plane results from Equation (15):

$$\frac{h_2}{h_1} = \left(\frac{\dot{m}_f + \dot{m}_a}{\dot{m}_a}\right)^{2/3} \left(\frac{T_i}{T_e}\right)^{1/3} = C. \tag{16}$$

Based on H according to Equation (14) for the height of the opening, height $h_1$ beneath the neutral pressure plane will be expressed as follows:

$$h_1 = \frac{H}{1 + C} \tag{17a}$$

and height $h_2$ above the neutral pressure plane by the relation

$$h_2 = \frac{HC}{1 + C}. \tag{17b}$$

If Equations (17a) and (17b) are inserted in the denominator and/or numerator of Equation (15) accordingly, mass flows of gases entering the compartment beneath the neutral pressure plane and/or flowing out of the compartment above this plane, may be calculated as shown under:

$$\dot{m}_a = \frac{2}{3} \alpha A_w H^{1/2} \left[2g\left(1 - \frac{T_e}{T_i}\right)\right]^{1/2} \rho_o \frac{T_o}{T_e} \left(\frac{1}{1 + C}\right)^{3/2} \tag{18a}$$

and

$$\dot{m}_f + \dot{m}_a = \frac{2}{3} \alpha A_w H^{1/2} \left[2g\left(\frac{T_i}{T_e} - 1\right)\right]^{1/2} \rho_o \frac{T_o}{T_i} \left(\frac{C}{1 + C}\right)^{3/2}. \tag{18b}$$

The mass loss of fuels per unit time in a fully-developed fire in a compartment, i.e. during the stationary phase of the fire depends either on the quantity of air available for the burning process or on the surface of the fuels where reactions of burning occur. Since the quantity of air entering the fire compartment depends on the number and size of openings available in this compartment provided that no other additional air supply by mechanical ventilation systems exists, the mass loss of fuels per unit time will be controlled by the ventilation where few and/or small openings are present only. If numbers and/or sizes of the openings increase, conditions in the fire compartment will approximate to those of a burning-off in the unconfined atmosphere which is only dependent on the surface of fuels.

Based on the results of a multitude of experiments carried out in laboratory scale and large scale tests, Harmathy [5] has been able to demonstrate the limits of both regimes. As shown in Fig. 5 which illustrates the mass loss of fuels related to the fuel surface per unit time above the ventilation parameter also related to the fuel surface, both regimes are separated by the critical value of the ventilation parameter

$$\phi = 0.263 \, A_f. \tag{19}$$

In this connection, Harmathy defines the ventilation parameter as follows:

$$\phi = \rho_e g^{1/2} A_w H^{1/2}. \tag{20}$$

Below this critical value of the ventilation parameter, the mass loss of fuels per unit time is controlled by the ventilation. For this regime the mass loss of fuels per unit time will be controlled by the following relationship:

$$\frac{\phi}{A_f} < 0.263 \; : \; \dot{m}_f = 0.0236 \, \phi. \tag{21a}$$

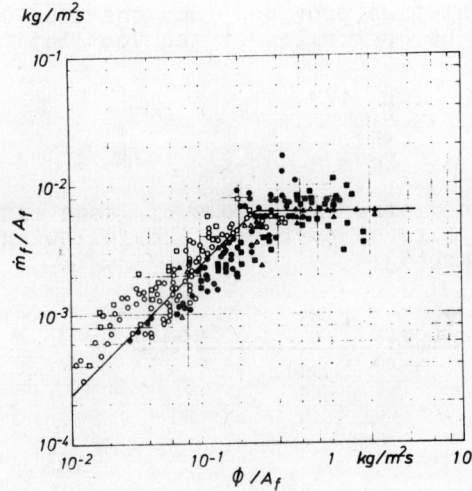

Fig. 5. Mass loss of fuels related to the fuel surface per unit time as a function of the ventilation parameter also related to the fuel surface.

Above this critical value, the mass loss of fuels per unit time depends on the fuel surface. For this regime the following relation applies for the mass loss of fuels per unit time:

$$\frac{\phi}{A_f} \geq 0.263 \; : \; \dot{m}_f = 0.0062 \, A_f. \tag{21b}$$

If adequate values are used in Equations (20) and/or (21a), it proves that the relation for the mass loss of fuels per unit time as specified by Harmathy is identical with the relation

$$\dot{m}_f = 5.5 \, A_w H^{1/2}. \tag{22}$$

established already previously by Fujita [6] and Kawagoe [7].

By means of the massflow of the air entering the fire compartment below the neutral pressure plane according to Equation (18a) and the mass loss of fuels per unit time according to Equation (22), Thomas et al. [4] found the following value for the ratio of these both quantities:

$$\frac{\dot{m}_f}{\dot{m}_a} = 0.185. \tag{23}$$

In this relation, the assumed values were: for the temperature $T_i = 1200$ K and $T_e = 300$ K, for density $\rho_o = 1.3$ kg/m$^3$, and for the discharge coefficient $\alpha = 0.7$.

Accordingly, the following expression is applicable for the ratio of the heights above and beneath the neutral pressure plane during a compartment fire provided that the mass loss of fuels per unit time will be controlled by the ventilation:

$$\frac{h_2}{h_1} = 1.185^{2/3} \left(\frac{T_i}{T_e}\right)^{1/3}. \tag{24}$$

For the ratio of the mass flows of gases entering and flowing out of the fire compartment, the following expression has been derived by John [8]:

$$\frac{\dot{m}_f + \dot{m}_a}{\dot{m}_a} = \frac{V_{fomin} + (\lambda - 1) V_{aomin}}{\lambda \, V_{aomin}} \tag{25}$$

where

$$\lambda = \frac{V_{ao}}{V_{aomin}}. \tag{26}$$

Also in this case it is presumed that the densities of both gas flows of the same temperature are equal to each other. If this expression according to Equation (25) is inserted in Equation (16), the following relation is obtained:

$$\frac{h_2}{h_1} = \left(\frac{V_{fomin} + (\lambda - 1) V_{aomin}}{\lambda \, V_{aomin}}\right)^{2/3} \left(\frac{T_i}{T_e}\right)^{1/3}. \tag{27}$$

For wood with a calorific value of 15.9 MJ/kg, the values as shown in Table 1 determine the ratio of mass flows according to Equation (25) for various air ratios.

Table 1. Ratio of mass flows according to Equation (25) for various air ratios with wood used as fuel.

| $\lambda$ | 0.8 | 1.0 | 1.2 | 1.5 | 2.0 | 10.0 |
|---|---|---|---|---|---|---|
| $\dfrac{V_{fomin} + (\lambda - 1) V_{aomin}}{\lambda V_{aomin}}$ | 1.199 | 1.159 | 1.132 | 1.106 | 1.079 | 1.016 |

As can be seen from Table 1, values for the ratio of the mass flows obtained by John [8] are of a similar order of magnitude as established by Thomas et al. in [4].

However, it must be emphasized here that the assumed air ratios $\lambda$ can only be obtained in furnaces where the shape of the combustion chamber and the flows of fuel and air are such as to guarantee a combustion which corresponds to the air ratio. But also in furnaces all concentrations may locally occur, ranging from pure fuel to pure air - provided that fuel and air have not yet mixed prior to entering the furnace -, i.e. the air ratio refers to the state prevailing when smoke gases are discharging from the furnace. Compartments where a fire breaks out, do not, in general, show those conditions which are required for ideal burning of fuels, since their geometry, distribution of fuels and ventilation will be influenced by other aspects, e.g. by those of architects, usage etc.. Therefore, in a fire compartment pyrolytic decompositions, complete and incomplete combustion phenomena will occur simultaneously and part of decomposition gases originating from the fuels will start burning only outside the fire compartment as shown by flames striking out of a fire compartment. Since, up to now, these processes and their share in the entire phenomena of burning have only been investigated to a small extent, it is at present not possible to specify the air ratios to be inserted in Equation (27) or in which manner the phenomena of pyrolytic decomposition and incomplete combustion have to be taken into account in the above mentioned equation.

FLOW CAUSED BY WIND EFFECTS ON A BUILDING

If a wind strikes a building, an overpressure will build up on the windward face while on the faces parallel to wind, above the roof and on the leeward side an underpressure will be created, as compared to the ambient pressure. The pressure distribution on the surface of a building due to wind is most nonuniform and depends on the direction of the wind, shape and height and on the environment (flat open country, urban areas) of the building. The local wind pressure on the surface of the building may be calculated by use of the velocity pressure in the stagnation point, the so-called dynamic pressure, and a pressure coefficient ß which shows the relation between the local pressure and the dynamic pressure, as follows:

$$q_w = \beta \frac{\rho v_w^2}{2} \tag{28}$$

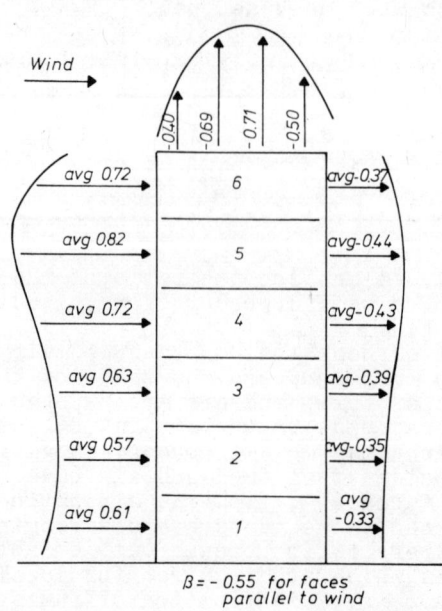

Fig. 6. Distribution of wind pressure on a high rectangular building.

Fig. 6 shows the distribution of wind pressure on a high rectangular building in an urban area which has been established according to results of envestigations [9]. The pressure coefficients ß therein refer to the dynamic pressure of the wind at the height of the top of the building. The positive symbol used in this connection indicates an overpressure, the negative symbol an underpressure and/or suction compared to the ambient pressure. Caused by the wind pressure, in particular, horizontal gas flows will be produced throughout the building flowing from the windward to the leeward side, but - due to underpressures at the roof and at the upper part of the building - also vertical flows may occur.

Wakamatsu [10] has found the following values for the pressure coefficient ß dependent on the angle $\gamma$ between wind direction and the normal on the wall of the building where the wind strikes:

$$\text{ß} = 0.75, \quad \text{if } 0° \leq /\gamma/ \leq 30°$$
$$\text{ß} = -0.021\gamma + 1.38, \quad \text{if } 30° < /\gamma/ \leq 90° \text{ and}$$
$$\text{ß} = -0.5, \quad \text{if } /\gamma/ \geq 90°.$$

## METHODS TO INFLUENCE AIR AND SMOKE FLOW IN A BUILDING

In order to influence the air and smoke flow in a building with the aim to keep escape routes free from smoke as far as possible, two methods may be applied [11, 12]. One method uses fresh air to be supplied to the fire floor while simultaneously smoke will be exhausted from the fire floor by means of two separated ventilation units. At the same time, the pressure on the fire floor will be regulated such as to be less than that on the escape routes within the fire floor. The magnitude of the fresh air flow will be choosen such as to make sure that the temperature of smoke and its optical density will be lowered to such an extent that also if penetrating escape routes no life risk to persons due to heat, toxic components of smoke or reduced sight will exist. In the second method, pressures in the building will be changed by natural venting or mechanical pressurization as far as to ensure that the pressure on the escape routes will be sufficiently high to prevent entering of smoke.

## METHOD OF AIR SUPPLY AND SMOKE EXHAUST OF THE FIRE FLOOR

The efficiency of the method of air supply and smoke exhaust of a fire floor has been tested by the Forschungsstelle für Brandschutztechnik an der Universität Karlsruhe in a 22-story high-rise dwelling-house with an internal staircase [13]. Final approval of the internal safety staircase by the building inspection authority was made conditional on the result of realistic fire tests in this high-rise building. As can be seen from the ground plan of a typical floor illustrated in Fig. 7, the four flats are grouped around the entrance hall with the lift shaft, around the lock chamber and around the internal staircase. In case of a fire, the entrance hall as well as the lock chamber can be vented by means of ventilation units via air supply and smoke exhaust shafts to be opened, however, only to the fire floor.

The main test was carried out in the 11th upper floor. For this purpose, the living room, the dining place and the kitchen in the fire flat were equipped with old furniture. The average fire load in those rooms was 419 $MJ/m^2$. The test lasted until, after approximately 70 minutes, all furnishing was completely burnt. At various points in the fire flat, in the entrance hall, in the lock chamber and in the staircase at the height of the fire floor, temperatures, concentrations of carbon monoxide and carbon dioxide, optical smoke density and the dynamic pressures corresponding to the supply air and exhaust smoke flows were measured. Some of the measuring points have been entered on Fig. 7. The development of the fire can be seen on Fig. 8 where the time curve for the temperatures at various measuring points is represented.

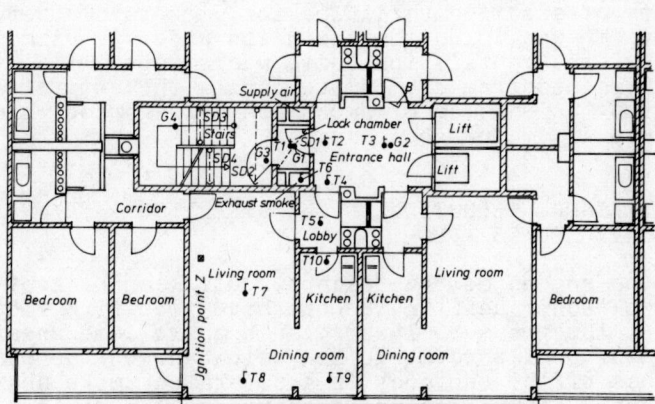

Fig. 7.  Ground plan of a residential floor with the arrangement of some of the measuring points.
        ⌠   Measuring point about 5 cm below the ceiling
        ⌡   Measuring point about 150 cm above the floor
        T   Temperature measuring point
        G   Gas extraction measuring point
        SD  Optical smoke density measuring section
        B   Observation window.

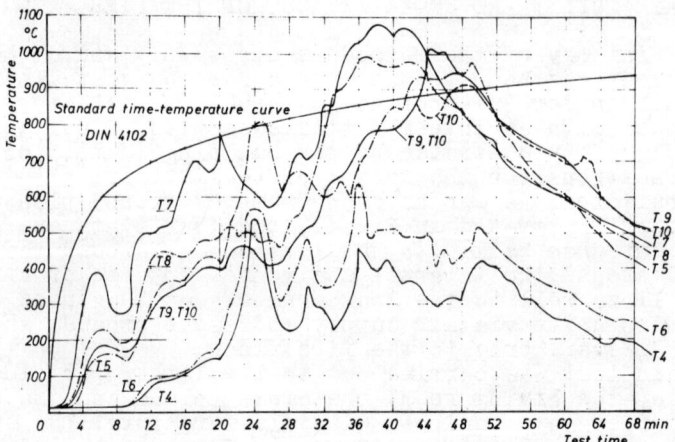

Fig. 8. Time curve for the temperature at measuring points T 4 to T 10.

The determination of the supply air and exhaust smoke flows by calculating from the measured dynamic pressures has shown that in the entrance hall the supply air flow was equivalent to the value of 323 for the hourly air exchange rate, whilst in the lock chamber this value amounted to 31, as provided for by the planning. At the same time an average underpressure of 40 N/m² was measured in the centre of the entrance hall, whereas in the centre of the lock chamber an average underpressure of 5 N/m² was recorded. Based on measurements of specific variables for smoke and observations carried out, it was found that with the understanding that the adjustment of the ventilation units was as specified, the internal staircase remained free from smoke during the entire test period, if very slight quantities of smoke are disregarded which, indeed, did not cause any major obstacle.

METHOD OF PRESSURE REDUCTION IN THE FIRE FLOOR AND/OR PRESSURE RISE ON ESCAPE ROUTES

With the second method, escape routes can be kept free from smoke by two techniques. On the one hand, the fire floor may be vented, i.e. the pressure lowered such as to come under the pressure prevailing on the floors above and below and on the escape routes, by making use of the buoyancy in a separated smoke shaft provided with an opening at the top and openings to the various floors sealed [11, 14, 15]. In case of a fire only that opening will be activated which is on the same level as the fire floor. On the other hand it is possible to increase the pressure existing in a shaft, e.g. stairwell shaft, by natural venting via an opening at the bottom of the shaft or by mechanical pressurization to such an extent as to make sure that the pressure on this escape route will, also in the event of a fire, be higher than in the fire floor [11, 12, 16].

Both possibilities are illustrated in Fig. 9. The full lines indicate the pressure pattern outside and inside the various floors and shafts for staircases, lifts etc. in a multi-storied building

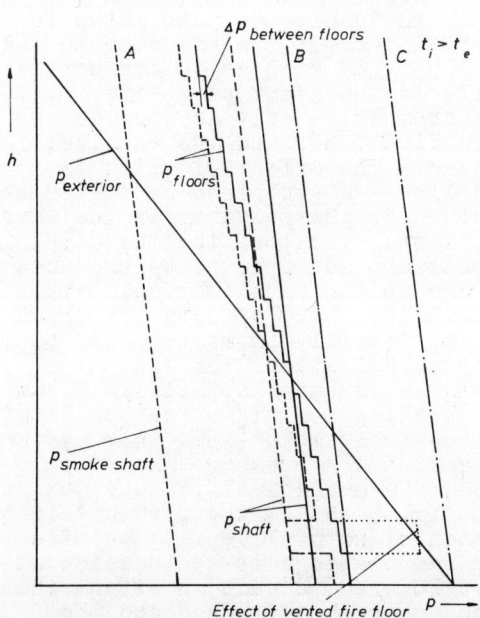

Fig. 9. Pressure pattern inside and outside a multi-storied building as a result of top venting via a smoke shaft and/or bottom venting or mechanical pressurization of the staircase

under ordinary conditions and on the presumption that the temperature in the building is higher than that of the ambient atmosphere. Due to pressure differences existing between adjoining building parts and the outside of the building, air will flow from the outside to the floors and to the shafts underneath the neutral pressure plane while the flow direction above the neutral pressure plane will be reversed, i.e. air will flow from shafts and floors to the outside. Due to leakages between floors, a limited quantity of air will also directly flow from one floor into the next one. A study by Tamura [17] has shown, however, that only a maximum of 5 % of the air entering at the lower part of the building and discharging from its top part were flowing directly from floor to floor whereas more than 95 % took its way through vertical shafts.

The pressure conditions occuring within the individual parts of and outside the building as a result of natural venting via a separated smoke shaft are represented on Fig. 9 as broken lines. As can be seen from this pressure pattern, the pressure (line A) in the smoke shaft - fitted with one opening to the outside only at the top - is everywhere lower than the pressure in the adjoining building areas. In the event of a fire, e.g. in the second floor, the fire floor will be vented by an additional opening between the fire floor and the smoke shaft, i.e. the pressure prevailing therein will be lowered. At the same time, the pressure in the other parts of the building will also decrease at all levels as shown by the dotted lines. By choosing adequate sizes for the openings at the top of the shaft and at the height of the fire floor, it is possible, to decrease the pressure in the fire floor so much as to be lower than in the neighbouring part of the stairwell shaft and in the floors above and below, i.e. in this area one flow direction only exists which runs from the neighbouring parts of the buildings into the fire floor. As this ventilation system, however, will break down as soon as an opening of a larger size is forming between the fire floor and the outside and, besides, this system strongly depends on the height and the characteristics of leakage of the building, natural venting of a fire floor will only be applicable to a limited extent [11, 14, 15].

The pressure pattern in various building parts and outside the building originating from pressure rise by natural venting or mechanical pressurization, e.g. in the staircase, are shown in Fig. 9 as full and/or dash-and-dot lines. If the pressure in the stairwell shaft will be increased only as much as to correspond to line B in Fig. 9, this shaft will, in the event of a fire, e.g. in the second floor, remain free from smoke only as long as no larger opening has formed between fire floor and the outside. In order to ensure that in such a case - the effect of which is illustrated in Fig. 9 as dotted line - a smokefree condition of the stairwell shaft will be maintained, the pressure in the shaft would have to be increased such as to correspond to line C in Fig. 9. Since, however, such a pressure could never be produced continuously by natural venting due to the fact that cold air entering at the bottom of the shaft adversely affects the pressure conditions prevailing in the shaft, it will, in most cases, be necessary to rely on mechanical pressurization [15, 16].

A combination of both techniques is being applied in Great Britain for some time past when erecting new high-rise buildings, in order to keep escape routes free from smoke [18]. This system works such that in the event of a fire the pressure on escape routes will be increased by means of a mechanical pressurization system as indicated by line C in Fig. 9 whilst the pressure in the fire floor will, by the application of natural venting as effected through a smoke shaft which only has openings to the outside at the top and to the fire floor, be lowered to such an extent that on the one hand - in spite of the expansion of gases and the buoyancy in the fire floor - the air will only flow from escape routes into the fire floor and/or to outside, and, on the other hand, it will be ensured that air which had entered the fire floor through existing leaks can be discharged.

The pressure differences originating from gas expansion, buoyancy and wind effects which must be overcome by such a pressurization system, have been investigated thoroughly by Hobson and Stewart [9]. The pressure difference occurring due to buoyancy at the upper edge of a closed, 2 m high door between a fire compartment with an uniform gas temperature of $800°C$ and the staircase can, theoretically, reach 17 $N/m^2$ if the neutral pressure plane is situated at floor level. In practice, however, the pressure difference will only reach half of this value since under normal circumstances, the neutral pressure plane will be near the centre of the door. This has been confirmed also by studies of Butcher et al. [19] who have found out, that the pressure difference at the upper edge of a door between the fire compartment and the staircase will not exceed the value of 7.5 $N/m^2$. As already mentioned before, pressure differences caused by expansion of gases may, in comparison to these pressure differences, be neglected. Pressure differences caused by buoyancy and wind effects calculated by Hobson and Stewart [9] by means of computer programs as to their dependence on the building height have been compiled in Table 2. Pressure increases on escape routes as a function of the building height recommended by aforesaid authors, are also listed in Table 2. In addition, Hobson and Stewart have demonstrated in [9] that with such pressure increases it will, in general, still be possible to open doors towards escape routes.

Table 2. Pressure differences due to buoyancy and wind effects in buildings and required minimum values of pressure increase on escape routes, according to [9].

| Height of building<br>m | Pressure difference between fire compartment and staircase<br>$N/m^2$ | Pressure difference due to buoyancy and wind effects<br>$N/m^2$ | Minimum pressure increase on escape routes<br>$N/m^2$ |
|---|---|---|---|
| 5 | 8.5 | 8 | 25 |
| 25 | 8.5 | 10.5 | 25 |
| 50 | 8.5 | 13 | 50 |
| 100 | 8.5 | 19.5 | 50 |
| 150 | 8.5 | 29.5 | 50 |

The efficiency of such a pressurization system has also been investigated recently by the Forschungsstelle für Brandschutztechnik in a 8-story office building [20]. Also in this case, the final approval of the internal staircase by the building inspection authority was made conditional on the result of a realistic fire test in this building. The groundplan of a typical floor in this building is shown in Fig. 10. The pressurization system has been designed such as to ensure that in case of a fire, the pressure in the entrance hall and in the anterooms in front of the offices as well as in the entire stairwell will be increased by 50 - 60 $N/m^2$, but, as far as the entire office area in the fire floor is concerned, will be kept at the level of the atmospheric pressure by venting this area via two smoke shafts to the roof. As can be seen on Fig. 10, for the test carried out in the second upper floor, one section of the office room has been used only and bricked up, according to the lines thick drawn out. In this so-called fire compartment with a total basis of 54 $m^2$, a fire load of approximate 365 kg - which related to the basis of the fire compartment equals about 120 $MJ/m^2$ - consisting mainly of wood and a few textiles and foamed plastics, was packed on to an area of about 10.5 $m^2$. The test lasted until, after some 40 minutes, all fuels was nearly burnt. Also during this test, temperatures, concentrations of carbon monoxide and carbon dioxide, optical smoke density and dynamic pressures corresponding to the supply air flows were measured at various points in the fire compartment, in the anteroom and in the staircase, at the height of the fire floor. The measuring points are also shown on Fig. 10. Since, at present, test results are being evaluated, final statements regarding this test cannot be made, yet. However, based on observvtions during the test, the following can be said already at this time: during the entire test period, the anteroom and the entire stairwell remained free from smoke with the exception of some smoke in the anteroom which, however, occurred for a short period only.

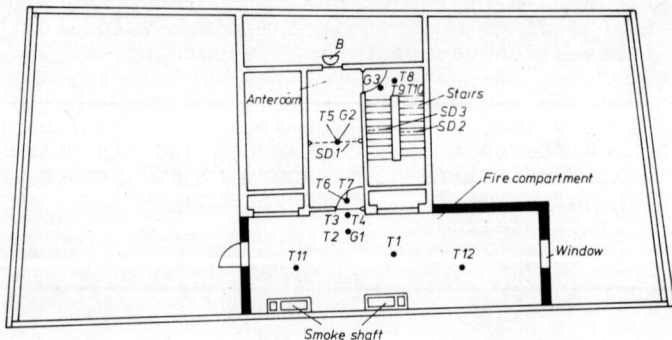

Fig. 10. Ground plan of a normal floor with the arrangement of the measuring points.
T   Temperature measuring point
G   Gas extraction measuring point
SD  Optical smoke deusity measuring section
B   Observation window.

REFERENCES

[1]  McGuire, J.H.: Smoke Movement in Buildings. Fire Technology, Vol. 3, 1967, No. 3, pp. 163-174.
[2]  McGuire, J.H.: Control of Smoke in Building Fires. Fire Technology, Vol.3, 1967, No. 4, pp. 281-290.
[3]  McGuire, J.H. and Tamura, G.T.: Simple Analysis of Smoke-Flow Problems in High Buildings. Fire Technology, Vol. 11, 1975, No. 1, pp. 15-22.
[4]  Thomas, P.H., Heselden,A.J.M.and Law, M.: Fully-developed Compartment Fires - Two Kinds of Behaviour. Fire Research Technical Paper No. 18, H.M. Stationery Office, London 1967.
[5]  Harmathy,T.Z.: A New Look at Compartment Fires, Part I. Fire Technology, Vol. 8, 1972, No. 3, pp. 196-217.
[6]  Fujita, K.: Characteristics of Fire Inside a Non-combustible Room and Prevention of Fire Damage. Building Research Institute Report 2(h), Japanese Ministry of Construction, Tokyo.
[7]  Kawagoe, K.: Fire Behaviour in Rooms. Building Research Institute Report No. 27, Japanese Ministry of Construction, Tokyo 1958.
[8]  John, R.: Rauchgas- und Luftströme durch Öffnungen in Brandräumen. 5th International Fire Protection Seminar, Vereinigung zur Förderung des Deutschen Brandschutzes e.V., Karlsruhe 1976.
[9]  Hobson, P.J. and Stewart, L.J.: Pressurisation of Escape Routes in Buildings. Fire Research Note No. 958, Joint Fire Research Organization, Borehamwood 1972.
[10] Wakamatsu, T.: Calculation of Smoke Movement in Buildings, Second Report. Building Research Institute Research Paper No. 46, Japanese Ministry of Construction, Tokyo 1971.

[11] Wilson, A.G. and Shorter, G.W.: The Smoke Problem and Its Control in High-Rise Buildings. 5th CIB Congress, Versailles 1971, pp. 205-207.
[12] Lie, T.T. and McGuire, J.H.: Control of Smoke in High-Rise Buildings. Fire Technology, Vol. 11, 1975, No. 1, pp. 5-14.
[13] Seeger, P.G. und John, R.: Brand- und Lüftungsversuche in einem Wohnhochhaus mit innenliegendem Treppenraum. VFDB-Zeitschrift, Vol. 21, 1972, No. 4, pp. 125-132.
[14] Tamura, G.T.: Analysis of Smoke Shafts for Control of Smoke Movement in Buildings. ASHRAE Transactions, Vol. 76, Part II, 1970, pp. 290-297.
[15] Tamura, G.T. and Wilson, A.G.: Natural Venting to Control Smoke Movement in Buildings via Vertical Shafts. ASHRAE Transactions, Vol. 76, Part II, 1970, pp. 279-289.
[16] McGuire, J.H., Tamura, G.T. and Wilson, A.G.: Factors in Controlling Smoke in High Buildings. ASHRAE Symposium "Fire Hazards in Buildings", San Francisco, Calif., 1970, pp. 8-13.
[17] Tamura, G.T.: Computer Analysis of Smoke Movement in Tall Buildings. ASHRAE Transactions, Vol. 75, Part II, 1969, pp. 81-92.
[18] Heselden, A.J.M. und Baldwin, R.: Rauchausbreitung in Rettungswegen in Gebäuden und deren Kontrolle. VFDB-Zeitschrift, Vol.24, 1975, No. 3, pp. 87-99.
[19] Butcher, E.G., Fardell, P.J. and Clarke, J.: Pressurization as a Means of Controlling the Movement of Smoke and Toxic Gases on Escape Routes. Fire Research Station Symposium No. 4 "Movement of Smoke on Escape Routes in Buildings", H.M. Stationery Office, London 1971, pp. 36-40.
[20] Seeger, P.G. und John, R.: Rauchfreihaltung eines innenliegenden Treppenraumes mittels eines Überdrucksystems. Research Report, Forschungsstelle für Brandschutztechnik an der Universität Karlsruhe, Karlsruhe 1976.

# ACCURACY OF THE FINITE DIFFERENCE COMPUTATION OF FREE CONVECTION

**JEAN J. PORTIER**

*Institut de Thermodynamique, University of Liège, Belgium*

**OZER A. ARNAS**

*Louisiana State University, Baton Rouge, Lousiana, USA*

ABSTRACT.

The upwind difference scheme of formulating differential expressions appears to be the most widely used in problems involving transport by simultaneous convection and diffusion, as it has been shown to have preferable convergence characteristics. It nevertheless remains that this approximation is only accurate to the first order in the grid spacing, and introduces important false transport mechanisms when dealing with increasing flow velocities. The purpose of the paper is to present a comparative study of the use of this unsymmetrical formulation for the convective terms, and of the central difference scheme with second order accuracy, for solving steady two-dimensional buoyant convection in a square cavity with horizontal temperature gradients. Use is made of the alternating direction implicit method for integrating the governing system, and a particular treatment of the equation, which defines the stream function field in terms of the vorticity, is suggested to promote the convergence of the numerical procedure.

NOMENCLATURE.

| | |
|---|---|
| $L$ | half the width of enclosure. |
| $x,y$ | cartesian coordinates. |
| $T_h, T_c$ and $T_m$ | uniform temperatures at hot and cold walls, and their arithmetic mean. |
| $x',y'$ | dimensionless space coordinates. |
| $u',v'$ | dimensionless velocity components. |
| $T'$ | dimensionless temperature. |
| $\omega$ | dimensionless vorticity. |
| $\psi$ | dimensionless stream function. |
| $g$ | acceleration of gravity. |
| $\nu$ | kinematic viscosity of fluid. |
| $\alpha$ | thermal diffusivity of fluid. |
| $\beta$ | coefficient of thermal expansion of fluid. |
| $k$ | coefficient in general form. |
| $i,j$ | subscripts for position in the space grid. |
| Pr | Prandtl number. |
| Gr | Grashof number. |
| Nu and $\overline{\text{Nu}}$ | local and mean Nusselt numbers. |

## INTRODUCTION.

The problem of accurately predicting the heat transfer due to natural convection within enclosed spaces by numerical means has not been solved definitively, although it has received increased attention recently. Besides the choice of the explicit or implicit integration procedure for the governing partial differential equations, the finite difference analogue of the non-linear convective derivatives appears to remain the most seriously deficient in accuracy. Excluding higher order schemes because of their inconvenience in use (1,2), the so-called upwind and central formula have been extensively utilized for their approximation. The convergence capabilities of the first one have undeniably been proven (3,4), and the use of three-point central differences often drastically limited the possible computable range of conditions for the undertaken studies (5,6). On the basis of very simple test problems, Spalding (7) and Runchal (8) also concluded its superior accuracy when the local Peclet number of the grid is large. From a strictly mathematical point of view, it nevertheless remains that the upwind scheme, because of its lack of symmetry, is a first order approximation of the convective terms, contrary to their central formulation which achieves the second order of accuracy.

This paper mainly concerns the differences in the flow patterns and temperature charts which arise from the use of these algorithms to predict two-dimensional natural convection in an enclosed square cavity whose vertical walls are held at different temperatures, and the top and bottom are perfect insulators. To emphasize the influence of the numerical formulation without interference of any other assumptions, the laminar flow regime has only been considered. A particular treatment of the Poisson equation, which defines the stream function field in terms of the vorticity, is also suggested to promote the convergence of the used iterative procedure, as it allows to overcome the usual restrictions on the Grashof number which were encountered in most of the previous studies of the same problem.

## THE MATHEMATICAL PROBLEM.

With reference to the physical model shown in figure 1, the governing equations are obtained by applying the Boussinesq approximation to the conservation laws to yield, in terms of the reduced variables $x' = x/L$ and $y' = y/L$,

$$\frac{\partial}{\partial x'}(u'\omega') + \frac{\partial}{\partial y'}(v'\omega') = \frac{\partial^2 \omega'}{\partial x'^2} + \frac{\partial^2 \omega'}{\partial y'^2} + \frac{Gr}{16}\frac{\partial T'}{\partial x'},$$

$$-\omega' = \frac{\partial^2 \psi'}{\partial x'^2} + \frac{\partial^2 \psi'}{\partial y'^2},$$

$$\frac{\partial}{\partial x'}(u'T') + \frac{\partial}{\partial y'}(v'T') = \frac{1}{Pr}\left(\frac{\partial^2 T'}{\partial x'^2} + \frac{\partial^2 T'}{\partial y'^2}\right),$$

and have to be solved together with the following boundary conditions

$$u' = \frac{\partial \psi'}{\partial y'} = 0 \text{ and } -v' = \frac{\partial \psi'}{\partial x'} = 0$$

$$\text{at } x' = \pm 1 \text{ and } y' = \pm 1,$$

# THE FINITE DIFFERENCE COMPUTATION OF FREE CONVECTION

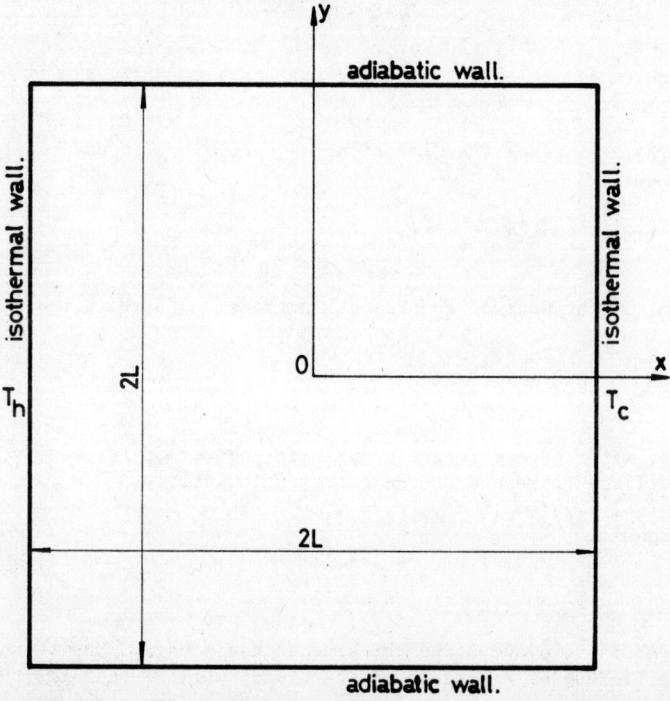

Figure 1. The Physical System.

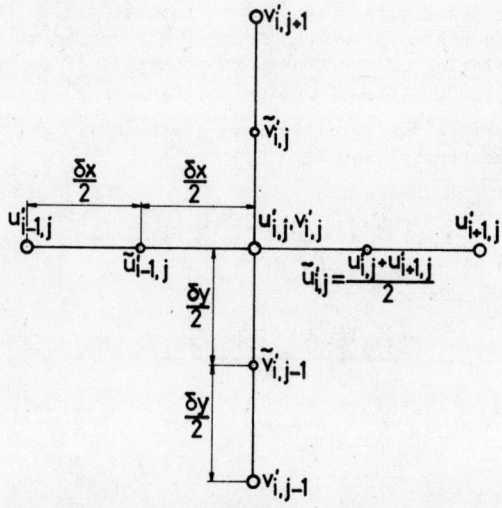

Figure 2. The Finite Difference Schemes.

$$T' = \mp 1 \qquad \text{at } x' = \pm 1 \text{ respectively},$$
$$\frac{\partial T'}{\partial y'} = 0 \qquad \text{at } y' = \pm 1.$$

This system defines the dimensionless vorticity, stream function and temperature fields

$$\omega' = \frac{L^2 \omega}{\nu}, \quad \psi' = \frac{\psi}{\nu} \quad \text{and} \quad T' = 2\frac{T - T_m}{T_h - T_c},$$

which depend on the Grashof and Prandtl numbers

$$Gr = \frac{8g\beta(T_h - T_c)L^3}{\nu^2} \quad \text{and} \quad Pr = \frac{\nu}{\alpha},$$

and exhibit, with respect to the geometrical center of the cavity, a symmetry which is utilized to reduce the numerical calculations

$$\omega'(x',y') = \omega'(-x',-y'),$$
$$\psi'(x',y') = \psi'(-x',-y'),$$
$$T'(x',y') = -T'(-x',-y').$$

The symbols $\nu$, $\alpha$ and $\beta$ are the kinematic viscosity, the thermal diffusivity and the coefficient of expansion of the fluid, while $T_m$ denotes the arithmetic mean of the hot and cold temperatures prescribed along the vertical walls of the enclosure.

THE NUMERICAL SOLUTION.

The steady-state distributions of the required variables are obtained as the asymptotic time-like limits of arbitrary initial fields through a sequential iterative process, each step of which successively considers the corresponding time-dependent formulation of the transport equations of energy and vorticity, and the evaluation of the stream function field.

With the appropriate finite difference approximation of the convective derivatives, namely the central scheme

$$\frac{\partial}{\partial x'}(u'T') = \frac{u'_{i+1,j}T'_{i+1,j} - u'_{i-1,j}T'_{i-1,j}}{2\delta x}$$

or the upwind representation

$$\frac{\partial}{\partial x'}(u'T') = \begin{cases} \dfrac{\tilde{u}'_{i,j}T'_{i,j} - \tilde{u}'_{i-1,j}T'_{i-1,j}}{\delta x} & \text{when } \tilde{u}'_{i,j} \text{ and } \tilde{u}'_{i-1,j} > 0 \\[2mm] \dfrac{\tilde{u}'_{i,j}T'_{i+1,j} - \tilde{u}'_{i-1,j}T'_{i,j}}{\delta x} & \text{when } \tilde{u}'_{i,j} \text{ and } \tilde{u}'_{i-1,j} < 0 \end{cases}$$

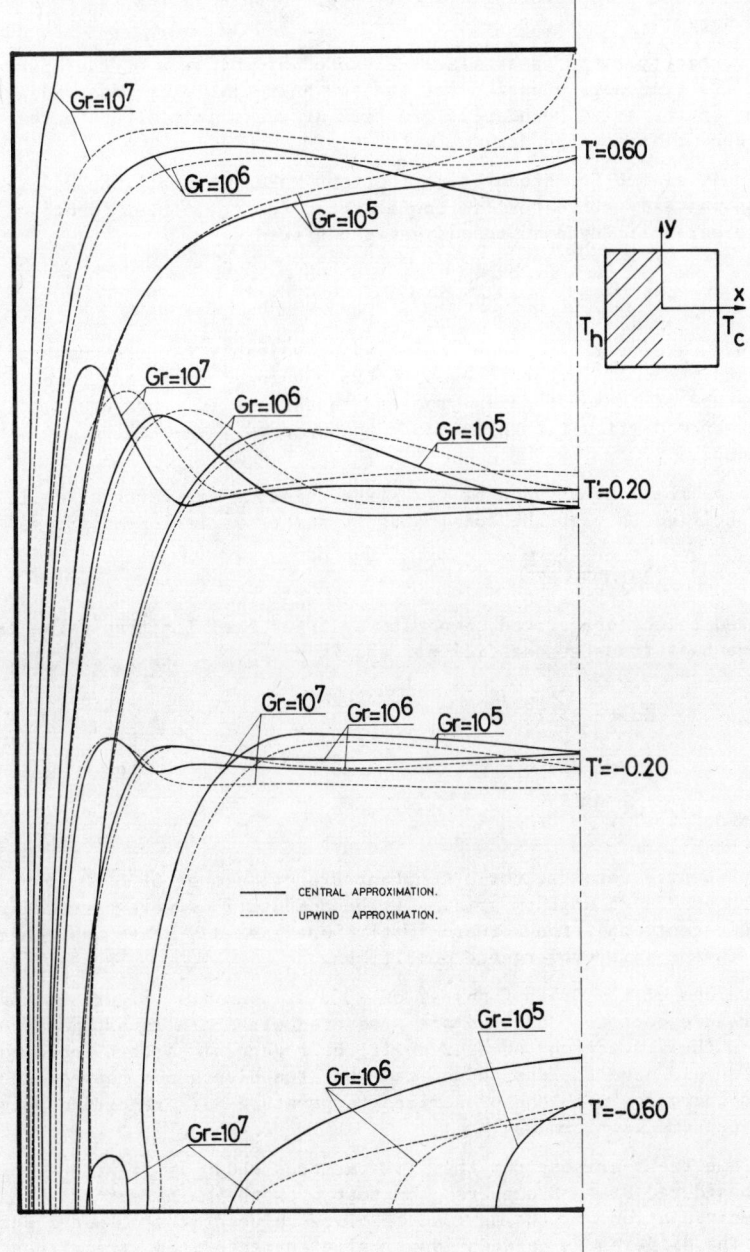

Figure 3. Isothermal Lines at the Considered Grashof Numbers, for $Pr = 0.70$.

according to the numbering of figure 2, and linearly extrapolated components of the velocity, the temperature field values are updated using the classical alternating direction implicit method.

The vorticity field is similarly advanced by solution of the discrete equivalent of its transport equation for the two half-time steps in tridiagonal form. Its values on the solid boundaries are left at their initial estimate from the boundary conditions on the stream function.

The only search for steady distributions makes possible the addition of a fictitious unsteady state term in the equation for the stream function, to convert it to parabolic type in the advantageous form of

$$k \frac{\partial \psi'}{\partial t'} = \frac{\partial^2 \psi'}{\partial x'^2} + \frac{\partial^2 \psi'}{\partial y'^2} + \omega'$$

The alternating direction scheme then remains suitable for its numerical integration, and the coefficient $k$, whose optimum value depends on the characteristic parameters of the undertaken problem, appears as an improvement to overcome the convergence difficulties previously encountered when dealing with increasing Grashof numbers.

Finally, the convective heat transmission along the vertical solid boundaries is specified through the local Nusselt number variation, expressed as

$$Nu_j = -\left(\frac{\partial T'}{\partial x'}\right)_{1,j}$$

when related to the prescribed temperature difference. Its mean value defines the average heat transfer coefficient, and is given by

$$\overline{Nu} = -\frac{1}{2}\int_{-1}^{+1}\left(\frac{\partial T'}{\partial x'}\right)_{x'=-1,y'} dy'$$

## THE OBTAINED RESULTS.

The presented results, for a constant Prandtl number of 0.70, were obtained with an $25 \times 25$ uniform grid. All the computations were carried from the initial quiescent condition forward in time until steady state was achieved, submitted to the same convergence conditions.

For values of the Grashof number of $10^5$, $10^6$ and $10^7$, figure 3 shows selected temperature contours, as they are predicted with both investigated difference schemes for the convection terms. Even if their general pattern looks quite similar, it should nevertheless be noticed that the divergence between corresponding lines increases with the prescribed temperature difference and mainly affects the extreme isothermals.

The same trend arises from the flow patterns shown in figures 4, 5 and 6 for the considered Grashof numbers, together with the main features of the appearing recirculation and the maximum absolute value of the computed stream function. The differences between identically characterized streamlines are still growing with higher values of the Grashof number and the stream function. One will notice the double centered counter circulation to the main buoyant tendency that is predicted through the use of the central difference scheme for the convective derivatives, and the increasing distance between the reported maxima.

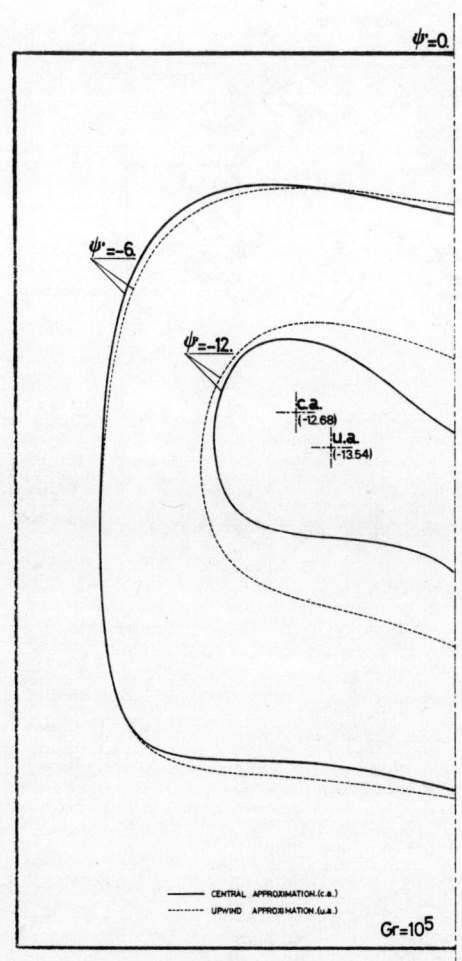

Figure 4. Selected Streamlines at a Grashof Number of $10^5$, for $Pr = 0.70$.

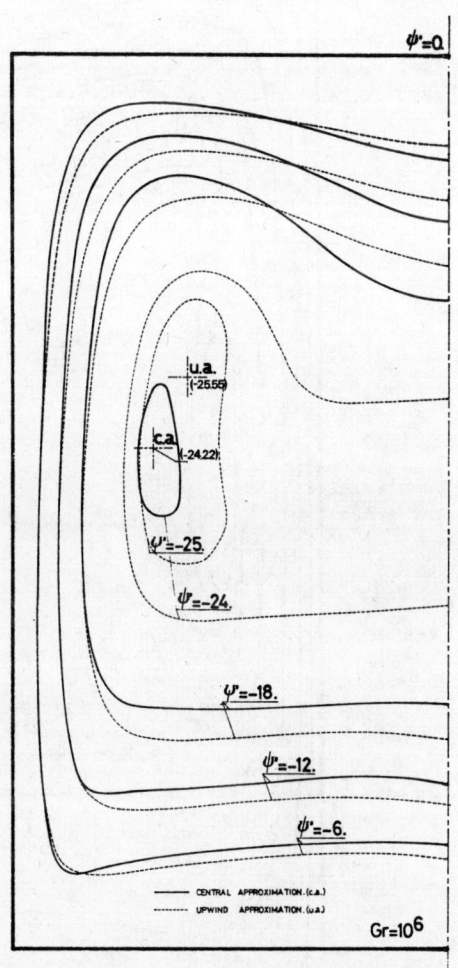

Figure 5. Selected Streamlines at a Grashof Number of $10^6$, for $Pr = 0.70$.

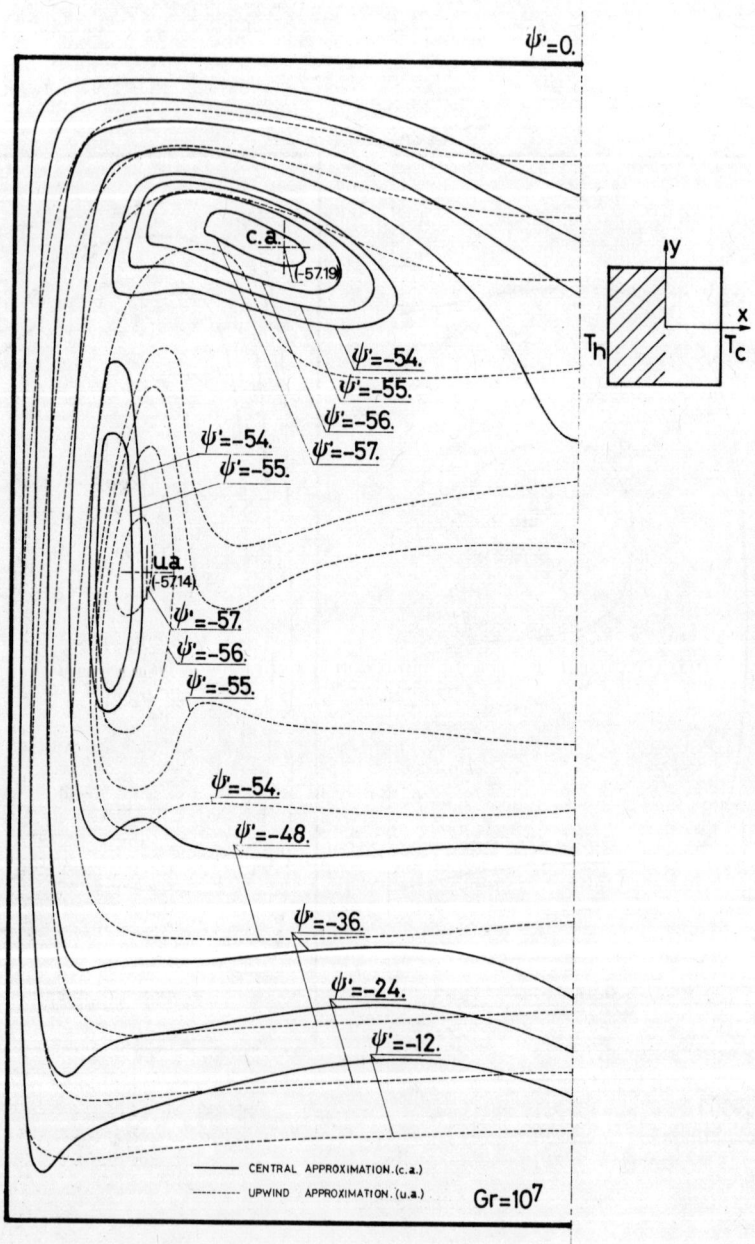

Figure 6. The Flow Pattern and the Predicted Recirculation at a Grashof Number of $10^7$, for $Pr = 0.70$.

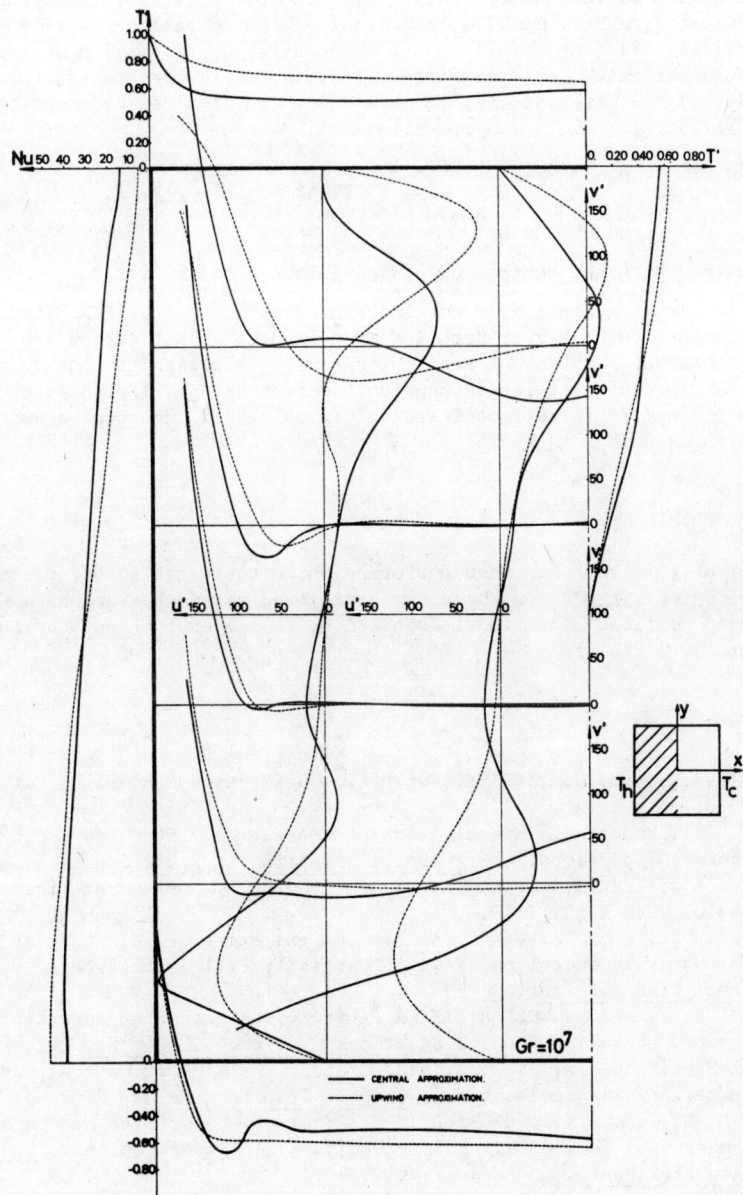

Figure 7. Some Characteristic Profiles in the Cavity at a Grashof Number of $10^7$, for $Pr = 0.70$.

Figure 7 at last compares some profiles of the horizontal and vertical components of the dimensionless velocity, details the temperature distribution along the adiabatic boundaries and the vertical axis of the cavity, and represents the local variation of the Nusselt number over the hot wall, for the highest investigated Grashof number. It will be added that the overall convective heat transfer coefficient is found smaller with the central approximation up to a value of the Grashof parameter just superior to $10^6$; the ratio changes afterwards. In this respect, the here reported results and those of other computations in the same range correlate with the power laws

$$\overline{Nu} = 0.163 \ Gr^{0.287}$$

or 

$$\overline{Nu} = 0.204 \ Gr^{0.270}$$

when obtained with the central or upwind formula, respectively.

If it cannot definitively conclude in favour of one or the other scheme without comparison to a very detailed experimental investigation, the present study has however pointed out an important problem related to the computation of natural convection fields by showing the increasing differences in the predictions of the two investigated and so commonly used numerical approximations, as the Grashof number is varied.

ACKNOWLEDGEMENTS.

Part of this work has been performed while the first author was at Louisiana State University, Baton Rouge, as a visiting researcher and as a Fellow of the Belgian American Educational Foundation. The support of both organizations is gratefully acknowledged.

REFERENCES.

1. ROBERTS,K.V., and N.O. WEISS,'Convective Difference Schemes.', Math. Comput. 20, 1966.
2. FROMM,J.E.,'Practical Investigation of Convective Difference Approximations of Reduced Dispersions.', High-Speed Computing in Fluid Dynamics, International Union of Theoretical and Applied Mechanics Symposium, American Institute of Physics, New York, 1969.
3. BARAKAT,H.Z., and J.A. CLARK,'Analytical and Experimental Study of Transient Laminar Natural Convection Flows in Partially Filled Containers.', Proc. 3rd Int. Heat Transfer Conf. 2, 1966.
4. RUNCHAL,A.K., and M. WOLFSHTEIN,'A Finite Difference Procedure for the integration of the Navier Stockes Equations', J. Mech. Engng Sci. 11, 1969.
5. NEWELL,M.E., and F.W. SCHMIDT,'Heat Transfer by Laminar Natural Convection within Rectangular Enclosures.', J. Heat Transfer, Trans. ASME, 32C. 1970.
6. CHU,H.H.-S., and S.W. CHURCHILL,'The Effect of Heater Size and Location on Two-Dimensional Laminar Natural Convection in a Square Channel.', National Heat Transfer Conf., San Francisco, 1975.
7. SPALDING,D.B.,'A Novel Difference Formulation for Differential Expressions Involving Both First and Second Derivatives.', Int. J. Num. Meth. Engng 4, 1972.
8. RUNCHAL,A.K.,'Convergence and Accuracy of Three Finite Difference Schemes for a Two-Dimensional Conduction and Convection Problem.', Int. J. Num. Meth. Engng 4, 1972.

# EXPERIMENTAL STUDY OF FREE CONVECTION IN A SQUARE CAVITY

G. BURNAY, J. HANNAY and J. PORTIER

*Institut de Thermodynamique, University of Liège, Belgium*

ABSTRACT.

Steady two-dimensional natural convection of air has been experimentally investigated within a long horizontal square enclosure. The vertical walls were held at different but uniform temperatures, and the top and bottom were to be adiabatic to heat and mass transfer. The presented results mainly concern the Grashof number range which includes the onset of laminar instability, and is of particular interest in the study of air circulation at environmental room temperatures. A relation for average heat exchange has been evaluated from direct measurements, after correction for the heat conduction losses and the radiative transfer.

INTRODUCTION.

The reported study concerns the steady convection motion of air contained within a long horizontal cavity of square section whose vertical walls are held at different constant temperatures, and the top and bottom are considered perfect insulators. The objective of the experimental investigation is to obtain a relationship between the convective heat transfer coefficient and the prescribed boundary conditions, for Grashof numbers at the occurring of the transition from laminar to turbulent flow regime.

Even if considerable attention has recently been devoted to convective flow within enclosed fluids, by theoretical, numerical and experimental means, very few studies have however reported detailed results for air circulation in conditions related to environmental temperatures of rooms. Among the available correlations for a square cavity, most have been derived from numerical solutions which were generally limited to values of the Grashof parameter less than $10^6$, and may therefore only be retained up to this upper bound. In this way, Han (1) suggested the relationship

$$\overline{Nu} = 0.0782 \; Gr^{0.3594},$$

between the mean Nusselt number and the Grashof one, when both are based on the temperature difference between the hot and cold vertical walls, for a Prandtl number of 0.733. Newell and Schmidt (2) indicated an expression of the form

$$\overline{Nu} = 0.0547 \; Gr^{0.397},$$

and Rubel and Landis (3) derived the relationship

$$\overline{Nu} = 0.082 \; Gr^{0.344}.$$

On the basis of a first experimental study and a subsequent numerical analysis, Elder (4,5) obtained the following approximation

$$\overline{Nu} = 0.231 \, Gr^{0.250},$$

and a recent numerical computation by Portier and Arnas (6) covered the Grashof number range from $10^5$ to $10^7$ and led to the correlations

$$\overline{Nu} = 0.163 \, Gr^{0.287}$$

or

$$\overline{Nu} = 0.204 \, Gr^{0.270}$$

according as the central or upwind difference scheme was utilized to simulate the convective derivatives.

No relationships for the Nusselt coefficient were unfortunately given in the very sophisticated study by Fromm (7), except an estimated value of 32.0 for $Gr = 10^8$ and $Pr = 1$.

THE TEST APPARATUS.

Figure 1 gives a sketch of the apparatus in its essential features. It consists of an horizontal square enclosure of 3 m. long to make possible the establishment of two-dimensional flow and heat transfer, which is controlled by repeating all the measurements in the two sections 1 m. apart from the central reference plane. The uniformity of the vertical walls temperature is achieved by conditioned air circulation with adequate output, while the horizontal boundaries, which are to be adiabatic to heat transfer, are covered with a 12.5 cm. thick insulating layer, which makes the respective thermal conductances of the upper and lower walls as low as 0.33 and 0.30 W./m$^2$.°C. The whole cavity is painted in black to prevent light reflection during the possible visualizing of the flow through one of the end sections of the apparatus, made of glass and covered with a removable insulating plate.

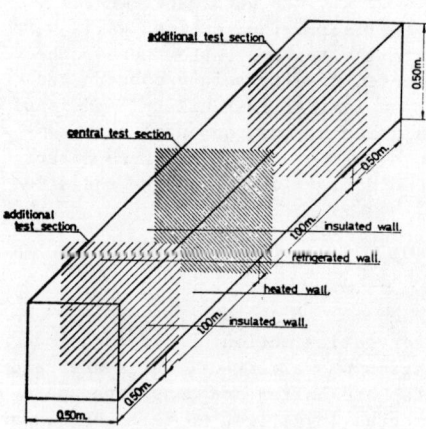

Figure 1. The Test Apparatus.

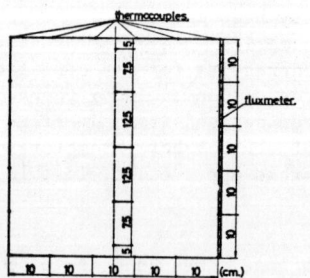

Figure 2. Probes Location in Each of the Test Sections.

Twenty thermocouples located as shown in figure 2 measure the surface temperatures in each of the three test planes, and five other ones, 0.1 mm. diameter and provided with radiation shields, control the air temperature along the vertical axis of the enclosure. An electrically driven carriage, with a platinum wire 3 μm. diameter sliding probe, also allows the scanning of both temperature and velocity fields in any section of the cavity. Several other thermocouples additionally measure the environmental conditions.

The overall heat flux from the hot plate is evaluated through a rectangular 25 cm. wide fluxmeter, covering the whole wall and made of the same material. Its stability is about 1 % of the measured quantity (8).

The steadiness of the flow regime is controlled every ten minutes during some two hours through systematic measurements the arithmetic means of which are the retained experimental results, with standard deviations of some 0.05 °C for the temperatures.

THE ANALYSIS OF THE EXPERIMENTAL RESULTS.

As an example, figure 3 reports the measured temperature profiles along the adiabatic walls in the central section of the cavity and along the vertical axis in the three mentioned test planes, for a Grashof number of 1.69 $10^8$.

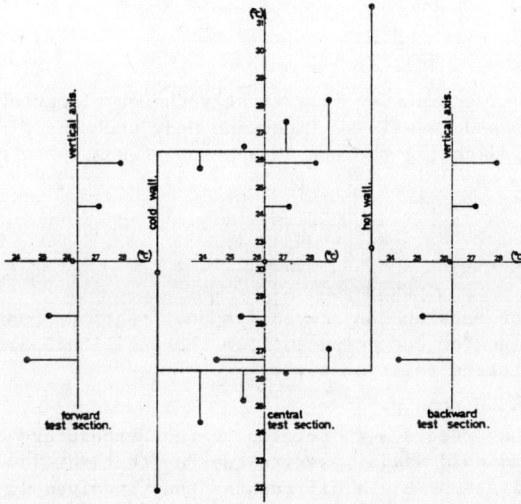

*Figure 3. Recorded Temperature Measurements at a Grashof Number of 1.69 $10^8$.*

This particular case will also serve to detail the derivation of the convective heat exchange from the available experimental data, as summarized in figure 4.

The total heat transmission through the hot plate is directly measured, and the knowledge of the thermal conductances of the horizontal walls, together with their surface temperatures and the environmental conditions, make possible the computation of their heat losses.

Considering the enclosure as a gray body divided into 22 zones with a constant emissivity of 0.925 and the uniform collected temperatures, its radiation balance results from the solution of the algebraic system of

$$R_i = \varepsilon \sigma T_i^4 + (1-\varepsilon) \sum_{j=1}^{22} R_j F_{i-j} \qquad i = 1, 2, \ldots, 22,$$

for the assumed constant radiosities. The net radiative heat flux densities at the zones are then determined from the equations

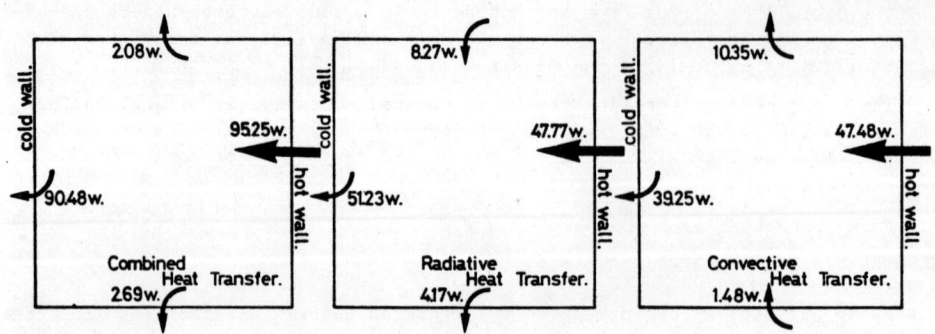

*Figure 4. Detailed Heat Exchange at a Grashof Number of $1.69 \; 10^8$.*

$$q_i'' = R_i - \sum_{j=1}^{22} R_j F_{i-j} \qquad i = 1,2,\ldots,22.$$

The net convective transfer is evidently the complementary part to the radiative components, and reveals an important heat exchange along the hotter horizontal boundary indicating there a strong interference between these two modes of heat transfer.

RESULTS AND CONCLUSION.

The correlation between the convective heat transfer coefficient and the prescribed Grashof number was computed from the individual data of figure 5, and led to the following relationship

$$\overline{Nu} = 0.0024 \; Gr^{0.540},$$

for dimensionless parameters both related to the temperature difference between the vertical hot and cold walls. Despite the few tests at low Grashof numbers, because of the small temperature differences they involved in the used test apparatus, the extrapolation of this law compares quite favourably with the results of reference 6.

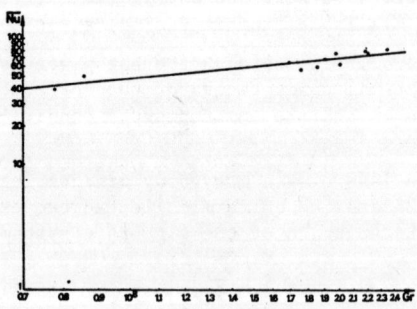

*Figure 5. The Experimental Data for the Mean Nusselt Number.*

The higher value of the exponent in the proposed expression, with respect to the previously reported correlations, is probably related to the different flow regimes they are concerned with. Emphasis will however be laid on that mentioned difficulty of separating convective and radiative effects in actual situations, which makes quite impossible the design of such an experiment with adiabatic walls to the convective transfer.

REFERENCES.

1. HAN,J.T.,'Numerical Solutions for an Isolated Vortex in a Slot and Free Convection Across a Square Cavity.', MSc thesis, Department of Mechanical Engineering, University of Toronto, 1967.
2. NEWELL,M.E., and F.W. SCHMIDT,'Heat Transfer by Laminar Natural Convection within Rectangular Enclosures.', J. Heat Transfer, Trans. ASME, 32C, 1970.
3. RUBEL,A., and F. LANDIS,'A Numerical Study of Natural Convection in a Vertical Rectangular Enclosure.', International Symposium on High-Speed Computing in Fluid Dynamics, The Physics of Fluids, Supplement II, Nov. 1969.
4. ELDER,J.W.,'Laminar Free Convection in a Vertical Slot.', J. Fluid Mechanics, Vol. 23, 1965.
5. ELDER,J.W.,'Numerical Experiments with Free Convection in a Vertical Slot.', J. Fluid Mechanics, Vol. 24, 1966.
6. PORTIER,J.J., and O.A. ARNAS,'Accuracy of the Finite Difference Computation of Free Convection.', ICHMT Int. Seminar on Turbulent Buoyant Convection, Dubrovnik Sept. 1976.
7. FROMM,J.E.,'Numerical Method for Computing Nonlinear, Time Dependent, Buoyant Circulation of Air in Rooms.', IBM J. Research & Development, May 1971.
8. NUSGENS,P.,'Sondes Utilisées pour la Mesure des Conditions de Confort et des Echanges Thermiques en Ambiance Climatisée.', Coll. Publ. Fac. Sc. Appl., U. Lg., 56, 1975.

# BUOYANT DIFFUSION FLAMES

JOHN DE RIS

Factory Mutual Research Corporation,
Norwood, Massachusetts
USA

ABSTRACT

The paper provides a general review of the current scientific understanding of turbulent buoyant diffusion flames - particularly from vertically directed fuel jets. Available experimental and theoretical results are surveyed with the view of suggesting future research in this challenging and important problem area. In the latter part of the paper we present a new Cascade Combustion Model for turbulent diffusion flames. The model follows the burning of eddies over their entire sequence of scales in the Kolmogorov cascade. At each cascade stage the burning at the molecular level is treated as a laminar diffusion flame driven by the strain rate associated with that scale. The burning takes place principally at the Kolmogorov microscale. The predicted mean volumetric burning rate versus fuel flow rate is consistent with Markstein's (10) buoyant diffusion flame measurements. Finally it is shown that the Kolmogorov microscale and associated Damköhler number varies only very slightly with flow rate; thus explaining why the radiant output of fully turbulent diffusion flames is proportional to the rate of combustion.

## I  INTRODUCTION

Over the past decade scientists have made great strides in predicting forced convection turbulent flows of engineering importance. Building on these successes scientists and engineers are now tackling more complex turbulent flow situations involving either buoyancy or combustion. It is the purpose of this article to review our fundamental understanding of turbulent flows involving both buoyancy and combustion. In particular we emphasize the fundamental problem of a turbulent diffusion flame from a vertical nozzle.

The problem of turbulent buoyant diffusion flames is of central importance to fire research. Hazardous fires are usually turbulent, they are almost always buoyancy dominated and they essentially always occur as diffusion flames. This problem is also of considerable importance to the field of furnace design. For both these applications engineers are principally concerned with the radiative interaction of the flames with their surroundings. For example, the burning rate of hazardous scale fires is principally controlled by the radiant heat feedback from the flames to the fuel surface. Also in most fire spread situations of practical interest the heat transfer to the unburnt surface takes place principally by radiation - this is true even when the buoyant flames directly impinge on the surface. Thus scientific studies of buoyant diffusion flames should if possible respond to the engineers need for predicting radiative interactions.

Only a small fraction of the radiation emitted by luminous flames is due to molecular emission; most of it comes from the soot. This is unfortunate, since little is known about soot formation. Instead of dealing with these problems engineers typically measure the effective flame radiation temperatures and absorption coefficients for the flames of interest and then calculate the radiative fluxes from the measured or predicted flame dimensions. Such calculations can be reasonably accurate provided one knows the flame shape and dimensions.

In considering the problem of radiation from turbulent buoyant diffusion flames fire researchers have long been puzzled by the common observation that the total flame radiant output for fully turbulent fires is proportional to the total burning rate of geometrically scaled fires. This result must surely be associated with some fundamental property of buoyant turbulent diffusion flames. Toward the end of this paper we provide a tentative explanation for this observation by considering the microscale burning processes associated with our cascade combustion model.

## II EXPERIMENTAL BACKGROUND

During the 1950's and early 1960's numerous studies were made on turbulent flame heights from: vertical fuel jets (1,2), wood crib fires (3,4) and liquid pool fires (5,6). These studies provided many useful dimensionless correlations as well as physical insight into the overall combustion processes. Thomas (3) reviewed many of these results and showed for example that the flame height, H, for a city gas buoyant fuel jet increases with volumetric flow rate $\dot{Q}$ according to

$$H/D = 29 \ (\dot{Q}^2/gD^5)^{1/5}$$

where D is the initial diameter. The formula implies that the diameter has no effect on the flame height. Experiments show this is true when $D \ll H$ and the initial jet momentum is much less than the total buoyancy induced momentum. In addition Thomas showed from Rasbash's (9) cine-photographic measurements that the characteristic flame tip velocity $V_t$ is given by

$$V_t = 0.36 \ (2g(T_t - T_\infty) \ H/T_\infty)^{1/2} \ .$$

Recent pitot-tube velocity measurements by Heskestad (7) are consistent with this result but suggest a somewhat higher coefficient. There is clearly a need for more such pitot-tube measurements throughout the flame plume.

Thomas et al. (4) made extensive measurements of the total air entrainment into the rising plume for a variety of fires. They found that the total entrainment below the flame tip is approximately a factor of ten greater than the air required for complete combustion. Some of this excess entrained air enters the flame envelope while the rest rises outside the flames. These total entrainment results were shown to be approximately predicted by the classical variable density entrainment model of Ricou and Spalding (8). Thomas et al. (4) also measured the oxygen concentration along the flame axis. This concentration increased from about 10% near the base of the fire to above 15% over the upper half of the flames. This suggests that the flames engulf several times their stoichiometric air requirement.

Rasbash et al. (9) in an earlier study photographically measured the flame volumes for a variety of liquid fuels burning in a 30cm diameter vessel. They found that for each fuel the flame volume remained proportional to the mass burning rate as the burning rate slowly increased with time. It turns out that the burning rate of stoichiometric air per unit flame volume was constant and the same for all the fuels (alcohol, benzene, gasoline, kerosine). Since in general the heat released per unit mass of air consumed is the same for all these fuels, we conclude that these fires all had the same heat release per unit volume - namely 1.8 $W/cm^3$. Recently Markstein (10), while studying the scaling characteristics of buoyant propane fuel jets, found an average volumetric heat release rate of about 2.0 $W/cm^3$ for a 100 cm high flame comparable in size to Rasbash's flames. However Markstein found that the volumetric heat release rate decreases with the negative 0.16 power of fuel flow rate.

In this study Markstein was principally interested in the distribution of radiant output from the flames. His flame lengths increased with the .452 power of flow rate, while his flame widths increased with the 0.355 power of flow rate. He also found that the total radiant output was precisely proportional to flow rate over a forty-fold range of flow rate. Later we will provide a tentative explanation for this result.

## III  THEORETICAL BACKGROUND

### 3.1  INTEGRAL THEORIES

From an engineering viewpoint one desires a theory which correctly predicts: flame height, flame width versus height and volumetric rate of burning. In addition it would be desirable to have the results in analytic form.

Steward (11) has made important progress in this direction. He assumed top hat profiles for velocity, temperature, density and species concentrations. He then formulated a set of integral boundary layer equations for mass, momentum, energy and species concentrations using the entrainment hypothesis of Ricou and Spalding (8). He treated the combustion by assuming that the ambient air burns instantaneously upon entrainment into the fire plume. This combustion assumption ignores both the unburnt air within the flame envelope and the air which rises in the plume outside the flame envelope. Nevertheless he obtained an analytical solution for both the flame height and the development of the thermal plume above the flame. When he compared his results with data for a wide range of different fuels, he found that the measured flame heights were considerably greater than those calculated; in fact the experimental flames reached the height at which the analytical model predicted that the thermal plume above the fire had entrained 400% excess air. The model has obvious shortcomings; however it seems to have the correct trends and from an engineering standpoint it is still the best theory we have available.

There is an urgent need for an improved integral theory which employs a more advanced combustion model and properly accounts for the air entrainment. Such a theory could provide deeper insight into buoyant flame combustion processes in addition to being of engineering importance.

### 3.2  $k$-$\varepsilon$-$g$ COMBUSTION MODEL

The $k$-$\varepsilon$-$g$ model recently developed by Launder and Spalding (24) is being intensively investigated by many researchers seeking a semi-empirical computer model for turbulent diffusion flames. Very recently Kent and Bilger (12) incorporated Favre averaging for variable density effects and showed that the model accurately predicted their extensive measurements of forced convection diffusion flames. Scientists at Imperial College have developed very general computer codes both for boundary layer (parabolic) and recirculating combustion situations. The model solves equations for mass and momentum as well as separate equations for:
- $k$ – the turbulent kinetic energy
- $\varepsilon$ – the dissipation rate of turbulent kinetic energy
- $f$ – the mean value of a conserved Shvab-Zeldovich scalar such as the fuel oxidant mixture fraction and
- $g = \overline{f'^2}$ – the mean square fluctuation of $f$.

The quantity $g$ is closely related to the molecular unmixedness of the turbulent flow. When $g$ is small the fluctuations of $f$ are small and the fluid is well mixed at the molecular level. When $g$ is large there are large local swings in concentration indicating that the molecular mixing has not caught up with the turbulent mixing.

This is a semi-empirical model. The structure of many of the terms in the governing equations is based on sound intuitive reasoning, while the coefficients appearing in such terms are determined by comparison with a wide range of forced convection experiments.

To guide the extension of the model to buoyant flow situations George et al. (13) at Factory Mutual made fundamental hot-wire turbulence measurements for the classical problem of a buoyant plume without combustion. They measured the mean and fluctuating temperatures and vertical velocities as well as their correlations across the plume. They found that the temperature and vertical velocity fluctuations are strongly correlated. Tamanini (14) at Factory Mutual used Rodi's concept of algebraic stress modeling (15) to describe the highly non-isotropic fluctuations; and using the k-ε-g model he obtained good agreement with experiment.

Tamanini (16) also used his algebraic stress model to study the flame shape and total burning rate versus height for the buoyant propane fuel jets measured by Markstein (10). He experienced several difficulties. First the k-ε-g model could not handle the laminar to turbulent transition phenomena occurring in the experiments. To overcome this problem he artificially introduced transition at appropriate heights. Apparently the k-ε-g model has difficulty in describing the rapid lateral transfer of turbulent kinetic energy and specie fluctuations just above transition. While the theory predicted the correct flame heights, it considerably underpredicted the flame widths. After adjustment of some of the coefficients Tamanini obtained considerably better agreement with the average instantaneous flame widths instead of the long-time averaged widths.

Apparently these buoyant fuel jet flames have a natural tendency to slowly meander from side to side, due possibly to a lateral instability mechanism associated with more intense burning and entrainment on the side toward which a flame is already leaning. Such meandering is less severe in thermal plumes. Thermal plumes decelerate in the vertical direction; and increased local lateral entrainment should have a cooling and consequently stabilizing effect. Emmons and Ying (17) experienced similar meandering problems for buoyant flames despite extensive efforts in eliminating laboratory disturbances. They experimentally overcame the problem by rapidly swinging an instrumented boom across the flames. Markstein's solution to the same problem was to electronically reject data when the flame meandered out of alignment. One should not expect the k-ε-g model to cope with this transient non-axisymmetric instability phenomenon. And even after experimentally correcting for gross meandering effects, some adjustment of constants in the theory may be necessary to correct for residual effects of any such instability mechanism.

## IV  FUNDAMENTAL COMBUSTION MECHANISMS

The previously mentioned turbulent diffusion flame theories are principally concerned with predicting overall mean parameters such as flame dimensions, velocities and primary reactant concentration distributions. These effects are controlled by macroscopic lengths and velocities. Even the turbulence quantities such as the turbulent kinetic energy, its rate of dissipation and mean square concentration fluctuations are principally controlled by these macroscopic flow variables. As a result these theories have so far been successful even though molecular processes were ignored.

However molecular processes are important in determining intermediate species and in particular soot concentrations which in turn control flame radiation and smoke generation. These molecular processes in turbulent diffusion flames are currently being intensively studied in several laboratories (18,19,20). Bilger (18) has rigorously shown that the local instantaneous generation of species "i" for a shifting-equilibrium flame with equal diffusivities is given by

$$\dot{m}_i''' = -\rho D (\nabla f)^2 d^2 Y_i / df^2$$

where f is the normalized fuel-oxidant concentration variable and $Y_i$ is the shifting-equilibrium concentration of species "i". This is an important result which has permitted Bilger (18) to estimate intermediate species concentrations. Magnussen and Hjertager (19) have recently developed a simplified turbulent combustion model including soot formation and combustion processes which provides excellent agreement with a wide range of turbulent premixed and diffusion flame experiments. Finally Spalding (20) is developing a new Eddy Break Up Model which describes the interleaving and subsequent combustion of fuel and oxidant eddies while their scale is decreased during the turbulent mixing.

V   TURBULENT CASCADE MODEL

In the remaining part of the paper we develop our turbulent cascade model. This model is related to Spalding's Eddy Break Up Model but differs from it in several significant respects - particularly in the dependence of the rate of eddy break-up on the eddy scale and in the use of an opposed-flow laminar diffusion flame model to describe the burning at the molecular level. The model follows the combustion between two large colliding fuel and oxidant eddies starting from their formation, on down the Kolmogorov turbulent cascade (23) to their final dissipation at the Kolmogorov microscale. The idea of describing the burning by a succession of strain driven flamelets was originally suggested to the author by the detailed discussion of Carrier, Fendell and Marble (21). They were in turn inspired by the entrainment visualizations of Brown and Roshko (22) shown schematically in the figure below for the mixing between two fluids.

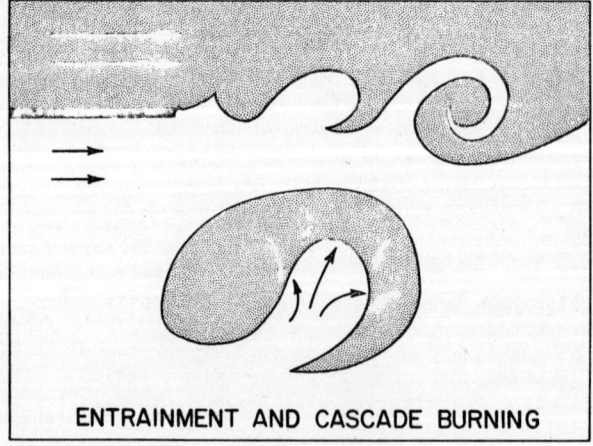

Figure 1

# BUOYANT DIFFUSION FLAMES

Notice the large scale motion sweeping across the entire mixing region. This is contrary to the common concept of "eddy diffusion" in use today. We see the lengthening of the interface between the fluids as the scale of the eddies is reduced. The burning itself is driven by the strain rate <u>along</u> the flame which brings the fuel and oxidant into close proximity. In the lower sketch we focus on the straining motion with most of the vorticity taken out for clarity. Vorticity without straining does not enhance the combustion.

With the turbulent combustion problem in mind Carrier et al. (21) have developed analytical solutions for a variety of transient and steady opposed flow situations including finite kinetic effects. Here we derive only the relatively simple rapid-kinetic laminar diffusion flame case. It is presumed that transient and finite kinetic effects could eventually be incorporated as the need arises.

## VI  OPPOSED FLOW LAMINAR BURNING

Consider the figure below. Fuel is supplied from below and oxidant from above with products leaving on the left and right. We assume a constant spatially uniform strain rate, s, with no vorticity so that the horizontal velocity u is given by

$$u = sx \tag{1}$$

with the coordinate system chosen such that $u = v = 0$ at the origin.

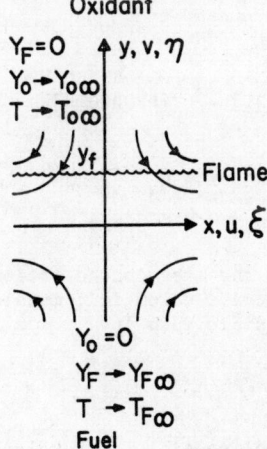

Figure 2

As shown below this assumed two-dimensional flow field results in a similarity solution with all parameters (except for u) being independent of x. Variable property effects are considered; but we note in passing that for variable properties the vertical momentum equation is satisfied only approximately except far downstream where the boundary layer approximation resolves the small variable property error in this momentum equation.

GOVERNING EQUATIONS

The usual Shvab-Zeldovich assumptions are made: unit Lewis number $\rho C_p D/\lambda$, equal specie diffusivities, low Mach number, the product $\rho^2 D = \rho\lambda/C_p = \rho_\infty^2 D_\infty =$ constant, a single global reaction going from fuel and oxidant to final products and heat. In addition it is assumed that a certain fraction, $\chi$, of the chemical heat release leaves the combustion zone in the form of radiation with the rest of the heat appearing as sensible enthalpy. Under these conditions the governing equations are:

$$\frac{\partial(\rho u)}{\partial x} + \frac{\partial(\rho v)}{\partial y} = 0 \tag{2}$$

$$\rho u \frac{\partial Y_F}{\partial x} + \rho v \frac{\partial Y_F}{\partial y} = \frac{\partial}{\partial x}\left[\rho D \frac{\partial Y_F}{\partial x}\right] + \frac{\partial}{\partial y}\left[\rho D \frac{\partial Y_F}{\partial y}\right] + \dot{m}_F''' \tag{3}$$

$$\rho u \frac{\partial Y_o}{\partial x} + \rho v \frac{\partial Y_o}{\partial y} = \frac{\partial}{\partial x}\left[\rho D \frac{\partial Y_o}{\partial x}\right] + \frac{\partial}{\partial y}\left[\rho D \frac{\partial Y_o}{\partial y}\right] + \dot{m}_o''' \tag{4}$$

$$\rho u \frac{\partial h_s}{\partial x} + v \frac{\partial h_s}{\partial y} = \frac{\partial}{\partial x}\left[\frac{\lambda}{C_p} \frac{\partial h_s}{\partial x}\right] + \frac{\partial}{\partial y}\left[\frac{\lambda}{C_p} \frac{\partial h_s}{\partial y}\right] + \dot{q}_s''' \tag{5}$$

$$\dot{q}_R''' = \chi(\dot{q}_s''' + \dot{q}_R''') \tag{6}$$

$$\nu_F[\text{FUEL}] + \nu_o[\text{OXIDANT}] \rightarrow \nu_p[\text{PRODUCTS}] + Q[\text{HEAT}] \tag{7}$$

$$\frac{-\dot{m}_F'''}{\nu_F M_F} = \frac{-\dot{m}_o'''}{\nu_o M_o} = \frac{\dot{q}_s''' + \dot{q}_R'''}{Q} = \frac{\dot{q}_s'''}{Q(1-\chi)} \tag{8}$$

Here $Y_F$ and $Y_o$ are respectively the mass concentrations of fuel and oxidant with $-\dot{m}_F'''$ and $-\dot{m}_o'''$ their respective volumetric mass consumption rates. Similarly $h_s$ is the gaseous sensible enthalpy

$$\int_{T_{o\infty}}^{T} C_p \, dt ,$$

and $\dot{q}_s'''$ is the volumetric rate of sensible heat release; whereas $\dot{q}_R'''$ is the net volumetric rate of radiant heat loss equal to $\chi$ times the total volumetric heat release rate $\dot{q}_s''' + \dot{q}_R'''$. For the presumed overall reaction, Eq. (7), $\nu_p$ moles of products are produced while Q units of heat are chemically released when $\nu_F$ moles of fuel and $\nu_o$ moles of oxidant are consumed. The final equation, (8), expresses this same chemical relation in terms of volumetric rates, where $M_F$ and $M_o$ are respectively the molecular weights of fuel and oxidant.

To solve the above equations we first assume that all the flow parameters (except u) are independent of x and later verify the consistency of the solution. Variable density effects are handled by the "Howarth" transformation

$$\eta = \sqrt{s/2D_\infty} \int_0^y (\rho/\rho_\infty) \, dy, \quad \xi = \sqrt{s/2D_\infty} \, x$$

so that

$$\frac{\partial}{\partial y}\bigg|_x = \sqrt{s/2D_\infty} \frac{\rho}{\rho_\infty} \frac{\partial}{\partial \eta}\bigg|_\xi \, ; \quad \frac{\partial}{\partial x}\bigg|_y = \sqrt{s/2D_\infty} \frac{\partial}{\partial \xi}\bigg|_\eta \qquad (9)$$

Using Eqs. (1) and (2) as well as the assumption that $\rho$ is independent of x one obtains

$$\rho v = -\rho_\infty \sqrt{2D_\infty s} \, \eta \qquad (10)$$

Again assuming invariance with x, the fuel, oxidant, and sensible enthalpy equations reduce to the form

$$-\rho s \eta \frac{dY_F}{d\eta} = \frac{\rho s}{2} \frac{d^2 Y_F}{d\eta^2} + \dot{m}_F'''$$

$$-\rho s \eta \frac{dY_o}{d\eta} = \frac{\rho s}{2} \frac{d^2 Y_F}{d\eta^2} + \dot{m}_o'''$$

$$-\rho s \eta \frac{dh_s}{d\eta} = \frac{\rho s}{2} \frac{d^2 h_s}{d\eta^2} + \dot{q}_s'''$$

where $\rho^2 D = \rho \lambda/C_p = \rho_\infty^2 D_\infty$ has been used.

Defining the Shavb-Zeldovich variables

$$\beta_1 = \frac{Y_o}{\nu_o M_o} - \frac{Y_F}{\nu_F M_F}, \qquad (11)$$

$$\beta_2 = \frac{Y_o}{\nu_o M_o} + \frac{h_s}{(1-\chi)Q} = \frac{Y_o}{\nu_o M_o} + \frac{\int_{T_{o\infty}}^T C_p \, dt}{(1-\chi)Q}, \qquad (12)$$

the above transformed equations become

$$-2\eta \frac{d\beta_i}{d\eta} = \frac{d^2 \beta_i}{d\eta} \qquad \text{for } i = 1,2$$

after eliminating the source terms with Eq. (8).

Using the boundary conditions shown in Fig. 2, $\beta_1$ and $\beta_2$ integrate to

$$\beta_1(\eta) = \frac{1}{2}\left\{\frac{Y_{o\infty}}{\nu_o M_o} + \frac{Y_{F\infty}}{\nu_F M_F}\right\} \text{erf}\eta + \frac{1}{2}\left\{\frac{Y_{o\infty}}{\nu_o M_o} - \frac{Y_{F\infty}}{\nu_F M_F}\right\} \qquad (13)$$

$$\beta_2(\eta) = \frac{1}{2}\left\{\frac{Y_{o\infty}}{\nu_o M_o} - \frac{\int_{T_{o\infty}}^{T_{F\infty}} C_p \, dT}{(1-\chi)Q}\right\} \text{erf}\eta + \left\{\frac{Y_{o\infty}}{\nu_o M_o} + \frac{\int_{T_{o\infty}}^{T_{F\infty}} C_p \, dt}{(1-\chi)Q}\right\} \qquad (14)$$

Under the assumption of rapid-kinetics fuel and oxidant cannot coexist at the same position. Thus above the flame, $\eta \geq \eta_f$, there is no fuel $Y_f=0$ and below the flame $\eta \leq \eta_f$, there is no oxidant $\overline{Y}_o=0$. With this knowledge and the definitions of $\beta_1$ and $\beta_2$, Eqs. (11) and (12), we can determine the complete distributions of $Y_F$, $Y_o$, and $h_s$. In particular at the flame $\eta=\eta_f$ both $Y_o=Y_f=0$; so that from Eq. (11) $\beta_1(\eta_f)=0$, yielding the dimensionless flame position $\eta_f$ from Eq. (13) as

$$\mathrm{erf}\,\eta_f = \frac{Y_{F\infty}/(\nu_F M_F) - Y_{o\infty}/(\nu_o M_o)}{Y_{F\infty}/(\nu_F M_F) + Y_{o\infty}/(\nu_o M_o)} = \frac{1 - Y_{o\infty}\nu_F M_F/(Y_{F\infty}\nu_o M_o)}{1 + Y_{o\infty}\nu_F M_F/(Y_{F\infty}\nu_o M_o)} \tag{15}$$

We note for later reference that $\eta_f$ is independent of the strain rate and depends only on the stoichiometric fuel to oxidant mass ratio $Y_{o\infty}\nu_F M_F / (Y_{F\infty}\nu_o M_o)$. For hydrocarbon fuels burning in air this ratio is usually less than 0.1 so that $\eta_f \gtrsim 1$, suggesting that the flame lies considerably above the stagnation plane $\eta=0$ as it seeks the oxidant source.

To determine the burning rate per unit flame area $-\dot{m}_F''$, we note that there is no oxidant on the fuel side of the flame $\eta \leq \eta_f$, so that from Eqs. (9), (11) and (13)

$$\frac{-\dot{m}_F''}{\nu_F M_F} = -\frac{\rho D}{\nu_F M_F} \left. \frac{\partial Y_F}{\partial y} \right|_{y_f} = \frac{\rho^2 D}{\rho_\infty} \sqrt{s/2D_\infty} \left. \frac{d\beta_1}{d\eta} \right|_{\eta_f}$$

$$= \left\{ \frac{Y_{o\infty}}{\nu_o M_o} + \frac{Y_{F\infty}}{\nu_F M_F} \right\} \rho_\infty \sqrt{sD_\infty/2\pi} \exp(-\eta_f^2) \tag{16}$$

where $\eta_f$ is given by Eq. (15) above in terms of stoichiometric parameters.

Thus we have arrived at the <u>important result</u> that the burning rate per unit flame area is proportional to $\rho_\infty\sqrt{sD_\infty/2\pi}$, that is proportional to the square root of the strain rate, s. From Equation (9) we see that the characteristic diffusion distance is proportional to $(\rho_\infty/\rho)\sqrt{2D_\infty/s}$. Both these results will be used later.

## VII  TURBULENT CASCADE COMBUSTION MODEL

Here we describe the turbulent combustion process as a sequence of opposed flow laminar diffusion flames of scales and strain rates provided by the Kolmogorov turbulence cascade. We assume that the turbulence Reynolds number is large and that the mixing is characterized by the collision of pure fuel and pure oxidant eddies. We also assume that from an overall standpoint the fuel and oxidant are introduced in their stoichiometric proportions. This stoichiometric assumption might eventually be relaxed by using Magnussen's (19) concept of ignoring any excess reactant and characterizing the burning rate by the availability of the deficit reactant.

### 7.1  BURNING OF THE INITIAL LARGE EDDIES

Consider the burning resulting from a large fuel clump or eddy penetrating into a previously undisturbed oxidant fluid as shown in Figure 1, or equivalently an oxidant eddy entering a fuel medium. Let the eddy scale be $a_o$ and the eddy velocity be $u_o$ so that the strain rate between these eddies is characterized by $u_o/a_o$. From Eq. (16) the burning rate per unit area between these eddies is

$$-\dot{m}_F''\bigg|_o = k\rho_\infty \sqrt{u_o D_\infty/a_o} \tag{17}$$

where

$$k = \frac{Y_{F\infty}}{\sqrt{2\pi}} [1 + \frac{Y_{o\infty} M_F \nu_F}{Y_{F\infty} M_o \nu_o}] \exp(-\eta_f^2) \tag{18}$$

which depends only on the reaction stoichiometry. To obtain the overall burning rate of this initial eddy we shall characterize its area by $\pi a_o^2$ corresponding to a sphere of diameter $a_o$ yielding an initial eddy burning rate of

$$-\dot{m}_F\bigg|_o = -\dot{m}_F''\bigg|_o \pi a_o^2 = \pi k\rho_\infty a_o^{3/2} \sqrt{u_o D_\infty} \tag{19}$$

To express this initial burning on a volumetric basis we divide the above expression by the eddy volume $\pi a_o^3/6$ and the total reactant to fuel stoichiometric ratio $\gamma = (1+Y_{F\infty}\nu_o M_o/Y_{o\infty}\nu_F M_F)$ obtaining the spatially averaged volumetric burning rate for the initial eddies as

$$\overline{-\dot{m}_F'''}\bigg|_o = 6(k/\gamma)\rho_\infty a_o^{-3/2} \sqrt{u_o D_\infty}$$

After a period of time these initial eddies will break up since the fluid motion is intrinsically unstable. It is not unreasonable to expect this break-up to occur at about the time the eddy has penetrated a distance $a_o$ into the undisturbed fluid. We therefore characterize the initial eddy lifetime as

$$t_o = a_o/u_o \quad , \tag{20}$$

obtaining the initial fractional mass burned during its lifetime as

$$-m_F'''\bigg|_o = t_o \dot{m}_F'''\bigg|_o = 6(k/\gamma)\rho_\infty \sqrt{D_\infty/a_o u_o} \tag{21}$$

## 7.2 INERTIAL CASCADE BURNING

The Kolmogorov turbulence cascade describes how the kinetic energy of turbulence is successively transferred to eddies of smaller and smaller scales before it is finally dissipated by viscosity. The turbulent kinetic energy is usually introduced by the mean flow so that its initial scale, $a_o$, is characterized by the mean flow. Subsequently these large eddies collide and deform producing eddies of smaller scale. This cascade continues until the scale becomes fine enough for the viscous effects to entirely consume the turbulent kinetic energy.

To model this process we assume that the initial eddy scale $a_o$ is much larger than the final Kolmogorov viscous dissipation scale, $\eta$, so that viscous effects are not involved during the inertial eddy break-up cascade. This assumption corresponds to a large turbulence Reynolds number $u_o a_o/\nu$. We also assume that at each stage of the cascade the eddy break-up process occurs by the interaction of eddies of similar scale. And finally we assume that all the turbulent kinetic energy is introduced from the mean flow. This latter assumption is strictly valid only for forced convection flows, since in buoyant flows gravity can directly inject turbulent kinetic energy by interacting with density fluctuations at finer scales. Here we ignore such effects.

With the above assumptions the rate of transfer of turbulent kinetic energy to eddies of smaller scale is independent of the particular eddy scale at each stage of the cascade since all the energy must be dissipated. This rate of transfer is commonly called the dissipation of turbulent kinetic energy, $\varepsilon$, which must equal the rate at which turbulence is initially introduced by the mean flow field, that is,

$$\varepsilon \sim u_o^2/t_o = u_o^3/a_o \tag{22}$$

For the sake of concreteness we shall model the cascade as a sequence of discrete stages with eddies of scale $a_n$ and rms velocity $u_n$ where the integer n identifies the stage number with the initial stage identified by n=o. Since viscous effects are not involved during the inertial cascade, by dimensionless analysis, the time, $t_n$, for the break-up of eddies of scale $a_n$, must be proportional to $a_n/u_n$. Here we define the separation of cascade stages such that

$$t_n \equiv a_n/u_n. \tag{23}$$

Accordingly the scale reduction for each successive stage can be expressed as

$$a_n = p \, a_{n-1}$$

where p is some number less than unity. The number p is independent of n by our assumption that the eddy break-up at each scale is governed by the interaction of eddies of similar scale. Consequently by iteration

$$a_n = p^n a_o \tag{24}$$

Since the dissipation of turbulent kinetic energy, $\varepsilon$, is invariant during the cascade, we must have $\varepsilon = u_n^3/a_n = u_o^3/a_o$ = constant yielding the cascade stage velocity

$$u_n = (\varepsilon a_n)^{1/3} = (\varepsilon a_o p^n)^{1/3} = u_o p^{n/3} \tag{25}$$

and the break-up time

$$t_n = a_n/u_n = (a_o/u_o) \, p^{2n/3} \tag{26}$$

To apply this cascade model to turbulent burning we shall follow the arguments developed in the previous section for the initial eddy burning. Once again we shall assume that the burning results from the collision of pure fuel and oxidant clumps impacting with strain rate $u_n/a_n$ and characterized by surface area $\pi a_n^2$ corresponding to a sphere of diameter, $a_n$. Replacing $a_o$ and $u_o$ in Eq. (19) by $a_n$ and $u_n$ respectively, given by Eqs. (24) and (25), the overall burning rate of a fuel eddy of scale $a_n$ is given by

$$-\dot{m}_F\big|_n = \pi k \rho_\infty \, a_o^{3/2} \sqrt{u_o D_\infty} \, p^{5n/3} \tag{27}$$

**BUOYANT DIFFUSION FLAMES**

Dividing this by the fuel eddy volume $\pi a_n^3/6$ and the total reactant to fuel stoichiometric ratio $\gamma = (1 + Y_{F\infty}\nu_O M_O / Y_{O\infty} \nu_F M_F)$ we obtain the spatially averaged volumetric burning rate for the cascade stage as

$$-\dot{m}_F'''|_n = 6(k/\gamma)\,\rho_\infty\, a_o^{-3/2}\, \sqrt{u_o D_\infty}\; p^{-4n/3}$$

Finally multiplying by the residence time, $t_n$, one obtains the spatially averaged mass burned during stage n as

$$-m_F'''|_n = 6(k/\gamma)\,\rho_\infty\, \sqrt{D_\infty/a_o u_o}\; p^{-2n/3} \qquad (28)$$

Since p is less than unity, we see that the additional mass burned increases at each successive stage of the cascade. This theory is approximate since at each stage we assume that the burning takes place between pure fuel and oxidant while ignoring the increasing presence of products. However these products will not be significant in the earlier cascade stages, if the flow Reynolds number $a_o u_o/\nu$ is large as seen from Eq. (28).

The mass burnt per unit volume at each stage of the cascade according to Eq. (28), increases with $a_n^{-2/3}$ as the eddy scale $a_n$ decreases. This increase is due to three factors: 1) the residence time $t_n$ at each stage decreases according to $a_n/u_n \sim a_n^{2/3}$; 2) the burning rate per unit interface area, $\dot{m}_F''$, increases with the square root of the strain rate $(u_n/a_n)^{1/2} \sim a_n^{-1/3}$; and 3) the interface area per unit volume increases according to $a_n^{-1}$.

This cascade burning model has made the implicit assumption that the boundary between fuel and oxidant eddies occurs where the strain rate is $u_n/a_n$ with complete separation of stages. This clearly is an oversimplification. There may well be some smaller scale straining within the pure fuel or oxidant eddies due to previous disturbances. Also small scale eddies from previous disturbances may eat into the boundaries of the larger scale eddies – this effect within the context of the present model should be ascribed to a previous cascade. Finally the straining rate will by no means be uniform along the boundaries of the eddies and the eddies of course are not really spherical. All these factors suggest that the model can at best provide only the correct overall trends.

### 7.3 FINAL DISSIPATION STAGE

To complete the turbulent burning model we must identify the final stage N for completion of burning. Such an identification can be accomplished in a variety of ways leading to similar results.

Consider first the total amount of fuel available for reaction. As before we assume the reactants are initially (but separately) introduced in their stoichiometric proportions so that the spatially averaged initial fuel per unit volume is

$$-m_F''' = \rho_\infty/\gamma$$

This quantity must be entirely consumed during the N stages of burning, or adding the terms from Eq. (28)

$$-m''_F = \rho_\infty/Y = 6k(\rho_\infty/Y)\sqrt{D_\infty/a_o u_o}\sum_{n=0}^{N} p^{-2n/3} \qquad (29)$$

Extracting $p^{-2N/3}$ from the sum and rearranging its terms we have

$$p^{2N/3} = 6k\sqrt{D_\infty/a_o u_o}\sum_{n=0}^{N}\bar{p}^{2n/3}$$

Noting that for typical fuels 6k is approximately one and assuming p is much less than unity we have approximately

$$N \simeq \frac{3}{4}\frac{\ln(u_o a_o/D_\infty)}{\ln(1/p)} \qquad (30)$$

A similar result can be obtained by considering the Kolmogorov-Batchelor microscale length $\eta_D$ for dissipation of concentration differences which is

$$\eta_D = D^{1/2}\nu^{1/4}\bar{\varepsilon}^{1/4}$$

for gases with constant properties, Ref. (23). Equating this to the final eddy scale $a_N$ from Eq. (24) and substituting $u_o^3/a_o$ for $\varepsilon$ one has

$$a_o p^N = D^{1/2}\nu^{1/4}a_o^{1/4}u_o^{-3/4}$$

or upon solving for N

$$N = \frac{3}{4}\frac{\ln(u_o a_o/D^{2/3}\nu^{1/3})}{\ln(1/p)}$$

for constant properties. This result is essentially identical to Eq. (30) in the case of gases which have $D \simeq \nu$.

Finally we can consider the cumulative dissipation of turbulent kinetic energy during the cascade. At stage n the bounding eddy surface area is characterized by $\pi a_n^2$, its volume by $\pi a_n^3/6$, the square of its boundary strain rate by $(u_n/a_n)^2$, and the thickness of its bounding strain layer (see Eq. (9)) is $\sqrt{\nu a_n/u_n}$; so that the total dissipation per unit volume at stage n is

$$\varepsilon_n = \nu(\pi a_n^2)(\pi a_n^3/6)^{-1}(u_n/a_n)^2\sqrt{\nu a_n/u_n}$$

$$= 6(u_n^3/a_n)(u_n a_n/\nu)^{-3/2}$$

Here we assume that this dissipation at each stage is driven by the main flow and that the viscous dissipation for each stage operates continuously during the entire cascade; so that the total dissipation rate $u_o^3/a_o$ is given by the sum of the $\varepsilon_n$, or using Eqs. (24) and (25)

$$u_o^3/a_o = 6(u_o^3/a_o)(u_o a_o/\nu)^{-3/2}\sum_{0}^{N} p^{-2n}$$

Factoring out $p^{-2N}$, rearranging the sum and ignoring the factor of 6 we have for small p

$$N \simeq \frac{3}{4}\frac{\ln(u_o a_o/\nu)}{\ln(1/p)}$$

We thus have arrived at a result similar to Eq. (30) in three different ways

## 7.4 MEAN BURNING RATE AND KOLMOGOROV MICROSCALES

According to this model for reactants introduced in their stoichiometric proportions the mean burning rate per unit volume $\overline{\dot{m}_F'''}$ is obtained by dividing the available fuel per unit volume $\rho_\infty/\gamma$ by the total burning time

$$\Sigma t_n = (a_o/u_o) \sum_{n=0}^{N} p^{2n/3},$$

or

$$-\overline{\dot{m}_F'''} = \frac{\rho_\infty u_o}{\gamma a_o} (\sum_{n=0}^{N} p^{2n/3})^{-1} \sim \frac{\rho_\infty u_o}{\gamma a_o} \qquad (31)$$

since p is a constant. This result says that the overall burning rate is controlled by the macroscopic variables, or more specifically by the characteristic time $a_o/u_o$ for the initial eddies to break up.

The result is consistent with experimental observations. For example, forced jet diffusion flame lengths are proportional to the jet diameter and independent of the initial jet velocity for sufficiently high Reynolds number. Consequently their total rate of fuel supply $\dot{M} = u_o \rho_\infty \pi D^2/4$ divided by their flame volume $V \sim \pi H D^2/4$ is

$$-\overline{\dot{m}_F'''} = \dot{M}/V \sim \rho_\infty u_o/H \sim \rho_\infty u_o/D.$$

In the case of buoyant fuel jet diffusion flames we consider the measurements of Markstein (10) who measured both flame widths and lengths. He found the flame length, H, and flame widths, D, were given by

$$H \sim \dot{Q}^{0.452}$$
$$D \sim \dot{Q}^{0.355}$$

respectively where $\dot{Q}$ is the volumetric fuel supply rate. Assuming that the upward buoyant velocities are characterized by $\sqrt{gH}$ the predicted mean burning rate should vary as

$$u_o/a_o \sim \sqrt{gH}/D \sim \dot{Q}^{-.129}$$

which is quite close to his measured volumetric burning rates which were proportional to $\dot{Q}^{-.161}$.

It is interesting to also examine the characteristic Kolmogorov microscale $\eta = \nu^{3/4} \varepsilon^{-1/4}$ for these buoyant fuel jets. Thus we have

$$\eta \sim \varepsilon^{-1/4} = a_o^{1/4} u_o^{3/4} \sim D^{1/4}/H^{3/8} \sim \dot{Q}^{-.0807}$$

which decreases slowly with flow rate. We anticipate that the flame chemistry and soot formation Damköhler number will be primarily determined by this characteristic diffusion length and thus are expected to vary only very slightly with flow rate or scale. We thus have a tentative explanation for the experimental observation of a constant fraction of heat release leaving in the form of radiation.

VIII  CONCLUSIONS

In this paper we develop a new analytic Cascade Combustion Model for turbulent diffusion flames. The model follows the burning of eddies over their entire sequence of scales in the Kolmogorov cascade. At each cascade stage the burning is treated as a laminar diffusion flame driven by the strain rate associated with the stage. The model shows that the residence time $t_n$ at each scale $a_n$ decreases according to $a_n^{2/3}$. Although most of the burning takes place at the final Kolmogorov dissipation microscale most of the time for burnout occurs during the initial large-scale cascade stages; so that the mean volumetric burning rate is governed by the macroscale time.

The predicted dependency of mean volumetric burning rate versus fuel supply rate agrees with Markstein's (10) buoyant turbulent diffusion flame measurements. Also the Kolmogorov microscale and the associated Damköhler number for burning is shown to be only very weakly dependent on fuel flow rate for these buoyant diffusion flame measurements. This provides an explanation for the observed constant fraction of heat leaving the flames in the form of radiation.

This review of buoyant diffusion flames also reveals several important research needs:

(1) First and foremost we need mean velocity and mean reactant concentration measurements for turbulent fuel jet diffusion flames to test new theories. Care should be taken to minimize the tendency of these flames to meander from side to side.

(2) There is an urgent engineering need for an improved integral theory for buoyant diffusion flames which incorporates a more advanced combustion model such as the cascade model presented here. The best model available today employs a 400% correction factor!!For the model to be useful for flame radiation calculations it must correctly predict flame shape and scale.

(3) Modern numerical turbulent combustion models developed for forced convection situations should be applied to buoyant diffusion flames to establish whether or not these flames have additional turbulence mechanisms. We are ultimately seeking a unified combustion model which can cope with both free and forced convection situations.

(4) There is a general need for improved turbulent diffusion flame models which relate Kolmogorov microscale burning to macroscale processes. In particular the "Cascade Combustion Model" presented here should be generalized to non-stoichiometric situations and should be tested experimentally. More attention should be placed on variable density effects and buoyancy generated turbulence.

(5) Soot concentration and temperature measurements should be made on small laminar diffusion flames to determine the sensitivity of soot to Damköhler number. Such measurements would provide fundamental insight into the influence of scale on radiation from turbulent diffusion flames.

(6) A transient laminar diffusion flame model including soot formation, soot oxidation as well as radiant extinction effects is needed to gain a fundamental insight into the smoke generation by buoyant turbulent diffusion flames.

## IX  NOMENCLATURE

| | |
|---|---|
| $a_o$ | initial eddy scale [L] |
| $a_n$ | scale of eddy at $n^{th}$ cascade stage [L] |
| $C_p$ | specific heat at constant pressure [E/M$\theta$] |
| D | orifice diameter [L] |
| D | species diffusivity [L$^2$/t] |
| f | conserved mixture variable [-] |
| $\overline{g = f'^2}$ | mean square fluctuation of f |
| g | acceleration of gravity [L/t$^2$] |
| $h_s$ | sensible specific enthalpy [E/M] |
| H | flame height [L] |
| k | see Eq. (18) [-] |
| $\dot{m}_i'''$ | volumetric mass generation rate of species "i" [M/L$^3$t] |
| M | mass per mole [M/mole] |
| N | number of final dissipation cascade stage [-] |
| n | cascade stage integer [-] |
| p | fractional decrease in eddy scale at each cascade stage [-] |
| $\dot{q}_s'''$ | volumetric rate of heat generation [E/L$^3$t] |
| $\dot{q}_R'''$ | volumetric generation of radiation [E/L$^3$t] |
| Q | heat of combustion for $\nu_F(\nu_o)$ moles of fuel (oxidant) [E/mole] |
| s | strain rate [1/t] |
| $t_n$ | residence time at cascade stage n [t] |
| T | temperature [$\theta$] |
| $u_n$ | velocity of eddy at $n^{th}$ cascade stage [L/t] |
| u,v | x,y components of velocity [L/t] |
| x,y | spatial coordinates [L] |
| Y | species mass fraction [-] |
| $\beta_1, \beta_2$ | Shvab-Zeldovich variables see Eqs. (11) and (12) [1/M] |
| $\gamma$ | mass of stoichiometric reactants per unit mass of fuel [-] |
| $\varepsilon$ | dissipation rate of turbulent kinetic energy [L$^2$/t$^3$] |
| $\eta$ | Kolmogorov microscale [L] |
| $\eta, \xi$ | transformed dimensionless coordinate [-] |
| $\lambda$ | thermal conductivity [E/Lt$\theta$] |
| $\rho$ | density [M/L$^3$] |
| $\nu$ | moles [mole] |
| $\nu$ | kinematic viscosity [L$^2$/t] |
| $\chi$ | radiative fraction [-] |

SUBSCRIPTS

F     fuel
o     oxidant
o     initial
n     cascade stage number
t     flame tip
$\infty$     infinity

SUPERSCRIPTS

–     mean
"'     volumetric
"     per unit area
•     per unit time

## X     REFERENCES

1. Hawthorne, W. R., Weddell, D. S., and Hottel, H. C.: Third Symposium on Combustion, Flame and Explosion Phenomena, p. 266, Williams and Wilkins, 1949.

2. Putnam, A. A. and Speich C. F.: Ninth Symposium (International) on Combustion, p. 867, Academic Press, 1963.

3. Thomas, P.H.: Ninth Symposium (International) on Combustion, p. 844, Academic Press, 1963.

4. Thomas, P.H.; Baldwin, R., and Heselden, A.J.M.: Tenth Symposium (International) on Combustion, p. 983, The Combustion Institute, Pittsburgh, Pennsylvania, 1965.

5. Blinov, V.I. and Khudyakov, G.N.: Diffusion Burning of Liquids, Izd. Nauk SSSR, Moscow, 1961; English Translation U.S. Army Engineering Research and Development Laboratories, T-1490 a-c, Astia AD 296 762.

6. Burgess, D.S., Strasser, A. and Grumer, J.: Fire Research Abstracts Revs. 3, 177 (1961).

7. Heskestad, G.: Optimization of Sprinkler Fire Protection Progress Report No. 9; FMRC Technical Report, March 1974. Factory Mutual Research Corporation, Norwood, Massachusetts.

8. Ricou, F.P. and Spalding D.B.: J. Fluid Mech 11, p.21, (1961).

9. Rasbash, D.J., Rogowski, A.W., and Stark, G.W.V.: Fuel 35, p. 94, (1956).

10. Markstein, G.H.: Sixteenth Symposium (International) on Combustion, (to appear 1977), also FMRC Technical Report No. 22361-4, Factory Mutual Research Corporation, Norwood, Massachusetts (1976).

11. Steward, F.R.: Comb. Sci. and Tech. 2, p. 203, (1970).

12. Kent, J.H. and Bilger, R.W.: Sixteenth Symposium (International) on Combustion, (to appear 1977), also T.N. F-82, Department of Mechanical Engineering, University of Sydney, Australia, (1976).

13. George, W.K., Alpert, R.L. and Tamanini, F.: FMRC Technical Report No. 22359-2, Factory Mutual Research Corporation, Norwood, Massachusetts, 1976.

14. Tamanini, F.: FMRC Technical Report No. 22360-4, Factory Mutual Research Corporation, Norwood, Massachusetts (in preparation).

15. Rodi, W.: ZAMM 56, p. 219, (1976).

16. Tamanini, F.: Sixteenth Symposium (International) on Combustion (to appear 1977), also FMRC Technical Report No. 22360-3, Factory Mutual Research Corporation, Norwood, Massachusetts. (1976).

17. Emmons, H.W. and Shuh-Jing Ying: Eleventh Symposium (International) on Combustion, p. 475, The Combustion Institute, Piitsburgh, Pa. (1967).

18. Bilger, R.W.: Prog. Energy Combust. Sci., 1, p. 87, (1976).

19. Magnussen, B.F. and Hjertager, B.H.: Sixteenth Symposium (International) on Combustion (to appear 1977).

20. Spalding, D.B.: Sixteenth Symposium (International) on Combustion, (to appear 1977) also HTS/76/1, Mechanical Engineering Department, Imperial College of Science and Technology, London SW7.

21. Carrier, G.F., Fendell, F.E. and Marble, F.E.: Technical Report TRW-5-PU, Project Squid, School of Mechanical Engineering, Purdue University, Lafayette, Indiana, (1973).

22. Brown, G. and Roshko, A.: J. Fluid Mech., 64, p. 775, (1974).

23. Tennekes, H. and Lumley, J.L.: A First Course in Turbulence, The MIT Press, 1972.

24. Launder, B.E. and Spalding, D.B.; Mathematical Models of Turbulence, Academic Press, 1972.

# INDEX

Air circulation in a room, 418, 421
Air jet apparatus, 304–305
Air neon, 504–506
Air and smoke flow, methods to influence, 788
Air warming, 423–425
Anisotropy, 4, 6
Argon, heat transfer measurements with, 504–506
Argon cover gas, 498
Argon-water vapor mass and heat transfer measurements, 503
Austausch phenomena, 95
Averaged temperature, 53

Benard cells, 733
Boiling in a slot, 639
Boundary-layer flow, 475
Boussinesq fluid, 79
Bouyancy
   and entrainment of fire, 458
   flow caused by, 779, 793
Buoyancy dominated turbulent flow, 683–685
Bouyancy-driven-convective flow, 473
Bouyancy-driven mixed layer, 65–66, 70–74
Buoyant flow calculation, 435–438
Buoyant plume and the similarity theory, 87
Bursting phenomena, 27–30, 34

Cascade Combustion Model, 813, 822
Cellular convection in a horizontal layer, 519
Center-line axial turbulence level distributions, 303–305
Center-line temperature distributions, 303–305
Center-line velocity distributions, 303–305
Conservation equation, 532–533, 573–574
Constant-density flow, 578–579
Convection, suppression of, 498
Convection heat flow densities, 417–427
Convection heat transfer, 451–452
Convective rollers, 520–521
Cooling pond, heat balance of, 143
Cooling of thermal electric power plants, 139–143
   with lake water, 141–142
"Corona" current, 673
Corridor entrance jet, 465–468
"Curl-like" flow, 487

Data processing equipment, 363
Dense plumes, 332
Density stratified flow, 625–631
Diffusion approximation and gaseous pollutants, 159
Diffusion simulation, 73–74
Dimensionless local jet volume, 357
Doorway mass flow rates, 468
Downstream step size, 169–170
Dry air and air plus water vapor heat transfer, 502–503
Duststream, passing through duct, 674

EBR-II, 545–553 passim
Eddy diffusivity, 408
Eddy distribution in lake surface, 97–100
Eddy viscosity parameter, 260
"Electric wind," 673
Energy consumption, 419

Energy dissipation, modeling of, 67
Entrainment function, 349
Entrainment hypothesis, 326, 329–330
Evaporation
   from lake and sea surfaces, 497
   from liquid pool, 498–500
Explosive boiling-up, 601–613

Fickian Law of Diffusion, 374
Film boiling, 616–617
Finite difference
   equations, 162
   schemes, 444–445
Fire induced flows in corridors, 457
Fire room configuration, effect of, 461–463
Fire size, effect of, 461–462
Fission heating, 548–549
Fission product decay heating, 547–548
Flow patterns
   at heat transfer surface, 489
   synchronous filming of, 523
Flows with density changes, 579
Flow visualizations, 225–229
Flue profiles, 216
Forced convection, dynamic field in, 512–514
Forced flow in a channel, 509
Forced ventilation of rooms, 430
Free convection
   with mass transfer, 498–500
   in vertical tube, 651
Free convection heat transfer, 587
Free-convection phenomena in gas-liquid mixtures, 570–576
Free-convective motion, 696
Free thermal convection, 488

Gaseous pollutants, 159–161
Gas expansion, flow caused by, 778–779
Gas and Liquid Analyser (GALA) technique, 569, 577–580
Gas-vapor mixtures, heat and mass transfer in, 497
Generalized velocity profile, 255
Gravitational acceleration vector, 487
Gravity vector, 487
Ground level deposition velocity, 166

Heated surface jets, prediction of, 281–285
Heat energy equation, 114
Heat flow density, transmission of, 417–427
Heat losses in a room, 417
Heat and mass transfer in environmental systems, 473
Heat transfer
   enhancement of, 771
   prediction of, 798
Heavy fluid jet, 529–532
Helium, heat and mass transfer in, 497
Helium plus water vapor, heat mass transfer in, 506–507
High-rate thermoconvective processes, 444
Homogeneous vapor nucleation theory, 611
Horizontal surface jet, 40, 43
   predictions of, 43–49

# INDEX

Hot plumes, 241
Hot wire anemometer, 363

Idealized three-layer-model, 200
Ideal thermal, 239
Ideal plume, 236–237
Inertial cascade burning, 823–825
Initial mixing, 400
Interface tracking, 579–580
Internal energy transport, 183
Internal Froude number, 400–407
Intermediate heat exchanger (IHX), 546
Internal heat generation simulation, 595–597
Internal hydraulics of turbulent jets, 626–628
InterPhase Slip Analyser (IPSA) technique, 569, 580–585
Inviscid central region, 510
Isothermal flow, 509
Isothermal flow calculations, 434–435, 439

Jet geometry, 253
Jet width parameter, 352

Kolmogorov microscales, 827

Lagrangian integral time scales, 289, 299
Lagrangian intensities of turbulence, 289, 297
Laminar flow, 468–469, 473, 487
Laminar free convection stagnation heat transfer, 663–670
Laminar to turbulent natural convection, transition from, 649
Large eddies, burning of, 822–823
Larvate boiling, 620–622
Laser Doppler anemometry, 349
Light-phase and dense-phase velocities, 584–585
Line impulse model, 325–336
Liquid-Metal-Cooled Fast Breeder Reactor (LMFBR), 529–531, 545
Liquid sodium
    eddy diffusivity of heat and momentum for, 534
    in reactor, 555, 565
Local equilibrium, 628–631
Local jet Richardson number, 352
Longitudinal jet turbulence, 256–258

Maximum transfer, 488–489
Mean burning rate, 827
Mean and fluctuating velocities, 512–514
Mean wave field, 8
Metastable liquid, 601
Microstructure, 4, 7–8
Mixing, effect on fire flow, 465
Mixing length parameter, 259
Modeling buoyant transport, 68–70, 72
Modeling of turbulence, 187
Momentum equation, 114
Multi-phase flow, 569

Nonisothermal bodies, heat transfer from, 663–664
Normal scram test, 536–542
Nusselt number-Rayleigh number relation, 84–87

Ocean, surface temperature of, 106–108
Oceanic fine structure, 221–222

Oceanic microstructure, 222–224
One-dimensional longitudinal mixing, 291
Open-channel flow, 290–291
Opposed flow laminar burning, 819–820
Optimal linearization, 108–112

Parallel flow, 475
Patankar and Spalding algorithm, 429, 431–432
Patankar-Spalding procedure, 690
Penetrative convection, 199–209
Perturbation equations, 476–477
Perturbation of forced convection by buoyancy, 514–517
Phenix nuclear reactor, 555
Physical model and coordinate system, 185
Pipe mixing, 395, 398
Plane heated vertical jet, 39
Plane vertical buoyant jet, 17
Plenum domain, 530–539
Plume
    definition of, 39, 235–238
    in stratified environment, 243–246
    in a cross flow, 246–248
Plumelike flow, 19
Pool-type reactors, 510
Potential core, 334–336
Pressure reduction on fire floor, 790
Pressure terms, 66–67
Pulsation heat fluxes, 53, 55

Quasi-Single-Phase model, 573–574

Radiation, calculation of, 421
Radiation heat flow densities, 417–427
Reactor coolant system, 546
"Ready centres" number, decrease of, 603
Real plumes, 239–240
Real thermals, 242
Recirculation, effect on fire flow, 465
Recirculation flows, 510
Recurrent flows, 519
Regions of plume, 361
Resistance thermometer, 363
Return-to-isotropy terms, 67
Reynolds stress tensor components, 53, 55
Room doorway configuration, effect of, 463
Rotating plug penetration, 564
Roughness geometry, 212
Rough surfaces, heat transfer data on, 657–659
Round heated turbulent jets, flow establishment for, 263–273

Scale, effect on fire induced flows, 457, 459–461
Seasonal thermocline, 113–121
Semi-empirical approximations, 127–128
Semi-Implicit-Method for Pressure-Linked Equations (SIMPLE) algorithm, 534, 583
Shallow cavity system, 475
Similarity theory and buoyant plume, 87–90
Single-phase thermosyphon, 639
Skin friction and heat transfer, 30
Slot
    vertical, 646
    inclined, 647–648
Slot thermosyphon, 641
Smoke control and extraction, 458
Smoke-Wire method, 27, 33
SNIP procedure, 577, 583

# INDEX

Sodium-cooled fast breeder reactors, 498, 545
Sodium-cooled-pool-type fast reactors, 500
Sodium vapor, transport of, 497
Source height, 167
Spreading surface plume, 177–180
Stability, 478–482
Steady rollers, 519
Steady-state code CONVECT, 552
Streak spacing, 35
Submerged jet, 173–175
Surface heat flux and fluid temperature, 591–593
Surface-layer meteorology, 340
"Sweep," 28

Temperature distribution in a natural lake, 93
Temperature fluctuations, 524–527
Temperature variance, destruction of, 67, 72
Theories of convection, 80–83
Thermal, definition of, 235–238
Thermal discharges
   rivers without, 189–193
   rivers with, 193–196
Thermal entrance region, 743-746
Thermistors, 349
Thermogravitation, 701, 719
Thermogravitational convection, 61
Thermosyphon model, 555, 560
Thermosyphon structure, 558,-561
Third-order closure model, 82–83
Three dimensional effects of fire flow, 463–465
Three-dimensional forced plumes, 333
Three-dimensional turbulent buoyant flow, 429, 433
Three-dimensional turbulent isothermal flow, 429, 433
Tracer mixing, 395
Transient flow coastdown, 550–551
Transition flow, 475
Transverse injection, 402
Tri-Diagonal Matrix Algorithm, 432
Turbulence, collapse of, 631–635
Turbulence energy equation, 18
Turbulence model, 115–116
Turbulent buoyant diffusion flames, 813, 828
Turbulent Cascade Model, 818–819
Turbulent convection, 487
Turbulent flows, effect of on buoyancy, 3–13
Turbulent fluctuations, phase characteristics, 23
Turbulent fluid flow equations, 276
Turbulent internal waves, 4–6
Turbulent mixed convection, 759
Turbulent mixed region, 123–135
Turbulent round jet, 362–368

Turbulent round plume, 363–368
Turbulent shear flows, 5, 28, 39
Turbulent thermal convection, 587
Two-dimensional forced plumes, 332
Two-dimensional free convective viscous imcompressible liquid flow, 447
Two-dimensional heated surface jet, 39
Two-dimensionality, tendency toward, 4, 7
Two-dimensional natural convection of air, 807

Unique phenomena, 4

Vapor-film from cooling-down, 615
Vapor-liquid interface, 617–619
Velocity
   inertial relaxation of, 576
   and temperature profile in a room, 426
Velocity vector components, 53
Vertical air cell, natural convection in, 695
Vertical buoyant jet, 40, 43
   predictions of, 43–49
Vertical eddy diffusivity, 166–167
Vertical fuel surface, burning problem of, 685–687
Vertical grid descretization and iteration procedure, 163–164
Vertical round buoyant jet, 354–358
Vertical thermocline in lake, 93
Viscous wall region, 510
Vortex rolls, 735, 743–744, 752, 756

Wall functions, 534, 688–689
Wall region in turbulent boundary layer, 33–34
Warm-water discharge, 152–153
Warm-water spreading in cooling lakes, 152–153
Washout coefficient, 166
Waste-heat rejection, 275
Waste water discharges, in lakes and oceans, 171–177
Water storage, 139
Water-surface heat transfer to atmosphere, 146–152
Wavelike turbulence, 8–9
Wind on building, flow caused by, 787–788, 793
Wind speed, 166
Winter thermal plume surveys, 313–323

Xenon, 504–506

Zero heat flux boudary conditon, 390

# ERRATA

In the list line $\overline{13}$ denotes the 13th line from the top; line $\underline{24}$ is the 24th line from the bottom.

| Page | Line | Reads | Should read |
|---|---|---|---|
| 3, 381, 457 | | Bouyant | Buoyant |
| 106 | $\underline{2}$ | ... it wust be retained | ... it must be retained |
| 112 | $\overline{7}$ | ... there is a possibiliy to subjet | ... there is a possibility to subject |
| 151 | page | | should be page 152 |
| 152 | page | | should be page 151 |
| 159 | $\underline{2}$ | ... factor of C | ... factor of c |
| 273 | $\overline{5}$ | Fan's [11] correlation ... | Hirst's [12] correlation ... |
| 320 | $\underline{22}$ | in winter re ealed that | in winter revealed that |
| 385 | Eq. (1b) | $\dfrac{DV}{Dt} = -\dfrac{1}{\rho}\widetilde{\nabla}p + \nu\widetilde{\nabla}^2 \vec{V} - \hat{e}_3 g$ | $\dfrac{\vec{DV}}{Dt} = -\dfrac{1}{\rho}\widetilde{\nabla}p + \nu\widetilde{\nabla}^2 \vec{V} - \hat{e}_3 g$ |
| 386 | Eq. (4b) | $\nabla^2 \phi + Ru =$ | $\nabla^2 \phi + Ru = 0$ |
| 388 | $\overline{9}$ | $q_{n+1}^2 + r_n^2 = R^{1/2}$ | $q_{n+1}^2 - r_n^2 = R^{1/2}$ |
| 389 | Eq. (15) | $R_c = \pi^4\left(\dfrac{1}{4} + \dfrac{1}{a^2}\right)$ | $R_c = \pi^4\left(\dfrac{1}{4} + \dfrac{1}{a^2}\right)^2$ |
| 390 | Table | 152.0 | 152.2 |
| 432 | $\overline{3}$ | $-\overline{\rho u_i \phi} = \dfrac{-\mu_t}{\sigma_\phi}\left(\dfrac{\partial \Phi}{\partial x_i}\right)$ | $-\overline{\rho u_i \phi} = \dfrac{\mu_t}{\sigma_\phi}\left(\dfrac{\partial \Phi}{\partial x_i}\right)$ |
| 433 | $\overline{10}$ | w = 243 mm | b = 243 mm |
| 471 | $\overline{3}$ | ... in figure 12 | ... in figure 11 |
| 471 | $\overline{20}$ | The undulation noted ... | The undulations noted ... |
| 472 | Ref. 4 | Mitta, K. | Nitta, K. |
| 497 | Abstract | | belongs on page 587 |
| 587 | Abstract | | belongs on page 497 |

4C
319.8
H42
v.2

FEB 12 1979